# Smart Innovation, Systems and Technologies

Volume 150

The Smart Innovation, Systems and Technologies book series encompasses the topics of knowledge, intelligence, innovation and sustainability. The aim of the series is to make available a platform for the publication of books on all aspects of single and multi-disciplinary research on these themes in order to make the latest results available in a readily-accessible form. Volumes on interdisciplinary research combining two or more of these areas is particularly sought.

The series covers systems and paradigms that employ knowledge and intelligence in a broad sense. Its scope is systems having embedded knowledge and intelligence, which may be applied to the solution of world problems in industry, the environment and the community. It also focusses on the knowledge-transfer methodologies and innovation strategies employed to make this happen effectively. The combination of intelligent systems tools and a broad range of applications introduces a need for a synergy of disciplines from science, technology, business and the humanities. The series will include conference proceedings, edited collections, monographs, handbooks, reference books, and other relevant types of book in areas of science and technology where smart systems and technologies can offer innovative solutions.

High quality content is an essential feature for all book proposals accepted for the series. It is expected that editors of all accepted volumes will ensure that contributions are subjected to an appropriate level of reviewing process and adhere to KES quality principles.

** Indexing: The books of this series are submitted to ISI Proceedings, EI-Compendex, SCOPUS, Google Scholar and Springerlink **

More information about this series at http://www.springer.com/series/8767

César Benavente-Peces ·
Sami Ben Slama · Bassam Zafar
Editors

# Proceedings of the 1st International Conference on Smart Innovation, Ergonomics and Applied Human Factors (SEAHF)

 Springer

*Editors*
César Benavente-Peces
Escuela Técnica Superior de Ingeniería y
Sistemas de Telecomunicación
Universidad Politécnica de Madrid
Madrid, Spain

Sami Ben Slama
Jeddah Community College
King Abdulaziz University
Jeddah, Saudi Arabia

Bassam Zafar
Department of Information Systems
King Abdulaziz University
Jeddah, Saudi Arabia

ISSN 2190-3018          ISSN 2190-3026   (electronic)
Smart Innovation, Systems and Technologies
ISBN 978-3-030-22963-4      ISBN 978-3-030-22964-1   (eBook)
https://doi.org/10.1007/978-3-030-22964-1

This Springer imprint is published by the registered company Springer Nature Switzerland AG
The registered company address is: Gewerbestrasse 11, 6330 Cham, Switzerland

# Preface

## Smart Innovation, Ergonomics and Applied Human Factors

Nowadays, technology surrounds every human activity supporting, monitoring and supervising equipment, facilities, commodities, industry, business and individuals' health. As technology evolves new applications, methods, techniques arise but also what citizens expect from technology.

In order to attend the insatiable demand for new applications, better performance and higher reliability, trustfulness, security, and power consumption efficiency, engineers and scientists must face those challenges based on **smart innovation**, i.e. developing the best techniques and technologies in a friendly way with individuals and environment.

This book contains the proceedings of the 1st edition of **International Conference on Smart Innovation, Ergonomics and Applied Human Factors** (SEAHF 2019) which targets different scientific fields and invites academics, researchers and educators to share innovative ideas and expose their works in the presence of experts from all over the world. The Conference is organized under the supervision of the General Chairman, Prof. César Benavente-Peces and the Co-Chairmen Prof. Sami Ben Slama (Tunisia) and Prof. Bassam Zafar (Saudi Arabia).

SEAHF focuses on original research and practice-driven applications encouraging the authors to extend and improve their research activity seeking for smart and innovative ideas while considering human factors as a road map to reach the excellence in research–development–innovation. SEAHF is aimed at providing a common link between a vibrant scientific and research community and industry professionals by offering a clear view of current real-life problems and how new challenges are faced supported by information and communication technologies. SEAHF is organized by setting a balance between innovative industrial approaches and original research work while it doesn't forget relevant issues affecting citizens:

individual and collective rights, security, pollution, health, applications and new technologies.

**Real-life issues** are the goal of this conference including industrial activity, energy generation, education, business and health. In order to cover such relevant areas on people life, the conference is organized into eight tracks: Smart technologies and Artificial Intelligence (SAI), Green Energy Production and Transfer Systems (GETS), Aerospace Engineering/Robotics and IT (AERIT), Information Security and Mobile Engineering (ISME), IT in Bio-Medical Engineering and smart agronomy (BESA), IT, Smart Marketing, Management and Tourism Policy (SMTP), Technology and Education (TE) and Hydrogen and Fuel cell energy technologies (HFCET).

Additionally, the conference is an opportunity for students developing their Ph. D. Thesis, given they will share their ideas with experts from different sectors: academics, researchers and the industry. In this way, they can discuss their ideas and get onsite relevant remarks to improve or re-conduct the investigation.

In order to guarantee the quality of the accepted and published papers, several committees have been created: Associated Editors Committee, Advisory Committee, Organizing Committee, International Program Committee, Local Committee, and Conference Support Members. Furthermore, every paper has been reviewed at least by two researchers whose expertise includes the topic os the assigned papers. A total of 142 papers were submitted to the SEAHF 2019 conference from a number of institutions all around the world. The review process was exhaustive and 85 papers were accepted for oral presentation and publication in the proceedings.

Madrid, Spain                                                    César Benavente-Peces
Jeddah, Saudi Arabia                                               Sami Ben Slama
Jeddah, Saudi Arabia                                                 Bassam Zafar
December 2018

# Organization

## General Chair

César Benavente-Peces      Universidad Politécnica de Madrid, Spain

## Co-chairs

Sami Ben Slama      King AbdulAziz University, Saudi Arabia
Bassam Zafar      King AbdulAziz University, Saudi Arabia

## Associated Editors Committee

Fariborz Haghighat      Canada
Luis Fernandez-Ramirez      Spain
David Luengo-Garcia      Spain
Balas Valentina Emilia      Romania
Farouk Kamoun      Tunisia
Kamal Mansour Jambi      KSA
Ayfer Veziroglu      USA
Maria Gabriella Xibilia      Italy
Riccardo Caponetto      Italy
Ali Sayigh      UK
Radhi Mhiri      Canada
Ali Eltamaly      KSA
Radhi Mhiri      Canada
Kort Bremer      Germany
Ruan Delgado Gomes      Brasil
Krim LOUHAB      Algeria
Abderrahman Khechekhouche      Algeria
Khaled Ghedir      Tunisia
Vikrant Sharma      Quantum University
Hafaifa Ahmed      Djelfa University, Algeria

| Nura Usman Danzaki | Tunisia |
| Kouzou Abdellah | Djelfa University, Algeria |
| Vincenzo Piuri | CAO IEEE Member |
| Vivek Kumar | Vice Chancellor Quantum University |

## Advisory Committee

| Ghada Amer | Egypt/UAE |
| Ahmed Abdu Alwahab | KSA |
| Ana Stevanovi | Serbia |
| Marcin Paprzycki | Poland |
| Vania V. Estrela | Brazil |
| Khaled Ghedira | Tunisia |
| Fatima Kies | Italy |
| Mimouni Mohamed Faouzi | Tunisia |
| Khalil Kassmi | Morocco |
| Francisca Segura | Spain |
| Adnane Cherif | Tunisia |
| Ashish Khanna | India |
| Dhaker Abbes | France |
| Josep M. Guerrero | Denmark |
| Mohamed NjibMansouri | Tunisia |
| Lotfi Hassine | Canada |
| Soumi Dutta | India |
| Jude Hemanth | India |
| Rachid Benifoughal | Algeria |
| Ahmed dhouib | Tunisia |
| Hadiyanto | Indonisia |
| Katarzyna Barbara Szymczyk | Poland |
| Maria Gabriella Xibilia | Italy |
| Riccardo Caponetto | Italy |
| Kort Bremer | Germany |
| David Luengo-Garcia | Spain |
| Ruan Delgado Gomes | Brasil |
| Ridha Bouallegue | |
| Donia Ammam | |
| Le Hoang Son | |
| Ekaterina Auer | |
| Ahmed Hfaifa | |
| Bright Keswani | |
| Jilali Antari | |

## Organizing Committee

| | |
|---|---|
| Sihem Nasri | Tunisia |
| Houda Jouini | Tunisia |
| Omar Jebali | Finland |
| Nouha Mansouri | Tunisia |
| Mubarek Rahmoun | KSA |
| Salsabil Gherairi | Tunisia |
| Jihen Bokri | USA |
| Fatima Kies | Italy |
| Thamer Ibrahim | KSA |

## International Program Committee

| | |
|---|---|
| Aditya Prasad Padhy | India |
| Lila Ghomri | Algeria |
| Sami Awani | Tunisia |
| Fatima Kies | Italy |
| Imene Ben Trad | France |
| Slim Ayadi | France |
| Fouzi Bellalem | UK |
| Abdulmuttalib T. Rashid | Iraq |
| Mankour Mohamed | Algeria |
| Gisele Ferreira Souza | Brazil |
| Nawel Souiss | Tunisia |
| Nabil Litayem | Tunisia |
| Kun Harismah | Indonesia |
| Abderrahmene Sellami | Tunisia |
| Abdulaziz Sahbani | Tunisia |
| Nabil Ben Slimane | Tunisia |
| Merieme Jaber | Maroc |
| Abdessamad Malaoui | Maroc |

## Local Committee

| | |
|---|---|
| Ana Belen Garcia-Hernando | Universidad Politécnica de Madrid |
| David Luengo-Garca | Universidad Politécnica de Madrid |
| Francisco Javier Ortega Gonzlez | Universidad Politécnica de Madrid |
| Jose Manuel Pardo-Martin | Universidad Politécnica de Madrid |

## Conference Support Members

| | |
|---|---|
| Friedhelm Steffen Hamann | Germany |
| Diana de Falco Alfano | Italy |
| Laura Spies | Gemany |
| Guillermo Ampudia Correa | Mexico |
| Franz Michelle Garcia Boada | Venezuela |
| Patricia Fratilescu | Rumania |

| Omar Jebali | Finland |
| Nouha Mansouri | Tunisia |
| Merieme Jaber | Morocco |
| Aymen Elamraoui | Tunisia |

## Additional Reviewers

Abdelilah Chtaini
Abdellah Kouzou
Abdessamad Malaoui
Abdessamad Tounsi
Aboulkas Adil
Abounada Abdelouahed
Adnen Cherif
Agrawal Rashmi
Ahrens Friedhelm
Alam Aftab
Alghazzawi Daniyal
Amer Ghada
Amir Soumia
Antari Jilali
Attia Nour
Auer Ekaterina
B. S. Abhigna
Bahi Lahoucine
Barguigua Abouddihaj
Barkouti Wahid
Baslam Mohamed
Belhouideg Soufiane
Ben Salem Anis
Ben Slama Sami
Benabdallah Ibrahim
Benavente Cesar
Berkani Mohamed
Bita Hassan
Bokri Jihen
Boudaia El Hassan
Bouissane Latifa
Boulbaroud Samira
Bremer Kort
Caponetto Riccardo
Castilla Mara del Mar
Chabchoub Abdelkader
Chakraborty Sudeshna
Chelangat Mercy
Chihab Yazough

Chikh Khalid
Cosovic Marijana
Danzaki Nura
De Souza Gisele
Derbal Kerroum
Devasia Reneesh
Dhaker ABBES
Dhouib Ahmed
Dutta Soumi
El Alaoui Aicha
El Amraoui Aymen
El harfi Khalifa
El Yousfi Hicham
Elegbede Isa
Elngar Dr. Ahmed
Fernndez-Ramrez Luis M.
Garba Ibrahim
Gomes Ruan
Guetari Ramzi
Hadiyanto Hadiyanto
Hafaifa Ahmed
Hamamouch Noureddne
Harismah Kun
Helaly Amira
Hicham Sahmouda
Ibrahim Tamer
Jouini Houda
Kassmi Khalil
Kebir Chaji
Keswani Dr. Bright
Khechekhouche Abderrahmane
Khlifa Nawres
Khorchani Abdelghaffar
Kies Fatima
Klilou Abdessamad
Krimissa Samira
Louhab Krim
Luengo David
Mahmood Muhammad Habib

Mehdaoui Youness
Mohamed Merzouki
Mostafa Bouzaid
Mouayn Zouhair
Moubarik Amine
Mueen Ahmed
Mustapha Namous
Nachaoui Mourad
Nasri Loubna
Nasri Sihem
Nouha Mansouri
Nouri Zineb
Outanoute Mohamed
Ozakturk Meliksah
Padhy Dr. Neelamadhab
Paprzycki Marcin
Piuri Vincenzo
Rabeh abbassi
Rahmoun Mbarek
Rayyan Mohammad

Rhazi Youssef
Saheb Djohra
Salsabil
Sboui Noureddine
Sharma Vikrant
Snchez Delgado Mnica
Srikanth Panigrahi
Stouti Abdelkader
Sultan Kiran
Szymczyk Katarzyna
Tahri Ali
Tanane Omar
Touati Rabeb
Xibilia Maria Gabriella
Yahyaoui Imene
Yousef Malik
Zafar Bassam
Zougagh Hicham
Zriouel Sanae

# Contents

**Control and Management of Hybrid Renewable Energy System** . . . . . .    1
Abderrahmane Djellouli, Fatiha Lakdja, and Meziane Rachid

**Motion Detection in Digital Video Recording Format with Static
Background** . . . . . . . . . . . . . . . . . . . . . . . . . . . . . . . . . . . . . . . . . . 11
Fahd Abdulsalam Alhaidari, Atta-ur-Rahman, Anas Alghamdi,
and Sujata Dash

**Automatic Text Categorization Using Fuzzy Semantic Network** . . . . . . . 24
Jamal Alhiyafi, Atta-ur-Rahman, Fahd Abdulsalam Alhaidari,
and Mohammed Aftab Khan

**An Empirical Approach on Absorption of EMI Radiations
by Designing Elliptical Microstrip Patch Antenna** . . . . . . . . . . . . . . . . 35
Mandeep Singh Heer and Vikrant Sharma

**Quantitative Analysis of Software Component's Reusability** . . . . . . . . . 42
Nisha Ratti and Parminder Kaur

**Improve the Tracking of the Target in Radar System Using
the Particular Filter** . . . . . . . . . . . . . . . . . . . . . . . . . . . . . . . . . . . . 55
Zarai Khaireddine, BenSlama Sami, and Cherif Adnane

**Robot Intelligent Perception Based on Deep Learning** . . . . . . . . . . . . . 63
Sehla Loussaief and Afef Abdelkrim

**Performances Study of Speed and Torque Control of an
Asynchronous Machine by Oriented Stator Flux** . . . . . . . . . . . . . . . . . 71
M. Benmbarek, F. Benzergua, and A. Chaker

**If You Want to Well Understand Today, You Have to Search History,
Babylon is a Good Example** . . . . . . . . . . . . . . . . . . . . . . . . . . . . . . . 85
Emad Kamil Hussein

**WhatsApp for Defamiliarizing Foundation Year English Learners: A Collaborative Action Research Evaluation** ..................... 91
Najat Alsowayegh and Ibrahim Garba

**Exploratory Study on SDI-ICMM Implementation: Case Study Tunisia** ......................................... 98
Abdelghaffar Khorchani, Fatima Kies, and Cesare Corselli

**Studying the Corrosion Behavior of Pd–Cu Alloy Used in Hydrogen Purification Systems** ...................................... 114
Hatem M. Alsyouri, Farqad F. Al-Hadeethi, and Sohaib A. Dwairi

**New Parametrization of Automatic Speech Recognition System Using Robust PCA** ......................................... 119
Sonia Moussa, Zied Hajaiej, and Ali Garsallah

**Voltage and Reactive Power Enhancement in a Large AC Power Grid Using Shunt FACTS Devices** .............................. 125
Mankour Mohamed and Ben Slama Sami

**Concept of Internal Momentum Absorber Structure to Reduce Impact During Accidents** ......................................... 144
Mandip Kumar Nar, Prabhjeet Singh, and Paprinder Singh

**Denoising Magnetic Resonance Imaging Using Fuzzy Similarity Based Filter** ............................................. 150
Bhanu Pratap Singh, Sunil Kumar, and Jayant Shekhar

**Augmented Reality in Children's Education in the Republic of Macedonia** .......................................... 165
Marija Minevska and Smilka Janeska-Sarkanjac

**Impact of Quality Management on Green Innovation: A Case of Pakistani Manufacturing Companies** ........................ 169
Tahir Iqbal

**Hamming Distance and K-mer Features for Classification of Pre-cursor microRNAs from Different Species** .................. 180
Malik Yousef

**Hybrid Feature Extraction Techniques Using TEO-PWP for Enhancement of Automatic Speech Recognition in Real Noisy Environment** ......................................... 190
Wafa Helali, Zied Hajaiej, and Adnen Cherif

**Investigation on the Behavior of Detonation Gun Sprayed Stellite-6 Coating on T-22 Boiler Steel in Actual Boiler Environment** .......... 196
Yogesh Kumar Sharma, Anuranjan Sharda, Rahul Joshi, and Ketan Kakkar

**ICI Reduction Using Enhanced Data Conversion Technique with**
**1 × 2 and 1 × 4 Receive Diversity for OFDM Systems** . . . . . . . . . . . . . 211
Vaishali Bahl, Vikrant Sharma, Gurleen Kaur, and Ravinder Kumar

**Architecture Analysis and Specification of the RacingDrones**
**Platform: Online Video Racing Drones** . . . . . . . . . . . . . . . . . . . . . . . 221
César Benavente-Peces, David Tena-Ramos, and Ao Hu

**Optimised Scheduling Algorithms and Techniques**
**in Grid Computing** . . . . . . . . . . . . . . . . . . . . . . . . . . . . . . . . . . . . . 231
Rajinder Vir, Rajiv Vasudeva, Vikrant Sharma, and Sandeep

**Deep Learning Algorithm for Predictive Maintenance**
**of Rotating Machines Through the Analysis of the Orbits Shape**
**of the Rotor Shaft** . . . . . . . . . . . . . . . . . . . . . . . . . . . . . . . . . . . . . . 245
R. Caponetto, F. Rizzo, L. Russotti, and M. G. Xibilia

**False Data Injection in Smart Grid in the Presence of Missing Data** . . . 251
Rehan Nawaz, Muhammad Awais Shahid,
and Muhammad Habib Mahmood

**A Predictive Real-Time Energy Management Control for a Hybrid**
**PEMFC Electric System Using Battery/Ultracapacitor** . . . . . . . . . . . . . 258
Nasri Sihem, Ben Slama Sami, Bassam Zafar, and Cherif Adnane

**Social Networkopia: Metamorphosing Minds with Hypnotic Halo** . . . . . 266
Neetu Vaid and Disha Khanna

**Design of the Furniture and Fixture Smart Automatic Batch**
**Simulation Mobile Application** . . . . . . . . . . . . . . . . . . . . . . . . . . . . . . 270
Jung-Sook Kim, Taek-Soo Heo, and Tae-Sub Chung

**Performance Evaluation of an Appraisal Autonomous System**
**with Hydrogen Energy Storage Devoted for Tunisian**
**Remote Housing** . . . . . . . . . . . . . . . . . . . . . . . . . . . . . . . . . . . . . . . . 274
Ben Slama Sami, Nasri Sihem, Bassam Zafar, Cherif Adnane,
and A. Elngar Ahmed

**A Technology Centric Strategic Approach as Decision Support**
**System During Flood Rescue for a Better Evacuation**
**and Rehabilitation Plan** . . . . . . . . . . . . . . . . . . . . . . . . . . . . . . . . . . . 282
Mohammad Nasim and G. V. Ramaraju

**Specification of the Data Warehouse for the Decision-Making**
**Dimension of the Bid Process Information System** . . . . . . . . . . . . . . . . 289
Manel Zekri, Sahbi Zahaf, and Faiez Gargouri

**Insights in Machine Learning for Cyber-Security Assessment** . . . . . . . . 296
César Benavente-Peces and David Bartolini

**Application of 3D Symbols as Tangible Input Devices** . . . . . . . . . . . . . . 306
Hirofumi Shishido and Rentaro Yoshioka

**Portable Braille Reading Device with Electromagnetic Actuators** . . . . . 310
Daniel Aparicio, Andhers Piña, and Alexis Bustamante

**Modeling and Simulation of Renewable Generation System: Tunisia
Grid Connected PV System Case Study** . . . . . . . . . . . . . . . . . . . . . . . . 316
Mansouri Nouha, Bouchoucha Chokri, and Cherif Adnen

**Privacy and Security Aware Cryptographic Algorithm for Secure
Cloud Data Storage** . . . . . . . . . . . . . . . . . . . . . . . . . . . . . . . . . . . . . . 323
B. Muthulakshmi and M. Venkatesulu

**Sequential Hybridization of GA and PSO to Solve the Problem
of the Optimal Reactive Power Flow ORPF in the Algerian Western
Network (102nodes)** . . . . . . . . . . . . . . . . . . . . . . . . . . . . . . . . . . . . . . 338
I. Cherki, A. Chaker, Z. Djidar, N. Khalfellah, and F. Benzergua

**Combined Use of Meta-heuristic and Deterministic Methods
to Minimize the Production Cost in Unit Commitment Problem** . . . . . . 347
Sahbi Marrouchi, Nesrine Amor, and Souad Chebbi

**Development and Experimental Validation of a Model to Simulate an
Alkaline Electrolysis System for Production of Hydrogen Powered
by Renewable Energy Sources** . . . . . . . . . . . . . . . . . . . . . . . . . . . . . . . 358
Mónica Sánchez, Ernesto Amores, David Abad, Carmen Clemente-Jul,
and Lourdes Rodríguez

**Building a Digital Business Technology Platform in the Industry
4.0 Era** . . . . . . . . . . . . . . . . . . . . . . . . . . . . . . . . . . . . . . . . . . . . . . . . 369
Maurizio Giacobbe, Maria Gabriella Xibilia, and Antonio Puliafito

**Single-Phase Grid Connected Photovoltaic System Using Fuzzy
Logic Controller** . . . . . . . . . . . . . . . . . . . . . . . . . . . . . . . . . . . . . . . . . 376
Abdelaziz Sahbani

**Synthesis of Sit-to-Stand Movement Using SimMechanics** . . . . . . . . . . . 386
Samina Rafique, M. Najam-l-Islam, and A. Mahmood

**Comparative Study of MPPT Algorithms of an Autonomous
Photovoltaic Generator** . . . . . . . . . . . . . . . . . . . . . . . . . . . . . . . . . . . . 393
Troudi Fathi, Houda Jouini, and Abdelkader Mami

**A Comparative Analysis of DPC and SMC-DPC-SVM Control
Approaches in Three-Phase Electrical Power Systems** . . . . . . . . . . . . . . 401
Maha Zoghlami and Faouzi Bacha

**Design of Intelligent Controllers, for Sun Tracking System Using Optical Sensors Networks** ................................... 414
Abdelaziz Sahbani, Ali Hamouda Ali Saeed, and Abdullah ALraddadi

**A Conceptual Model of Cyberterrorists' Rhetorical Structure in Protecting National Critical Infrastructure** ................... 421
Khairunnisa Osman, Ala Alarood, Zanariah Jano, Rabiah Ahmad, Azizah Abdul Manaf, and Marwan Mahmoud

**MP3 Steganalysis Based on Neural Networks** ................... 428
Marwan Mahmoud and Alaa Abdulsalam Alarood

**Author Index** ............................................... 437

# Control and Management of Hybrid Renewable Energy System

Abderrahmane Djellouli[1(✉)], Fatiha Lakdja[2], and Meziane Rachid[3]

[1] Electro-Technical Engineering Laboratory, Faculty of Technology, University of Saida, Saida, Algeria
djellouli7abderrahmane@gmail.com
[2] ICEPS Laboratory, University of Sidi-Bel-Abbes, Sidi Bel Abbès, Algeria
flakdja@yahoo.fr
[3] Faculty of Technology, University of Saida, Saida, Algeria

**Abstract.** The importance of energy from renewable resources, such as wind and solar, is increasing and their penetration rate in power increases each year due to several factors. Firstly, the perpetual rise in demand, particularly because of population growth and economic development. Second, pledges made by many governments to increase their reliance on renewable sources of energy, with a view to reducing the devastating consequences of climate change on the environment. The multiplication of decentralized production connected to the low-voltage power grid causes the appearance of a bidirectional energy flow. This is at the origin of many electrical phenomena that are increasingly difficult to manage it by distribution system operators. An innovative solution consists in controlling the integration of renewable energies and managed the flow of the powers for a different source. This work covers integration of renewable energy into the public grid for hybrid system. Or more precisely, a microgrid that contains two renewable energy sources (PV + wind), battery and public network all the system is connected in a residential charge. Furthermore, the control of all devices for this integration, management of load, renewable energy and public grid.

**Keywords:** Renewable energy · Wind · Solar · Management · Microgrid

## 1 Introduction

Today, in many countries, there is a sharp increase in decentralized generation sources (solar photovoltaic, wind, hydroelectric, heat-power coupling, etc.) with the addition of the storage system and the diesel generator that are connected to the utility grid. The multiple combination of these sources that are called hybrid power systems.

Although positive, the multiplication of these sources of energy connected to the low voltage (LW) electrical network also has limits. In fact, the decentralized injection of electricity into the grid (by fossil and/or renewable sources) is a factor of instability.

This is due to the presence of a final consumer who, having become both a producer and a user of electricity, causes the appearance of a bidirectional energy flow

© Springer Nature Switzerland AG 2019
C. Benavente-Peces et al. (Eds.): SEAHF 2019, SIST 150, pp. 1–10, 2019.
https://doi.org/10.1007/978-3-030-22964-1_1

using an infrastructure that was not designed, at the same time, originally, to welcome him.

Paradoxically, a good coordination of these different sources of injection would better manage power flows and have better oversight of the quality of electricity [1]. The general trend is therefore to move towards the use of smart grid or namely micro-grid. Research in the field of microgrid system has been reviewed in the preview studies using the computer tool and control strategy [2].

This work aims to design a simple micro-grid that includes two renewable energy sources (PV + Wind) and a storage battery, this system is connected in a public grid and do not forget the control system. The study is done in MATLAB interface.

## 2  Micro-grid

With the progress of human society, electric power has been the symbol of modern civilization. Power supply reliability and power quality have become more and more important. Facing pressures from traditional resource depletion and environmental pollution, power generation methods based on fossil fuel and the centralized power supply mode have been difficult to meet the requirements of economic and social development. So, to facilitate access to energy, local-scale power generating and consumption systems called microgrids are gradually being introduced [3].

Microgrid idea is generally developed in countries such as the United States, Canada, Japan and the United Kingdom. It has been investigated and implemented [4]. Micro-grid is defined as a system that consisting of renewable sources integrate in the electrical grid, small in size, it comprises low voltage (LV) system with distributed energy resources (DERs) together with storage devices and flexible loads, which can be operated in either on-grid (grid connected) or off-grid (islanded) mode of operation [5, 6].

Design of a microgrid can be differentiated according to the applications and expectations. Some microgrids are developed to achieve high end use reliability and stability. Reliability is a common property for a microgrid. Microgrids are able to use more reliable and controllable energy sources like diesel sets, natural energy sources etc. to ensure reliability and stability of the system. Most of these types are implemented as a backup system for large industries and military stations located in remote areas. Some others will integrate more renewable and waste energy resources. Overcoming limitations introduced by these kinds of resources is a challenging task. It is seen that there is no clear common architecture for microgrids. So, design, implementation and operation of a microgrid are differentiated according to the purpose [7].

A microgrid is typically made up of:

- Renewable energy sources (solar, wind or biomass).
- Fossil fuel energy sources to ensure grid stability.
- Energy storage solutions (batteries, hydrogen storage, mechanical storage, etc.).
- A low-voltage supply grid regulated by a smart control system.

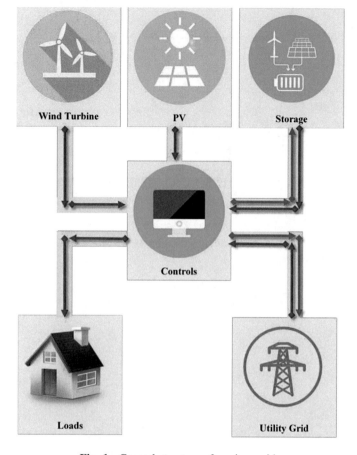

**Fig. 1.** General structure of a micro-grid.

"Figure 1," represents the general design of a micro-grid system.

## 3   Photovoltaic System

Photovoltaic solar energy comes from the conversion of sunlight into electricity due to the photovoltaic effect [8, 9]. When the photons of sunlight come into contact with the semiconductor materials, they are given the necessary energy for the electron to move from one band to another, this movement produces an electric current [10]. This continuous micropower current calculated in watt peak (Wc) can be converted into alternating current due to an inverter.

The equivalent circuit of a photovoltaic panel is shown in Fig. 2. It includes a current source, a diode, a series resistor and a shunt resistor [11].

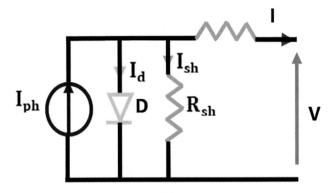

**Fig. 2.** Electrical model of a PV panel.

Based on the circuit Above and using Kirchhoff law, the current generated by the solar panel can be determined [12], as in (1).

$$I = I_{ph} - I_0 \left( \exp \frac{q(V + R_S I)}{aKTN_S} - 1 \right) - \frac{(V + R_S I)}{R_P} \tag{1}$$

where

$$I_{ph} = (I_{SC} + K_i(T - 298.15)) - \frac{G}{1000} \tag{2}$$

$$I_0 = \frac{I_{SC} + K_i(T - 298.15)}{\exp\left(\frac{q(V_{oc} + K_v(T - 298.15))}{aKTN_s}\right) - 1} \tag{3}$$

where $I_{pv}$ and $I_0$ are the photovoltaic and saturation currents of the array; $I_m$ and $V_m$ represent the voltage and the current at the terminals of the module; $R_s$ and $R_p$ are series and shunt resistance; T mean the ambient temperature in Kelvin; G [W/m$^2$] is the irradiation on the device surface; q represents the charge of the electron; a is the factor of ideality.

## 4   Wind Power System

Wind energy is a renewable source energy that transforms the wind's kinetic energy into electrical energy. "Figure 3," shows the operation of this energy.

The model is based on the steady-state power characteristics of the turbine. The output power of the turbine is given by the following (4) [13].

$$P_m = \frac{1}{2} C_P \times S \times \rho \times V^3 \tag{4}$$

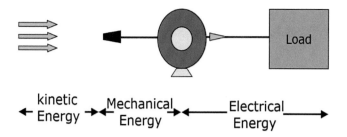

**Fig. 3.** The operation principle of wind energy.

where

$P_m$ : Mechanical output power of the turbine (W).
$C_P$ : Performance coefficient of the turbine.
$\rho$ : Air density (kg/m$^3$).
$S$ : The Total Blade Area swept by the rotor blades (m$^2$).
$V$: The wind velocity (m/s).

## 5 Model Proposed

The proposed model of this work using MATLAB is illustrated in Fig. 4. The micro-grid contains a photovoltaic system, wind, public grid and a battery that are connected to a load.

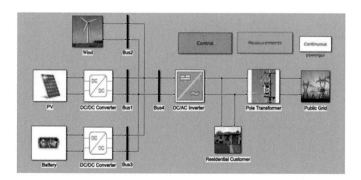

**Fig. 4.** Proposed micro-grid model.

The photovoltaic system is characterized by a power of 4.5 kW and a voltage of 290 V for a temperature of 25 °C and an irradiance of 1000 W/m$^2$. The Maximum Power Point Tracking (MPPT) control takes maximum power from the solar panel using a unidirectional DC/DC converter.

The wind turbine in this model generates a power of 3 kW and a voltage of 440 V, this source is connected in an AC/DC rectifier.

A controller consisting of a power conditioning system and a battery control system is also connected directly to the system.

The power grid is connected to the system by a pole-mounted distribution transformer it can absorb or demand energy from the system.

The previous three elements (PV + Wind + Battery) that are putting in cascade, they are connect to a bidirectional DC/AC inverter which connected in series with a public grid.

Three loads of 1.5 kW can be connected to the system. They consume a total of 6 kW of electrical power when All of them are connected.

## 6 Results and Discussion

### 6.1 Results

After the execution of the proposed model shown in Fig. 4. We obtained the following results:

"Figure 5," shows the operating time of each load, such that load 1 operates between {0.25–0.75 s and 1.25–1.75 s}, load 2 operates between {0.5–0.75 s and 1.5–1.75 s} and load 3 works between 0.75 and 1.25 s.

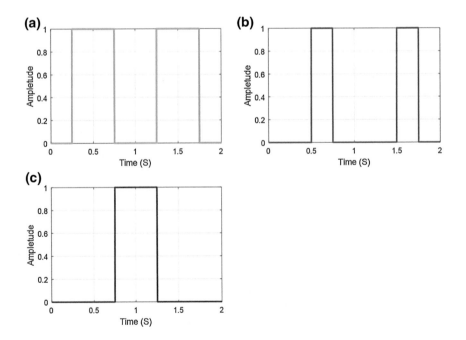

**Fig. 5.** The operation time for the three loads: **a** Load1, **b** Load2, **c** Load3.

"Figure 6," shows the operating time of the grid and battery. We managed the time so that if the grid works, the batterie is disconnected and vice versa. This management makes it possible to maintain the continuity of service.

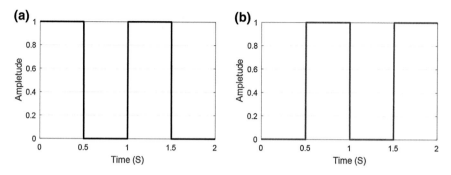

**Fig. 6.** The operation time for: **a** The grid, **b** The battery.

"Figure 7," shows the voltage and current profile for the grid in the left and current in the right.

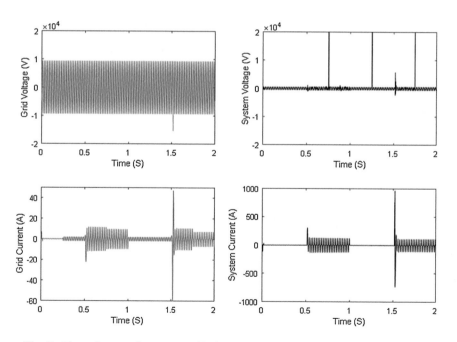

**Fig. 7.** The voltage and current profile for: the grid (blue), the hybrid system (dark).

"Figure 8," shows the power profile of the battery in right and the grid in left. "Figure 9," shows the power profile of the three loads.

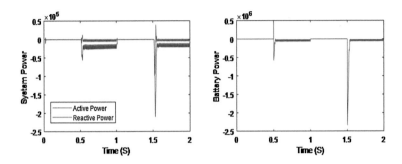

**Fig. 8.** The power profile of the battery (right) and the hybrid system (left).

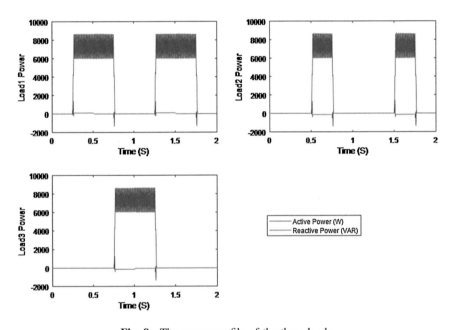

**Fig. 9.** The power profile of the three loads.

## 6.2   Discussion

1. Grid connected

- The battery is not connected to the system and the charges are not activated.
- The power generated by the hybrid system is sent to the public grid.
- The power generated by the hybrid system is kept approximately constant at the nominal power due to the integration of the control system.

2. Battery connected

- The grid not yet connected.
- The power generated by the system is negative because all three loads are connected. The system requires an additional energy from the grid.
- The power generated by the system hybrid rest constant.

Load 1 et 2 then disconnected, load 3 is disconnected at time (t = 1.25 s). Then the battery is connected to the system. Therefore, we can disconnect the public grid from management system because the hybrid system (PV + Wind + Battery) is sufficient to provide the total load.

# 7   Conclusion

In this article we started by giving a generality on the micro-grids system. Then we present the two renewable energy sources (PV + Wind) that we used in this work.

This system consists of a photovoltaic source, a small wind source, a storage unit that is the fuel cell and a load of 6 kW, the whole is connected to a DC bus.

We realized this global system in MATLAB Simulink, with the setting of the initial variables of each component. In addition, we have installed a process to control all system variables, such as DC bus voltage control, MPPT control, control of the DC/AC converter for the synchronization of the AC output with the public grid. As we have installed a power management system et de control from all sources that is in the charge, battery and public grid.

Finally, the simulation is launched in a duration of 2 s, the results obtained are satisfactory fact that, the following functions are provided:

- The control of system components and variables facilitates the integration of renewable energies in public grid and guarantees operation even in adverse cases.
- The installation of a power management system plays a very important role not only to promote the use of renewable energies but also the reduction of the use of the public network (consumer become both a producer and a user of electricity).

In perspective, we propose to work for a system of great power, the integration of other communication devices and energy storage such as batteries and super capacity. The medium-term realization of this study by the creation of real management models of production of the powers.

# References

1. Jaton, J., Besson, G., Carpita, M., Ducommun, D.: Gestion intelligente et autonome du réseau électrique basse tension. Electrosuisse, Technopark, Zürich, 17.Oktober 2013
2. Nehrir, M.H., Wang, C., Strunz, K., Aki, H., Ramakumar, R., Bing, J., Miao, Z., Salameh, Z.: A review of hybrid renewable/alternative energy systems for electric power generation: configuration, control, and applications. IEEE Trans. Sustain. Energy 2(4), 392–403 (2011)

3. Xu, Y., Shi, Z., Wang, J., Hou, P.: Discussion on the factors affecting the stability of microgrid based on distributed power supply. Energy Power Eng. **5**(4B), 1344–1346 (2013)
4. Moradian, M., Tabatabaei, F.M., Moradian, S.: Modeling, control and fault management of microgrids. Smart Grid Renew. Energy **4**(1), 99–112 (2013)
5. Luu, N.A.: Control and management strategies for a microgrid, Saint-Martin-d'Hères, Grenoble University, 18 Dec 2014
6. Vinayagam, A., Swarna, K.S.V., Khoo, S.Y., Oo, A.T., Stojcevski, A.: PV based microgrid with grid-support grid-forming inverter control-(simulation and analysis). Smart Grid Renew. Energy **8**, 1–30 (2017)
7. Ariyasinghe, M., Hemapala, K.: Microgrid test-beds and its control strategies. Smart Grid Renew. Energy **4**(1), 11–17 (2013)
8. Venkateshkumar, M., Indumathi, R., Poornima, P.: Photovoltaic cell power generation for stand-alone applications. In: IEEE-International Conference on Advances in Engineering, Science and Management (ICAESM-2012), pp. 57–62
9. Indumathi, R., Venkateshkumar, M., Raghavan, R.: Integration of D-Statcom based photovoltaic cell power in low voltage power distribution grid. In: IEEE-International Conference on Advances in Engineering, Science and Management (ICAESM-2012), pp. 460–465
10. Venkateshkumar, M., Raghavan, R.: Hybrid Photovoltaic and Wind Power System with Battery Management System Using Fuzzy Logic Controller, vol. 5(2), pp. 72–78 (2016)
11. Villalva, M.G., Gazoli, J.R., Filho, E.R.: Modeling and circuit-based simulation of photovoltaic arrays. In: Power Electronics Conference, COBEP'09, Brazilian, pp. 1244–1254 (2009)
12. Bellia, H., Youcef, R., Fatima, M.: A detailed modeling of photovoltaic module using MATLAB. NRIAG J. Astron. Geophys. **3**(1), 53–61 (2014)
13. Kariyawasam, K.K.M.S., Karunarathna, K.K.N.P., Karunarathne, R.M.A., Kularathne, M.P.D.S.C., Hemapala, K.T.M.U.: Design and development of a wind turbine simulator using a separately excited DC motor. Smart Grid Renew. Energy **4**(3), 259–265 (2013)

# Motion Detection in Digital Video Recording Format with Static Background

Fahd Abdulsalam Alhaidari[1], Atta-ur-Rahman[2(✉)], Anas Alghamdi[2], and Sujata Dash[3]

[1] Department of Computer Information System (CIS), Imam Abdulrahman Bin Faisal University, 1982, Dammam, Kingdom of Saudi Arabia
faalhaidari@iau.edu.sa
[2] Department of Computer Science (CS), College of Computer Science and Information Technology (CCSIT), Imam Abdulrahman Bin Faisal University, 1982, Dammam, Kingdom of Saudi Arabia
{aaurrahman,2150009467}@iau.edu.sa
[3] Department of Computer Science, North Orissa University, Baripada, India
sujata238dash@gmail.com

**Abstract.** Nowadays, security of valuable and secret assets is very important for large organization. Need of an efficient and reliable security system arises due to limitation of human resources and manpower. In such systems, video surveillance systems have their own significance. Although some people don't like the idea of being monitored, surveillance systems improve the public security, allowing the system to detect dangers. Multimedia data typically means digital imagesaudio, video, animation and graphics together with text data. The acquisition, generation, storage and processing of multimedia data in computers and transmission over networks have grown tremendously in the recent past. In video surveillance, motion detection refers to the capability of the surveillance system to detect motion, identify the activity and capture the events while preventing the negligible activities. Motion detection is usually a software-based monitoring algorithm which, when it detects motions will signal the surveillance camera to begin capturing the event. Also called activity detection. An advanced motion detection surveillance system can analyze the type of motion to see if it warrants an alarm. In this research, we propose a novel motion detection algorithm for digital video recording (DVR) system using a Fuzzy Rule Based System that detects the motion in the running video intelligently.

**Keywords:** Motion detection · DVR · Fuzzy rule based system · Surveillance

## 1 Introduction

Due to increase of criminal activities CCTV cameras are deployed on various locations to monitor the activities of people, now a day's malicious activities are detected manually, these cameras are not intelligent to detect malicious activity and notify operator therefore a person (operator) is also looking on people activities through these cameras in control room.

© Springer Nature Switzerland AG 2019
C. Benavente-Peces et al. (Eds.): SEAHF 2019, SIST 150, pp. 11–23, 2019.
https://doi.org/10.1007/978-3-030-22964-1_2

In the literature, it is found out that there are increasing research interests in the field of Motion Detection. As described by Kadam (2015) For motion estimation by using dynamic camera, Blob detection algorithm used on binary image from thresholding, which detect the moving object in that image. After the object detection object is registered and track [1]. Zalevsky and Garcia (2014) A method is presented for imaging an object. The method comprises imaging a coherent speckle pattern propagating from an object, using an imaging system being focused on a plane displaced from the object [2].

Boregowda and Ibrahim (2006) A moving object is detected in a Video data stream by extracting color information to estimate regions of motion in two or more sequential Video frames, extracting edge information to estimate object shape of the moving object in two or more sequential Video frames; and combining the color information and edge information to estimate motion of the object [3]. According to Jain and Rajagopal (2007) a system detects motion in video data. In an embodiment, a difference frame is created by comparing the pixels from a first frame and a second frame. The difference frame is divided up into blocks of pixels, and the system calculates standard deviations on a block basis. A threshold value is calculated based on the standard deviation, and the presence or absence of motion is determined based on that threshold value [4].

Living (2008) A video motion detection apparatus includes a mechanism applying a motion test by comparing test areas of an image with respective sets of candidate areas in that or another image and generating motion vectors in dependence on a displacement between each test area and a candidate area giving a greatest similarity between the test area and that candidate area; and a mechanism applying an integrity test to test the motion vectors. For a motion vector failing the integrity test, the corresponding test area is divided into two or more smaller test areas and the motion test is applied again in respect of the two or more smaller areas, and a motion vector passing the integrity test is made available as an output of the apparatus [5].

According to Kavita et al. (2014), recent research in computer has persuaded more studies in human motion observation as well as analysis. Visual analysis of human motion is more active component in computer vision. With the help of video surveillance, we are able to work in this area conveniently. Human motion analysis concerns the detection, tracking, recognition of human activities and human behaviors. This paper helps to comprehend multiple techniques of human motion detection and behavior understanding [6].

From the literature review, it is observed that, currently detecting malicious activities and firing alarm on malicious activities are not mature enough. Various research has been carried out by different researchers to make this system mature so that it could help in reducing unwanted activities and reduce disastrous results [7]. The proposed system is smart enough to identify the malicious activity among other activities. The algorithm will use differential analysis of the video frames for this purpose. Fuzzy rule-based systems (FRBS) are getting popular due to their promising nature for handle the situation that are unclear, wage and missing some of the information [8–15]. In this paper, a FRBS is proposed to estimate the threshold for differential analysis of the video frames to identify the activity with accuracy.

Rest of the paper is organized as follow: Sect. 2 contains proposed technique, Sect. 3 presents the results and discussion while Sect. 4 concludes the paper.

## 2 Proposed Approach

The section contains the proposed approach for motion detection in the DVR systems.

### 2.1 Obtaining Foreground

Motion detection plays a fundamental role in any object tracking or video surveillance algorithm, to the extent that nearly all such algorithms start with motion detection. The reliability with which potential foreground objects in movement can be identified, directly impacts on the efficiency and performance level achievable by subsequent processing stages of tracking and/or recognition. However, detecting regions of change in images of the same scene is not a straightforward task since it does not only depend on the features of the foreground elements, but also on the characteristics of the background such as, for instance, the presence of vacillating elements.

In this section, we will find the motion detection on static scenes, that is, the only elements in movement will be the targets. In that way, it is possible to analyze the activity more precisely. Moreover, our focus is on CCTV footages where most of the cases background remains static.

In Fig. 1, it is shown the difference of two relative frames for sake of activity detection. This is mainly done by taking the difference of two successive frames with ideally a static background.

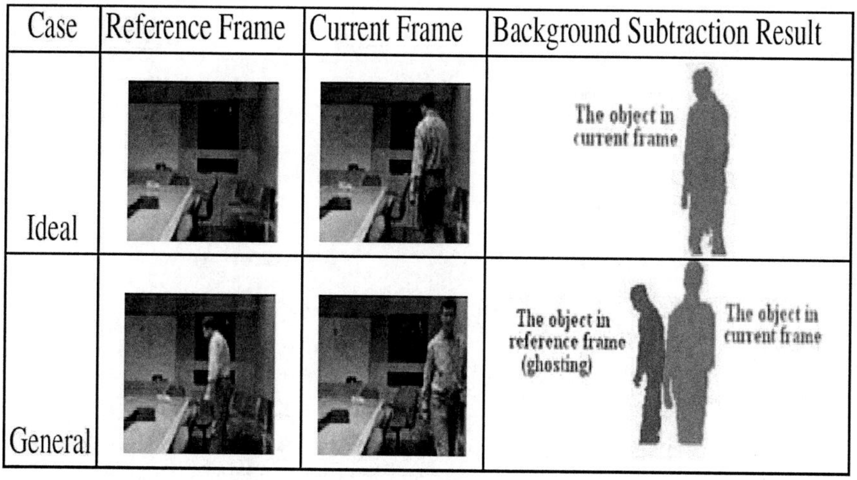

| Case | Reference Frame | Current Frame | Background Subtraction Result |
|------|-----------------|---------------|-------------------------------|
| Ideal | | | The object in current frame |
| General | | | The object in reference frame (ghosting)     The object in current frame |

**Fig. 1.** Background difference results

In figure shows two scenes with and without object in an ideal situation. By taking the difference, all the details are subtracted except the object of interest. In second scenario, there is one more snapshot usually called ghost frame. Usually exists in sharp videos with high frame rate otherwise its effect is negligible.

On the other hand, techniques based on temporally adjacent frames could be considered. Basically, this time-differencing approach suggests that a pixel is moving if its intensity has significantly changed between the current frame and the previous one. That is, a pixel x belongs to a moving object if $|It (x, y) - It - 1(x, y)| < \tau$ where It (x, y) represents the intensity value at pixel position (x, y) at time t and $\tau$ corresponds to a threshold describing a significant intensity change. It − 1(x, y) is the previous frame's pixel's intensity. Usually to avoid the daylight effects this threshold may vary based on the time at the capturing video.

The intensity values are actually the pixel's value as an integer between 0 and 255 if the image is in grayscale. Where intensity value zero corresponds to black and 255 corresponds to white value in terms of intensity. This is mainly comprised as a byte value or an 8-bit value to represent the corresponding pixel's intensity in the frame.

In case of colored images this range may be increased to two bytes (16-bit) or three bytes (24-bit) also called RGB images where R corresponds to Red, G corresponds to Green and B corresponds to Blue. This can also be considered as three separate frames of 8-bit each or one big frame with the pixels having values of 24 bits, ranging between 0 and 224.

Figure 2 shows the difference taking technique considered in this research. In the difference three frames are considered, current, previous and next frame, represented as t, t − 1 and t + 1 respectively.

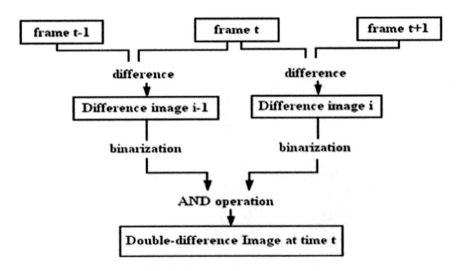

**Fig. 2.** Double difference detection

Three frames are taken into consideration. Previous, middle and next frame. So, in this regard, two differences appeared. To find the optimum value AND operation is carried out. Here AND operation precisely obtains the desired outcome because of its nature.

Algorithm 1 (Migliore 2006), given below (Fig. 3) exploits the same philosophy of detection based on joint difference [16].

---

**Algorithm 1** Joint Difference Algorithm

---

**if** $((|F_t(x) - B_{t-1}(x)| > \tau_B)$ AND $(|F_t(x) - F_{t-1}(x)| > \tau_A))$ **then**
    Foreground Pixel;
**else if** $((|F_t(x) - B_{t-1}(x)| > \tau_B)$ AND $(|F_t(x) - F_{t-1}(x)| < \tau_A))$ **then**
    Collect pixels in blobs;
    **if** ($\sharp$ Foreground Pixels $\geq (\gamma * (\sharp$ Total Pixels)))$ **then**
        Foreground Pixel; //foreground aperture problem solution
    **else**
        Background Pixel; //a background object suddenly starts moving at time $t$
    **end if**
**else if** $((|F_t(x) - B_{t-1}(x)| < \tau_B)$ AND $(|F_t(x) - F_{t-1}(x)| > \tau_A))$ **then**
    Background Pixel; //ghosting problem solution
**else**
    // $|F_t(x) - B_{t-1}(x)| < \tau_B$ AND $|F_t(x) - F_{t-1}(x)| < \tau_A$;
    Background Pixel;
**end if**

---

**Fig. 3.** Algorithm 1

However, the thresholds being used are still to be determine and their values depend on the nature of the application. Moreover, they are static in nature. To improve the accuracy, a fuzzy rule based system is designed to find these thresholds precisely.

## 2.2   Fuzzy Rule Based System for Finding Optimum Thresholds

To obtain the optimum threshold for our proposed algorithm, a fuzzy rule based system is designed (FRBS). The FRBS decides the value of threshold based on three factors.

- **Daylight time**

This factor is important because there are patches of daylight when the change in pixel's intensity is more frequent and it may lead to a misunderstanding. That is why during those hours threshold guess may be different than those hours where day light change is less frequent. That is why it is divided into three classes. Day, night and day-night.

- **Frame Rate**

That is another important factor that may impact the threshold value. If the framerate is significantly high, this means the interframe differences will be less, which must result in a lower threshold and vice versa. In this regard, three classes are suggested that are low medium and high.

- **Neighboring Average**

This is another very important factor that can help us deciding the optimal threshold difference. Neighboring average is an average intensity of the adjacent pixels of a pixel. Say pixel is $(x, y)$ then four connected neighbors are $(x + 1, y), (x - 1, y), (x, y + 1), (x, y - 1)$. Now aggregating the corresponding intensities and after normalizing, if the there is a change that means the pixel $(x, y)$ must also possess that change. In this regard, the neighboring average is also divided into three classes namely low, medium and high.

Based on the values of factors mentioned above, the output value (threshold) is intuitively decided. This is to reduce the false alarms that may occur due to noise and other factors. The classes for threshold are also three named as low, moderate and high in fuzzy world.

### 2.2.1   Input/Output Variables

The proposed fuzzy rule-based system is designed in such a way that it takes values of all factors and provides a suggested threshold. Every factor contributes in its own way which is independent from the other. This is to make the relationship unbiased (Fig. 3).

This can be written as.

Threshold = FRBS(daylight, framerate, neighboring-average)

Here the input to the proposed fuzzy rule based system are the above-mentioned factors that contributes to obtain the final threshold of the given frames, given in Figs. 4, 5 and 6. similarly, the output variable is given in Fig. 7 respectively.

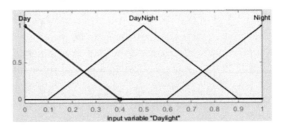

**Fig. 4.**  Input variable "Daylight"

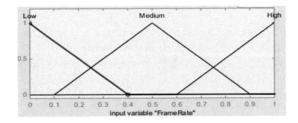

**Fig. 5.**  Input variable "Framerate"

The range of all the variables is between 0 and 1 where 1 corresponds to highest and 0 corresponds to the lowest value. The sets in all the input variables are kept three that are low, medium and high while the number of fuzzy sets in output variable are three as well. Triangular fuzzifiers are used in all the input output variables and

Mamdani Inference Engine (MIE) is selected for the design of underlying proposed Fuzzy Rule Based System.

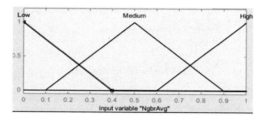

**Fig. 6.** Input variable "Neighboring average"

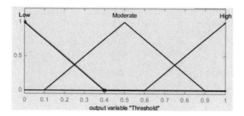

**Fig. 7.** Fuzzy output variable

Figure 8 shows the overall picture of fuzzy rule-based system that is comprised on all input and output variables. This shows that there are three input variables and one output variable. Moreover, the inference engine used is Mamdani Inference Engine (MIE) that infers to the output based on the given set of inputs at any given time.

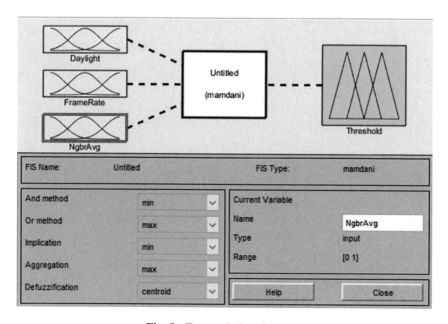

**Fig. 8.** Fuzzy rule based system

## 2.2.2    Inference System

The parameters being used in the proposed Fuzzy Rule Based System are given in the Table 1 along with their values. This table shows that which type of fuzzy operations are used during the execution of fuzzy inference engine. These parameters and their values may vary from application to application.

**Table 1.**  FRBS parameters

| Sr. | Parameter | Value |
|---|---|---|
| 1 | Number of inputs | 3 |
| 2 | Number of output | 1 |
| 3 | Type of fuzzifiers | Triangular |
| 4 | Inference engine | Mamdani inference engine |
| 5 | AND method | Min |
| 6 | OR method | Max |
| 7 | Implication | Min |
| 8 | Aggregation | Max |
| 9 | Defuzzifier | Center average defuzzifier (CAD) |
| 10 | Cardinality of input output | $3 \times 1$ |
| 11 | Size of rule base | $3 \times 3 \times 3 = 27$ |
| 12 | Rule base complete? | Yes |

## 2.2.3    Rule Base

The rule base contains all the rules pertaining to the output threshold by given input factors discussed in the previous section. The three input variables, after fuzzification, map to three possible categories in their respective universe of discourse. That corresponds to 27 rules in total.

Figure 9 shows rule editor, which provides way to add rules based on all possibilities. For example, in the given diagram,

IF (daylight = 'day' & Framerate = 'low' & NeighborAverage = 'low') THEN (threshold = 'low')

It is just a sample single rule. Similarly, rules may be added for all possible combinations of values of the input variables based on the intuitive approach. In this regard, the rule base is complete.

Figure 10 shows the rule surface built by comparing the relationship of neighbor average and Framerate with the output variable threshold. Figure 11 shows the rule surface built by comparing the relationship of neighbor average and daylight with the output variable threshold. Both impacts are almost same. Figure 12 shows the rule surface built by comparing the relationship of framerate and daylight with the output variable threshold. That impact is different from the previous relationships. This is mainly because both input and the output variable are in directly proportion to each other.

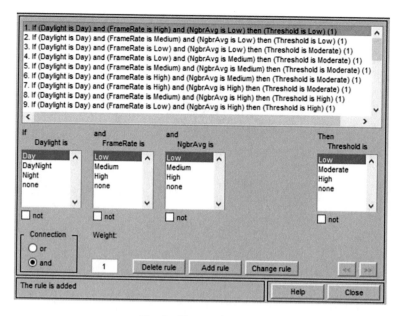

Fig. 9.  Fuzzy rule editor

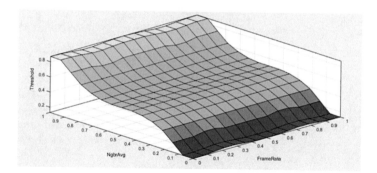

Fig. 10.  Fuzzy rule surface 1

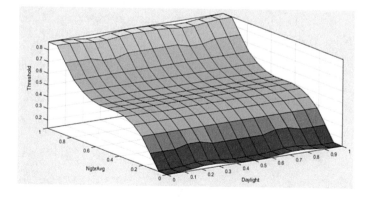

Fig. 11.  Fuzzy rule surface 2

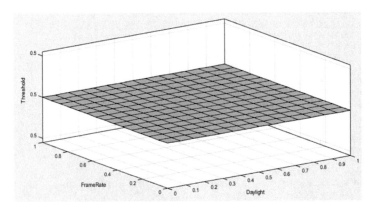

**Fig. 12.** Fuzzy rule surface 3

# 3 Results and Discussion

This section contains the results of the proposed Fuzzy Rule Based System based threshold detection scheme for precise object detection. This corresponds to the updated version of Algorithm 1. The Algorithm 2 is given below.

Algorithm 2 (Fig. 13) looks very much similar to the Algorithm 1 except the thresholds $T_A$ and $T_B$ become $T_{A_f}$ and $T_{B_f}$ that are the fuzzy versions of their original counter parts. Form the algorithm it is apparent that it is highly based on these thresholds for sake of finding the joint difference.

Once the thresholds are the finest the differentiation and detection are the most precise. The results of this accuracy can be seen in the subsequent section where it is compared with the Algorithm 1.

## 3.1 Example

For experiment purpose, we took few samples of a DVR video frames. For sake of detection three successive frames are chosen. Namely frame at time t, the reference frame (the default static background) and the frame at time t − 1 (previous frame).

From the visual effects (Fig. 14) it can be seen that by keeping frame at t and the reference frame static and varying the previous frame, the algorithm can precisely detect the object. Not only the object is detected but the noise is also reduced compared to the results of Algorithm 1 shown in the previous section, Fig. 1 that contains a significant noise factor.

That noise factor may lead to misdetection of the later detection in terms of motion later.

---

**Algorithm 2** Joint Difference Algorithm (Proposed)

---

**if** $((|F_t(x) - B_{t-1}(x)| > \tau_{B_f})$ AND $(|F_t(x) - F_{t-1}(x)| > \tau_{A_f}))$ **then**
  Foreground Pixel;
**else if** $((|F_t(x) - B_{t-1}(x)| > \tau_{B_f})$ AND $(|F_t(x) - F_{t-1}(x)| < \tau_{A_f}))$ **then**
  Collect pixels in blobs;
  **if** ($\sharp$ Foreground Pixels $\geq (\gamma * (\sharp$ Total Pixels)))$ **then**
    Foreground Pixel; //foreground aperture problem solution
  **else**
    Background Pixel; //a background object suddenly starts moving at time $t$
  **end if**
**else if** $((|F_t(x) - B_{t-1}(x)| < \tau_{B_f})$ AND $(|F_t(x) - F_{t-1}(x)| > \tau_{A_f}))$ **then**
  Background Pixel; //ghosting problem solution
**else**
  // $|F_t(x) - B_{t-1}(x)| < \tau_{B_f}$ AND $|F_t(x) - F_{t-1}(x)| < \tau_{A_f}$;
  Background Pixel;
**end if**

---

**Fig. 13.** Algorithm 2

Figure 15 shows another continuation of the same example by just changing the reference frame while keeping other settings as it is. Means same frame at t and frames at t − 1. The results remain same. That mean any noise factor appearing in the reference frame will be discarded. No impact of ghost can be observed which was a significant part in Algorithm 1 and was quite misleading.

From the experiment, it is clear that the proposed fuzzy rule based threshold algorithm performs way better in term of precise object detection and noise

| Frame at t | Reference frame | Frame at t − 1 | Result |
|---|---|---|---|

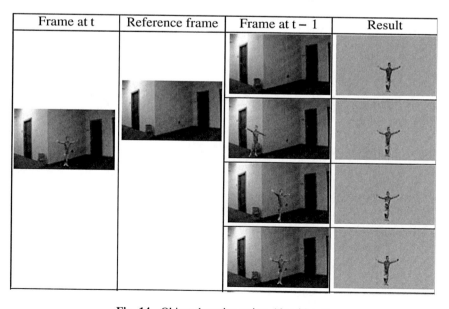

**Fig. 14.** Object detection using Algorithm 2

| Frame at t | Reference frame | Frame at t − 1 | Result |
|---|---|---|---|

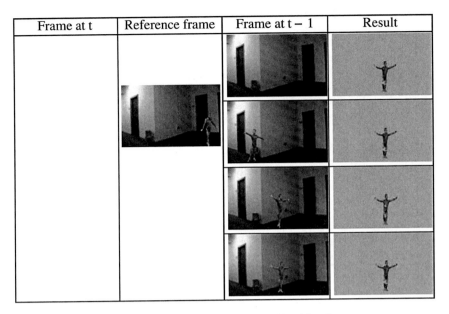

**Fig. 15.** Object detection using Algorithm 2

cancellation. This is the basic requirement of any advanced video signal processing algorithms that the successive frame differences must be precisely obtained and false alarms (noise, ghosts) must be eliminated.

It is also shown that the algorithm can tolerate a significant drift in the successive frames. This happens when the DVR camera is mounted outdoor and it might be staggered due to air pressure and other atmospheric factors.

## 4 Conclusion

The main purpose of this research was motion detection and finding the malicious activities from the videos. For this the motion detection is the first and key thing to perform because the existing algorithm exhibits a significant noise factor in detection that may lead to misdetection. The proposed FRBS helps in terms of finding the optimum thresholds based on which detection becomes more precise and noise free. This consequently ended up in the finest motion detection that leads to finest activity detection. For sake of obtaining the best thresholds, proposed FRBS took three factors into account namely the daylight effect, the framerate of the video and the neighboring pixel's average threshold. Here we considered four connected neighbors of the pixel to make the system less complex and suitable for real-time. In future, more factors can be incorporated based on video type and resolution. Instead of four-connected, eight-connected neighbors of the pixels may be investigated for further precision. Moreover, hybrid intelligent systems may be employed for further intelligent digital image processing.

# References

1. Zalevsky, Z., Garcia, J.: Motion detection system and method. Int. J. Comput. Appl. (IJCA) **44**(12), 5550–5557 (2014)
2. Kadam, V.: Real time motion detection using dynamic camera: a case study. Int. J. Comput. Appl. (IJCA) **114**(7), 8875–8887 (2015)
3. Boregowda, L.R., Jain, M.D., Rajagopal, V.: Video motion detection using block processing. Int. J. Comput. Appl. (IJCA) (2007)
4. Bharadwaj, B.K., Pal, S.: Motion detection in a video stream. Int. J. Comput. Appl. (IJCA) (2006)
5. Living, J.: Video motion detection. Int. J. Comput. Appl. (IJCA) (2008)
6. Bhaltilak, K.V.: Human motion analysis with the help of video surveillance. Int. J. Comput. Sci. Eng. Technol. (IJCSET) **4**(9), 245–249 (2014)
7. García-Valls, M., Basanta-Val, P., Estévez-Ayres, I.: Adaptive real-time video transmission over DDS. In: Conference on Industrial Informatics (INDIN), vol. 9(1), pp. 130–180 (2013)
8. Atta-ur-Rahman, Qureshi, I.M., Malik, A.N., Naseem, M.T.: QoS and rate enhancement in DVB-S2 using fuzzy rule base system. J. Intell. Fuzzy Syst. (JIFS) **30**(1), 801–810 (2016)
9. Atta-ur-Rahman, Qureshi, I.M., Malik, A.N., Naseem, M.T.: Dynamic resource allocation for OFDM systems using DE and fuzzy rule base system. J. Intell. Fuzzy Syst. (JIFS) **26**(4), 2035–2046 (2014). https://doi.org/10.3233/IFS-130880
10. Atta-ur-Rahman, Qureshi, I.M., Malik, A.N., Naseem, M.T.: A real time adaptive resource allocation scheme for OFDM systems using GRBF-neural networks and fuzzy rule base system. Int. Arab J. Inf. Technol. (IAJIT) **11**(6), 593–601 (2014). http://ccis2k.org/iajit/PDF/vol.11,no.6/6305.pdf
11. Atta-ur-Rahman, Qureshi, I.M., Malik, A.N.: Adaptive resource allocation in OFDM systems using GA and fuzzy rule base system. World Appl. Sci. J. (WASJ) **18**(6), 836–844 (2012). https://doi.org/10.5829/idosi.wasj.2012.18.06.906
12. Atta-ur-Rahman, Qureshi, I.M., Malik, A.N.: A fuzzy rule base assisted adaptive coding and modulation scheme for OFDM systems. J. Basic Appl. Sci. Res. **2**(5), 4843–4853 (2012)
13. Atta-ur-Rahman, Qureshi, I.M., Muzaffar, M.Z., Naseem, M.T.: A fuzzy rule base aided rate enhancement scheme for OFDM systems. In: IEEE Conference on Emerging Technologies (ICET'12), Islamabad, Pakistan, pp. 151–156, 8–9 October 2012
14. Atta-ur-Rahman, Qureshi, I.M., Naseem, M.T., Muzaffar, M.Z.: A GA-FRBS based rate enhancement scheme for OFDM based HYPERLANS. In: 10th IEEE International Conference on Frontiers of Information Technology (FIT'12), Islamabad, Pakistan, pp. 153–158, 17–19 December 2012
15. Atta-ur-Rahman, Qureshi, I.M.: Comparison of coding schemes over FRBS aided AOFDM systems. J. Netw. Innov. Comput. (JNIC) **1**, 183–193 (2013)
16. Migliore, D., Matteucci, M., Naccari, M.: A revaluation of frame difference in fast and robust motion detection. In: 4th ACM International Workshop on Video Surveillance and Sensor Networks (VSSN), Santa Barbara, California, pp. 215–218 (2006)

# Automatic Text Categorization Using Fuzzy Semantic Network

Jamal Alhiyafi[1], Atta-ur-Rahman[1(✉)], Fahd Abdulsalam Alhaidari[2], and Mohammed Aftab Khan[1]

[1] Department of Computer Science (CS), Imam Abdulrahman Bin Faisal University, 1982, Dammam, Saudi Arabia
{jalhiyafi, aaurrahman, mkhan}@iau.edu.sa
[2] Department of Computer Information System (CIS), College of Computer Science and Information Technology (CCSIT), Imam Abdulrahman Bin Faisal University, 1982, Dammam, Saudi Arabia
faalhaidari@iau.edu.sa

**Abstract.** Knowledge Representation (KR) is an art of representing knowledge in such a form that later it can be retrieved, searched, queried and inferred with a precision and along with all possible associations. That is why it can readily be said that KR is one of the key areas of research in Semantic web, Artificial Intelligence Data mining and big data analytics. In this regard, several techniques have been investigated in the literature. In this paper, the fuzzy semantic network (FSN) has been investigated for automatic text categorization in religious domain. The network is consisted of religious entities and their relationships obtained from various religious book and scriptures. This state of the art network is built and investigated for various search criteria. The results show that the scheme is promising in terms of better categorization.

**Keywords:** Knowledge representation · Semantic networks · Fuzzy semantic network · Information retrieval first section

## 1 Introduction

For knowledge representation, Semantic Networks are the basic and initial structures that were proposed and utilized effectively over the years. In fact, they set the basis of semantic web which is one of the hottest areas in today's web technologies [1, 2] one of the oldest and effective-most technique for KR.

Early, Semantic networks were investigated for the computers [3] at Cambridge Language Research Unit (CLRU), for the early stage of natural language processing (NLP). However, they were further enhanced by [4] for KR and other purposes. Later, the idea was further intervened by [5–7]. Almost over two decades that is 1960–1980 the semantic networks were considered as a mesh of hyperlinks in a hypertext format.

C. Benavente-Peces et al. (Eds.): SEAHF 2019, SIST 150, pp. 24–34, 2019.
https://doi.org/10.1007/978-3-030-22964-1_3

In this regard, several software has been implemented for sake semantic networks. In [8], authors described six common types of Semantic Network [8]. Two examples of semantic networks are shown in Figs. 1 and 2. Over here the famous link "is a", is used to represent animals' relationships along with their characteristics. This form also helps a lot in terms of deduction. Hence logical rules of inference like Modus Tollens, Modus Ponens etc., can easily be inferred and implemented.

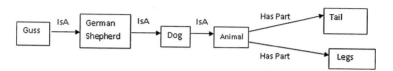

**Fig. 1.** A typical semantic network example [21]

Figure 2 shows a stronger semantic network that not only emphasizes on "is a" relationship but also represents property list. Property list means the characteristics related to any concept other than object's relationships.

Fuzzy systems have several applications in various fields of studies other than data mining like control and telecommunication [9–12]. This paper focuses on the application of Fuzzy Semantic Networks for knowledge representation and automatic text classification in the religious repository.

Rest of the paper is organized as follows. Section 2 presents types of semantic relations. Section 3 presents the fuzzy semantic network and motivation behind it. Section 4 presents the application of FSN for automatic text categorization in a religious domain while Sect. 5 concludes the paper.

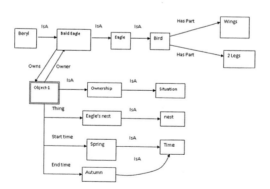

**Fig. 2.** Semantic network with object properties [21]

## 2  Types of Semantic Relations

### 2.1   Set-Membership Relation.... ∈

It can be read as 'member of'. It refers to typical crisp set membership with membership value exactly '1'. In this way a concept is wholly belongs to a class or another concept. It is depicted in Fig. 3.

**Fig. 3.**  Set membership relation

### 2.2   Subset Relationship .... IsA

It can also be read as 'belongs to'. It is depicted in Fig. 4. Any object can be expressed as a subset of another object.

**Fig. 4.**  Subset relation

### 2.3   Events by "Case Frames"

This relation is a connection between two objects via some event and event can also have a type. It is depicted in Fig. 5.

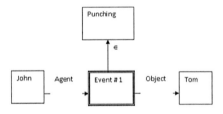

**Fig. 5.**  Case frames

### 2.4   N-arg Relations (with N > 2)

A relation in semantic network not necessarily binary (one node connected with exactly one more node), it could be tertiary or more. It is represented by creating an entity or case frame to stand for n-arg (n number of arguments) relationship e.g. machine uses 20 parts. It is depicted in Fig. 6. As the relationship name is depicting, the attribute could be of any type. This is like the class-relationship in an Object-Oriented Paradigm (OOP) [21]. That may be classified as:

- Association (weak)
- Composition (strong association) and so on

To view these relations in one network, the example is given in Fig. 7. This incorporates all the above-mentioned relations.

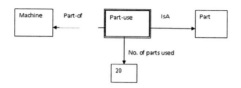

**Fig. 6.** N-arg relation

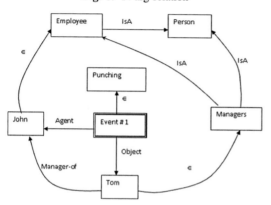

**Fig. 7.** Semantic Network with all relations

## 3   Fuzzy Semantic Networks

Semantic Networks provide the means of KR. Though the concept of Semantic Networks is not new, and it has been widely used, however, there exist certain limitations in it. To overcome these limitations a modified version is proposed in this paper that is named as Fuzzy Semantic Networks (FSN) in which the relationship between two concepts can be provided more realistically by means of fuzzy membership function instead of just an 'is a' link. In this way, a true relationship between any pair of entities can be represented. Moreover, as better representation promises a better retrieval, in [21], it is projected that this concept can play a vital role in numerous applications. Certain application areas are highlighted where FSN can be used and the working order and algorithm structures with example are also presented. [13]

According to [21], there are various application areas of FSN such as:

- Classification [16]
- Document classification
- Behavior classification

- Disease classification
- Natural Language Processing [17]
- Semantic web [18]
- Data/text mining [22]

## 4   Results and Discussions

In this section, the proposed Fuzzy Semantic Network are investigated for story building using automatic text classification and categorization in religious repository (corpora) that contains religious material from different Holy sources like Holy Quran, Holy Ahadith and Holy Bible etc. The process is shown in Fig. 8.

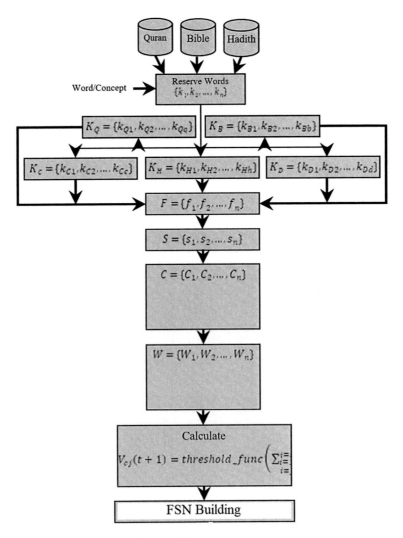

**Fig. 8.** FSN building process

The steps FSN building are given below.

Word or reserve word basically belong to a main story, that a user wants to build. First, a user finds the words/ reserve words from relevant holy book. Or combine holy books namely holy corpus (Holy Quran, Holy Bible, Holy Hadith). Words/ reserve words are represented by $K = \{k_1, k_2, \ldots, k_n\}$ This set of reserve words is further divided into five subsets given as follows:

$$K_Q = \{k_{Q1}, k_{Q2}, \ldots, k_{Qq}\}$$
$$K_B = \{k_{B1}, k_{B2}, \ldots, k_{Bb}\}$$
$$K_H = \{k_{H1}, k_{H2}, \ldots, k_{Hh}\}$$
$$K_C = \{k_{C1}, k_{C2}, \ldots, k_{Cc}\}$$
$$K_D = \{k_{D1}, k_{D2}, \ldots, k_{Dd}\}$$

where $K_Q$, $K_B$, $K_H$ is the set of reserve words is taken only from Holy Quran, Holy Bible, Holy Hadith respectively. $K_C$ is the set of common reserve words, those are taken from Holy Corpus. $K_D$ is the set of different reserve words, those are not common in Holy Corpus.

After defining/ extracting reserve words, find the frequency of those reserve words. Number of occurrences of a reserve words is called the frequency of that word, then find the state value of reserve word automatically. Every reserve word has its own state value $S = \{s_1, s_2, \ldots, s_n\}$ the meaning of state value is own worth of reserve word. After finding this co-occurrence is required as completion of FSN $C = \{C_1, C_2, \ldots, C_n\}$ where $C_i = \{C_{i1}, C_{i2}, \ldots, C_{in}\}$ and in combine fashion, $C$ can be represented as;

$$C = \begin{bmatrix} C_{11} & C_{12} & \cdots & C_{1n} \\ C_{21} & C_{22} & \cdots & C_{2n} \\ \vdots & \vdots & \vdots & \vdots \\ C_{n1} & C_{n2} & \cdots & C_{nn} \end{bmatrix}. \tag{1}$$

Fuzzy semantic weightage of that co-occurrence of reserve word is required $W = \{W_1, W_2, \ldots, W_n\}$ where $W_j = \{w_{j1}, w_{j2}, \ldots, w_{jn}\}$ and in matrix form is;

$$W = \begin{bmatrix} w_{11} & w_{12} & \cdots & w_{1n} \\ w_{21} & w_{22} & \cdots & w_{2n} \\ \vdots & \vdots & \vdots & \vdots \\ w_{n1} & w_{n2} & \cdots & w_{nn} \end{bmatrix} \tag{2}$$

Now, all material is ready for calculation, according to the formula [19, 20] for automatic building of FSN with thrash-holding function in Eq. 3.

$$V_{cj}(t+1) = threshold\_func\left(\sum_{\substack{i=1 \\ i \neq j}}^{i=n} V_{ci}(t)w_{ij}\right) \tag{3}$$

Table 1 contains the keywords searched along with their frequencies and state values. Table 2 contains the co-occurrence of the word "Adam" from Holy bible.

**Table 1.** Keywords, frequencies and respective state values (Holy Bible)

| Keyword | Frequency | State value |
|---|---|---|
| Adam | 33 | 0.999999999 |
| God | 3883 | 1 |
| First | 13 | 0.999655595 |
| Tell | 27 | 0.99999997 |
| Satan | 55 | 1 |
| Heaven | 5 | 0.931109609 |
| Special | 6 | 0.96402758 |
| Gathering | 12 | 0.9993293 |
| Together | 436 | 1 |
| Creation | 6 | 0.96402758 |
| First man | 3 | 0.761594156 |
| Eve | 153 | 1 |

**Table 2.** Co-Occurrence from Holy Bible

| | Adam | God | First | Tell | Satan | Heaven | Special | Gathering | Together | Creation | First man | Eve |
|---|---|---|---|---|---|---|---|---|---|---|---|---|
| Adam | 0 | 8 | 1 | 0 | 0 | 0 | 0 | 0 | 0 | 0 | 1 | 3 |
| God | 8 | 0 | 6 | 5 | 10 | 1 | 3 | 3 | 51 | 4 | 1 | 25 |
| First | 1 | 6 | 0 | 0 | 0 | 0 | 0 | 0 | 0 | 0 | 0 | 0 |
| Tell | 0 | 5 | 0 | 0 | 0 | 0 | 0 | 0 | 0 | 0 | 0 | 0 |
| Satan | 0 | 10 | 0 | 0 | 0 | 0 | 0 | 0 | 1 | 0 | 0 | 2 |
| Heaven | 0 | 1 | 0 | 0 | 0 | 0 | 0 | 0 | 0 | 0 | 0 | 0 |
| Special | 0 | 3 | 0 | 0 | 0 | 0 | 0 | 0 | 0 | 0 | 0 | 0 |
| Gathering | 0 | 3 | 0 | 0 | 0 | 0 | 0 | 0 | 2 | 0 | 0 | 0 |
| Together | 0 | 51 | 0 | 0 | 1 | 0 | 0 | 2 | 0 | 1 | 0 | 3 |
| Creation | 0 | 4 | 0 | 0 | 0 | 0 | 0 | 0 | 1 | 0 | 0 | 0 |
| First man | 1 | 1 | 0 | 0 | 0 | 0 | 0 | 0 | 0 | 0 | 0 | 0 |
| Eve | 3 | 25 | 0 | 0 | 2 | 0 | 0 | 0 | 3 | 0 | 0 | 0 |

Now the Fuzzy Semantic Weightages of the respective experiment can be shown in the table-3. For example, the keyword Adam with God has 0.0021, Adam with First has 0.0769 as fuzzy membership value or the weightage.

Finally, Fig. 9 shows the obtained Semantic Fuzzy Network of the word "Adam" in Holy Bible. In Fig. 9, word Adam is semantically connected with other words according to the Corpus of Holy Bible with corresponding degree of relationship is given in Table 3. There are some links that are no directly connected with "Adam". For example, "Satan" and "Heaven", they are not connected directly but indirectly they are connected. To compliment these gaps, other corpora like Holy Quran and Ahadith can be incorporated to complete the knowledge representation.

**Table 3.** Fuzzy Semantic weightages (degree of relationship)

|  | Adam | God | First | Tell | Satan | Heaven | Special | Gathering | Together | Creation | First man | Eve |
|---|---|---|---|---|---|---|---|---|---|---|---|---|
| Adam | 0.0000 | 0.2424 | 0.0303 | 0.0000 | 0.0000 | 0.0000 | 0.0000 | 0.0000 | 0.0000 | 0.0000 | 0.0303 | 0.0909 |
| God | 0.0021 | 0.0000 | 0.0015 | 0.0013 | 0.0026 | 0.0003 | 0.0008 | 0.0008 | 0.0131 | 0.0010 | 0.0003 | 0.0064 |
| First | 0.0769 | 0.4615 | 0.0000 | 0.0000 | 0.0000 | 0.0000 | 0.0000 | 0.0000 | 0.0000 | 0.0000 | 0.0000 | 0.0000 |
| Tell | 0.0000 | 0.1852 | 0.0000 | 0.0000 | 0.0000 | 0.0000 | 0.0000 | 0.0000 | 0.0000 | 0.0000 | 0.0000 | 0.0000 |
| Satan | 0.0000 | 0.1818 | 0.0000 | 0.0000 | 0.0000 | 0.0000 | 0.0000 | 0.0000 | 0.0182 | 0.0000 | 0.0000 | 0.0364 |
| Heaven | 0.0000 | 0.2000 | 0.0000 | 0.0000 | 0.0000 | 0.0000 | 0.0000 | 0.0000 | 0.0000 | 0.0000 | 0.0000 | 0.0000 |
| Special | 0.0000 | 0.5000 | 0.0000 | 0.0000 | 0.0000 | 0.0000 | 0.0000 | 0.0000 | 0.0000 | 0.0000 | 0.0000 | 0.0000 |
| Creation | 0.0000 | 0.2500 | 0.0000 | 0.0000 | 0.0000 | 0.0000 | 0.0000 | 0.0000 | 0.1667 | 0.0000 | 0.0000 | 0.0000 |
| Together | 0.0000 | 0.1170 | 0.0000 | 0.0000 | 0.0023 | 0.0000 | 0.0000 | 0.0046 | 0.0000 | 0.0023 | 0.0000 | 0.0069 |
| Creation | 0.0000 | 0.6667 | 0.0000 | 0.0000 | 0.0000 | 0.0000 | 0.0000 | 0.0000 | 0.1667 | 0.0000 | 0.0000 | 0.0000 |
| First man | 0.3333 | 0.3333 | 0.0000 | 0.0000 | 0.0000 | 0.0000 | 0.0000 | 0.0000 | 0.0000 | 0.0000 | 0.0000 | 0.0000 |
| Eve | 0.0196 | 0.1634 | 0.0000 | 0.0000 | 0.0131 | 0.0000 | 0.0000 | 0.0000 | 0.0196 | 0.0000 | 0.0000 | 0.0000 |

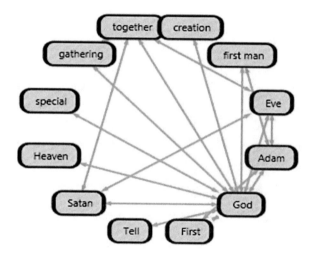

**Fig. 9.** FSN of word "Adam" in Holy Bible

In Fig. 10, it is apparent that there is a semantic link between "Adam" and "Satan". Similarly, the Quranic FSN has more number of links compared to Holy Bible for the word "Adam".

Just imagine that if entire Holy Corpora is used for building FSN for the same word, then network will be more precise and well connected in terms of link and relationships.

This is given in Fig. 11. The network is so dense that it may not be fit in the space allowed here. Just as glimpse it can be seen here and it contains countless collections of concepts and objects. It is a fuzzy semantic network of the word "Adam" in the entire Holy Corpora with immense number of relationships that cannot be thought as just "is a" or any arbitrary subset relationship in simple Semantic Network. Moreover, more than one words and the concepts can also be viewed in the network but on a higher resolution display and cannot be displayed here.

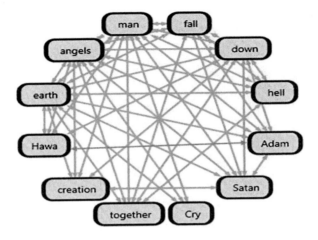

**Fig. 10.** FSN of word "Adam" in Holy Quran

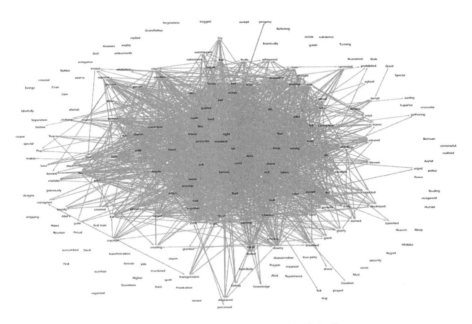

**Fig. 11.** FSN of word "Adam" in entire Holy Corpora

## 5    Conclusions

This paper presents a brief description of Semantic networks and Fuzzy Semantic Network (FSN) for knowledge representation. FSN makes use of basic Semantic Networks structure while incorporating fuzzy logic and fuzzy set theory to express the relationships among different objects in the real life. Fuzzy systems have proven their supremacy among the rest in terms of accuracy and precision. In this paper, the FSN are investigated for automatic text categorization in a huge religious repository that contains various concepts in the religious domain. From the results, it is apparent the categorization is excellent in terms of associations. The weakness and strength of associations can also be evidenced.

In future, the same idea can be investigated for other domains like social sciences and humanities. Because these areas fall in the category where the concepts are not linked scientifically but in a more realistic way. For better inferencing among multiple concepts, fuzzy inference engines can also be incorporated.

# References

1. Berners-Lee, T., Hendler, J., Lassila, O.: The semantic web. Scientif. Am. (2001)
2. Hendler, J., van Harmelen, F.: The Semantic Web: Webizing Knowledge Representation, pp. 821–839 (2008)
3. Richens, Richard H.: Preprogramming for mechanical translation. Mech. Trans. 3(1), 20–25 (1956)
4. Klein, S., Simmons, R.F.: Syntactic dependence and the computer generation of coherent discourse. Mech. Transl. 7
5. Allan, M., Collins, A., Quillian, M.R.: Retrieval time from semantic memory. J. Verbal Learn. Verbal Behav. 8(2), 240–248 (1969)
6. Allan, M., Collins, A., Quillian, M.R.: Does category size affect categorization time? J. Verbal Learn. Verbal Behav. 9(4), 432–438 (1970)
7. Allen, J., Frisch, A.: What's in a semantic network. In: Proceedings of the 20th. Annual Meeting of ACL, Toronto, pp. 19–27 (1982)
8. Sowa, J.F., Borgida, A.: Principles of Semantic Networks: Explorations in the Representation of Knowledge (1991)
9. Atta-ur-Rahman, Qureshi, I.M., Malik, A.N., Naseem, M.T.: Dynamic resource allocation for OFDM systems using differential evolution and Fuzzy rule base system. J. Intell. Fuzzy Syst. (JIFS), 26(4), 2035–2046 (2014)
10. Atta-ur-Rahman, Qureshi, I.M., Malik, A.N.: Adaptive resource allocation in OFDM systems using GA and fuzzy rule base system. World Appl. Sci. J. (WASJ), 18(6), 836–844 (2012)
11. Atta-ur-Rahman, Qureshi I.M., Malik, A.N., Naseem, M.T.: A real time adaptive resource allocation scheme for OFDM systems using GRBF-neural networks and Fuzzy rule base system. Int. Arab J. Informat. Technol. (IAJIT) 11(6), 593–601 (2014)
12. Atta-ur-Rahman, Qureshi I.M., Malik A.N.: A Fuzzy rule base assisted adaptive coding and modulation scheme for OFDM systems. J. Basic Appl. Sci. Res. 2(5), 4843–4853 (2012)
13. Wang, L.X.: A Course in Fuzzy Systems and Controls (Chapter 12). Prentice Hall Publications (1997)
14. Shahzadi, N., Atta-ur-Rahman, Shaheen, A.: Semantic network based semantic search of religious repository. Int. J. Comput. Appl. (IJCA) 36(9), 1–5 (2011)
15. Shahzadi, N., Atta-ur-Rahman, Sawar, M.J.: Semantic network based classifier of holy Quran. Int. J. Comput. Appl. (IJCA) 39(5), 43–47 (2012)
16. Bates, M. (1995). Models of natural language understanding. Proceedings of the National Academy of Sciences of the United States of America, Vol. 92, No. 22 (Oct. 24, 1995), pp. 9977–9982

17. Antoniou, G., Van Harmelen, F.: A semantic web primer, 2nd edn. The MIT Press. ISBN 0-262-01242-1 (March 31, 2008)
18. Sebastiani, F.: Machine learning in automated text categorization. ACM Comput. Surv. **34** (1), 1–47 (2002)
19. D. Kardaras and N. Karakostas, E-Service adaptation using fuzzy cognitive maps. In: Proceedings of 3rd International IEEE Conference on Intelligent systems (2006)
20. Luo, X., Fang, N., Hu, B., Yan, K., Xiao, H.: Semantic representation of Scientific documents for the E-Service Knowledge Grid, in Wiley InterScience, Shanghai University, Shanghai 200072, China, 31 October 2007
21. Atta-ur-Rahman: Knowledge Representation: A Semantic Network Approach", Chapter 4, in Handbook of Research on Computational Intelligence Applications in Bioinformatics, IGI Global (2016)
22. Faisal, H.M., Ahmad, M., Asghar, S., Atta-ur-Rahman: Intelligent Quranic Story Builder. Int. J. Hybrid Intell. Syst. preprint, 1–8 (2017)

# An Empirical Approach on Absorption of EMI Radiations by Designing Elliptical Microstrip Patch Antenna

Mandeep Singh Heer$^{(\boxtimes)}$ and Vikrant Sharma

GNA University, Phagwara, India
`er.mannheer86@gmail.com`

**Abstract.** In this paper a novel design and analysis of an Antenna with broadband behavior has presented. Antenna is designed and fabricated on an inset fed Microstrip Antenna and calculate the effect of antenna dimensions Length (L), Width (W) and substrate parameters relative Dielectric constant (εr), substrate thickness (h), which will be useful in radiation parameters of calculating band width. Also discovery of CNT & Nanomaterials, it is possible to use various techniques for absorption of EMI radiations. Our main aim is to design an antenna which is capable to absorb harmful radiations and obtain a suitable material which is used for making a patch or substrate. Proposed antenna frequency bands for communication purposes which will be called the low band (2.495–2.695 GHz), the medium band (3.25–3.85 GHz) and the high band (5.25–5.85 GHz). The design of required patch antenna is simulated by using software named "IE3D v-12 simulator Zeland Inc." and required patch antenna and its parameters can be realized as per design requirements.

**Keywords:** MSA (Microstrip Patch Antenna) · Dielectric constant (εr) · Resonant frequency ($f_r$)

## 1 Introduction

The simple dictionary meaning of an antenna is that it is usually metallic device (as a rod or wire) for radiating or receiving radio waves [2, 8]. The IEEE standard definitions in terms for antennas is defines the as "a means for transmitting or receiving radio waves." In other words, the antenna is the transitional structure between free space and a guiding device [10]. The antenna is also referred to as aerial. So in simple words antenna is defined as "a transducer which convert electromagnetic waves into required electrical signal at receiver and also a metallic device used for transmitting or receiving electromagnetic waves which acts as the transition region between free spaces and guiding structure like a transmission line in order to communicate even in a longer distance."

By using various Nanomaterials and CNTs we will fabricate a patch on which by suitable probe feed we obtain maximum bandwidth and efficiency of antenna. Also use suitable substrate on which patch is fabricated by masking and lithography process. Masking can be done by software (like Coral, Autocad) simulations. For compactness

© Springer Nature Switzerland AG 2019
C. Benavente-Peces et al. (Eds.): SEAHF 2019, SIST 150, pp. 35–41, 2019.
https://doi.org/10.1007/978-3-030-22964-1_4

in advanced communication system, may be it wired or wireless communication, miniaturization is the basic needs. Microstrip patch antenna is most commonly used now a days for mobile applications or Wi-Fi, Wi-Max applications [9]. So antennas can be easily compatible with the application areas of embedded and fabricated antenna used in handheld wireless devices such as cellular phones, pagers etc. [4]. The telemetry and communication antennas on missiles need to be thin and conformal and are often in the form of Microstrip patch antennas. Another application is Radar communication and Satellite communication. In this communication rectangular patch is chosen for better response and ease of analysis rather than any other shapes.

## 2   Antenna Design

This given design having these dimensions: Ground dimension = 100 × 100. Ground cut at point (0, 50). The geometrical representation of proposed antenna [1] is shown in Fig. 1.

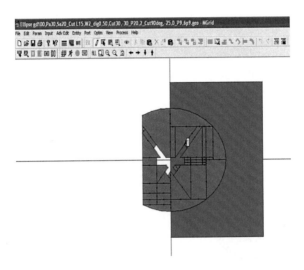

**Fig. 1.** Ellipse design having ground cut and patch cut

At patch, at axis (−25, 0), having length L = 50 mm and width W = 20 mm, cut at angle 90°. Ellipse major axis a = 30 mm, Ellipse Minor Axis b = 20 mm. $\in r$ = 4.4, tangent loss = 0.025, h = 1.59.

There are few considerations or parameters for the design of a rectangular Microstrip Patch Antenna which will be taken care of: A substrate with a high dielectric constant has been selected since it reduces the dimensions of the antenna according to our application. The resonant frequency ($f_r$) of the antenna must be selected appropriately. The Mobile Communication Systems uses the frequency range from 2 to 6 GHz. By doing this we fabricate or design an antenna which will be able to operate in the desired frequency range [5, 6]. The resonant frequency selected for my design is 3–8 GHz.

For the microstrip patch antenna to be used in cellular phones, it is essential that the antenna is not bulky. So, the dielectric substrate is selected as 1.59 mm height (h).

Hence, the essential parameters for the design are: $fo = 3$–8 GHz, $\varepsilon r = 4.4$ and $h = 1.59$ mm.

**Fig. 2.** A fabricated design of elliptical microstrip patch antenna with coral draw model

Zeland Inc's IE3D is the software we used to model and simulate the Microstrip patch antenna. IE3D is a full-wave electromagnetic simulator based on the method of moments. It analyzes 3D and multilayer structures of general shapes. It has been widely used in the design of MICs, RFICs, patch antennas, wire antennas, and other RF/wireless antennas [2]. It can be used to calculate and plot the $S11$ parameters, return losses, smith chart etc. as well as the radiation patterns (Fig. 2).

## 3   Result

The return loss for the patch antenna can be measured on a network Analyzer [7, 10]. The pattern created in a far field test can be shown as E-plane and H-plane patterns [13], (preferably in an anechoic chamber) calculated with a standard gain antenna as a transmitting antenna and the Antenna under test as a receiving antenna mounted on a pedestal [7]. Layout generation can be done in coral draw or autocad software for preparing the mask. Once the mask is printed on a transparent sheet, the patch can be fabricated using conventional photolithography process [10]. Once the dimensions are obtained the antenna can be simulated on a commercially available 2.5D or a 3D EM simulator [7, 10]. Further optimization and fine tuning of dimensions, other parameters like axial ratio, smith chart, 2D pattern etc. can be carried out to bring the resonance back at the desired frequency with acceptable return loss (Fig. 3).

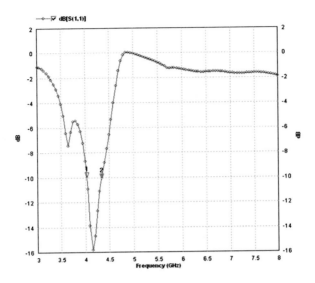

**Fig. 3.** Return loss in dB

The radiation pattern of antenna represents the radiation property of antenna as a function of space coordinates. The pattern gives the normalized field (power) value w.r. t. the maximum value [13]. In patch antenna various radiation properties include directivity, gain, power flux density, radiation intensity, radiation resistance, field strength, phase polarization. It will resonate at the frequency of 4.15 GHz, it means antenna give maximum radiation at this frequency (Fig. 4).

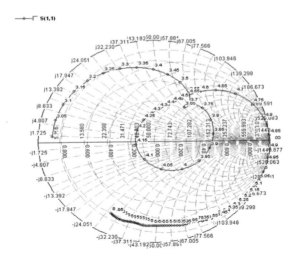

**Fig. 4.** Smith chart

Impedance calculated is 36.19–2.27j, by showing these values we see that the antenna impedance is not properly matched so there is losses occur in these design. So to rectify this problem we will shift or change the probe position by proper dimension and check the result on IE3D simulator (Figs. 5 and 6).

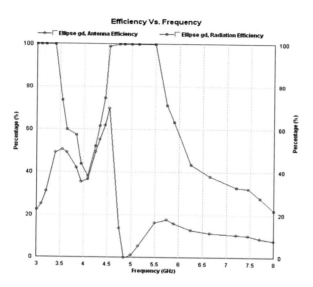

**Fig. 5.** Graph between efficiency versus frequency of elliptical patch antenna

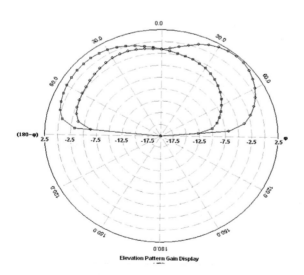

**Fig. 6.** 2-D pattern

Here we see that the antenna efficiency and radiation efficiency is 40.77 and 42.75% respectively. As we see in figure, by increase in the frequency we obtain the maximum efficiency. We generally use S band frequency range applications by this proposed design. The maximum bandwidth is 7.47 Hz at maximum resonant frequency i.e., 4.15 GHz (Fig. 7).

### 3.1   Simulation Setup and Results

| Freq. | Return loss | Impedance | Bandwidth | Antenna efficiency | Radiation efficiency | Axial ratio (0, 90) |
|---|---|---|---|---|---|---|
| 4.15 | −15.81 | 36.19-2.27j | 7.47 | 40.77 | 42.75 | 15.21, 25.38 |

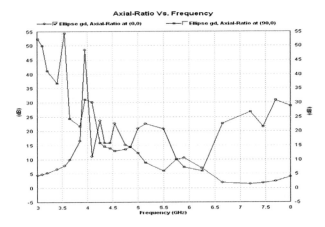

**Fig. 7.**   Axial ratio at 0° and 90°

## 4   Conclusion

We propose a design and fabrication of elliptical microstrip patch antenna in this paper. Our main objective is work on the allocation of frequency bands for communication purposes which will be called the L & S band (ranging from 1 to 4 GHZ). The 2.45–2.484 and 5.15–5.825 GHz frequency bands have been adopted by IEEE 802.11a5 for wireless local area networks (WLAN) [13]. The maximum absorption value will be achieved by designing that antenna with the composition of nanomaterial as base, maximum absorption ratio will be achieved. The study explicitly showed that the nanomaterial or CNT [11] with patch material i.e., copper, this compound show possibility of EMI shielding for various applications up to X band frequency range [12]. So by using this design in elliptical microstrip patch antenna we find various results by changing the position of probe at same ground cut, patch cut etc. Without ground cut we obtain better gain but poor bandwidth. As we change probes in a constant ground cut, its bandwidth and efficiency is improved.

# References

1. Kushwah, V.S., Tomar, G.S.: Size reduction of microstrip patch antenna using defected microstrip structures. In: 2011 International Conference on Communication Systems and Network Technologies. https://doi.org/10.1109/CSNT.2011.50. 978-0-7695-4437-3/11/$26.00 ©2011 IEEE
2. Pu, W., Xiao-Miao, Z., Zhen-Lin, L.: A novel elliptical UWB antenna with dual band-notched characteristics. 978-1-4244-8443-0/11/$26.00 ©2011 IEEE
3. Nath, M., Chakravarty, T., De, A.: An experimental study on the resonant frequency of shorted elliptical patch. 1-4244-0123-2/06/$20.00 ©2006 IEEE
4. Sharma, V., Saxena, V.K., Sharma, K.B., Saini, J.S., Bhatnagar, D.: Probe feed elliptical patch antenna with slits for WLAN application. In: Proceedings of International Conference on Microwave–08. 978-1-4244-2690-4444/08/$25.00 ©2008 IEEE
5. Bhattacharyya, A.K., Shafai, L.: Elliptical ring patch antenna for circular polarization. CH2563-5/88/0000-0022$1.00 ©1988 IEEE
6. Bhattacharyya, A.K.: Effects of ground plane truncation on the impedance of a patch antenna. In: IEEE Proceedings-II, vol. 138(6) December 1991
7. Behdad, N.: Simulation of a 2.4 GHz patch antenna using IE3D. EEL 6463, Spring 2007
8. Garg, R., Bhartia, P.: Microstrip Antenna Design Handbook. Antennas and Propagation Library. Artech House, London
9. Balanis, C.A.: Antenna Theory - Analysis and Design, 3rd edn.
10. Singh, S.P., Singh, A., Upadhyay, D.: Design and fabrication of microstrip patch antenna at 2.4 GHz for WLAN application using HFSS. IOSR J. Electron. Commun. Eng. (IOSR-JECE). e-ISSN: 2278-2834, p-ISSN: 2278-873, Special Issue-AETM'16
11. Caudillo, R., Troiani, H.E., Miki-Yoshida, M., Marques, M.A.L., Rubio, A., Yacaman, M.J.: A viable way to tailor carbon nanomaterials by irradiation-induced transformations. Radiat. Phys. Chem. **73**, 334–339 (2005). 0969-806X ©2004 Elsevier Ltd
12. Reshi, H.A., Singh, A.P., Pillai, S., Para, T.A., Dhawan, S.K., Shelke, V.: X Band frequency response and electromagnetic interference shielding in multiferroic $BiFeO_3$ nanomaterials. Appl. Phys. Lett. **109**, 142904 (2016). AIP publishing
13. Dong, J., Wang, A., Wang, P., Hou, Y.: A novel stacked wideband microstrip patch antenna with U-shaped parasitic elements. 978-1-4244-2193-0/08/$25.00 ©2008 IEEE

# Quantitative Analysis of Software Component's Reusability

Nisha Ratti[⊠] and Parminder Kaur

Guru Nanak Dev University, Amritsar, India
nisharatti@gmail.com

**Abstract.** This paper aims at nurturing the component-based development paradigm by quantitatively analyzing the component's reusability. The study intends to resolve the issues revolving around compatibility of the units to be incorporated. With the help of case study of open source software, multiple versions are studied. Metrics are one of the methods to analyze the growth of software quantitatively. A metric suite has been proposed for this purpose. These metrics can be helpful in making our decision regarding whether any new component can be deployed merely or former component can be easily removed. These metrics contribute to check the reusability properties of the component.

**Keywords:** Open source software · Software engineering · Software maintenance · Software metrics · Software reusability

## 1 Introduction

Components are composed to develop a system. The component may be commercial off-the-shelf (COTS) or free/open source Off-the-shelf components. The first step in the composition of the system is to procure the component. The selection of the component depends on many factors, like its cost, reviews, past experiences and the documentation given by the vendor. If the component is COTS, then it means the vendor is supposed to provide all the necessary information. Moreover, if it is the free/open source, then only that component is selected which is white-box so that the complete source code is available.

Component identification and selection process has to be elaborated before continuing with development process [5]. The process involves four steps. The first step is to look for the possible options/alternatives which can fulfill the requirements posed by the user. When the search process is complete, the next step is to evaluate all the alternatives following the guidelines laid down earlier. The component which matches the criteria is selected. However, after the criterion is matched, there is a possibility of negotiation required concerning requirements. After all the alterations in negotiation are done, the component is finally selected for composition. The following diagram better depicts this whole process.

Once the component is selected, its compatibility is required to be verified. If the component is found to be compatible by the dependencies of the component and the system, then the installation process proceeds. Otherwise, that component is rejected.

© Springer Nature Switzerland AG 2019
C. Benavente-Peces et al. (Eds.): SEAHF 2019, SIST 150, pp. 42–54, 2019.
https://doi.org/10.1007/978-3-030-22964-1_5

After the selection is finalized, the component has to be integrated into the system. The addition of a new component results in the revision of the software or we can say, a new version is generated. Figure 2 shows the flowchart of component procurement and deployment process.

## 2 Component Integration

Component Integration systems assume an essential part in the improvement of programming with segments. The modules, which are to be incorporated, may have an impact on numerous other modules effectively working with the present framework. Another issue is that that officially utilized component may rely upon the recently included components for different errands. With a specific end goal to deal with all such sort of vulnerabilities, there is a need to recognize the influenced parts at the season of incorporation of the new module/component. Two significant undertakings are related to the incorporation of the new module in the framework:

- To distinguish the influenced module alongside its different versions.
- To recognize the relationship of the influenced module.

Before the component configuration, services associated with the components need to be considered. The services may be the required services or the prohibited services. So, during Component configuration, if a component is providing all the required services, then only it is eligible to be installed. In addition to this, there are services which have conflicts with the system, so it should not provide those services. Prohibited services may be accompanied by some prohibited components. So, all these conditions are to be tested before installation of a component [1].

The interconnection between the components facilitates the information flow among components. Henry and Kafura [3] have proposed two measurements while discussing the information flow among procedures in traditional programming. The metrics are Fan-In and Fan-Out. Fan-In is essentially the data spill out of a specific segment, and Fan-Out is data from a specific part alongside the databases kept up or required.

For analyzing the component-based systems, Fan-In is usually termed as Afferent Coupling (AC) & Fan-Out as Efferent Coupling (EC). Afferent Coupling can be better explained as the number of classes which are dependent on the candidate class. Efferent Coupling can be defined as the number of classes on which the candidate class is dependent.

For example: If we have five components in a system. The graph in Fig. 1 showing the dependencies of the system. A → B shows that B is dependent on A for its functionality & change in A will affect B & E directly & indirectly D, C will also be affected. Consider the AC & EC of component B (Fig. 3).

This scenario is true if we consider only direct dependencies. If the user considers the indirect dependencies too then, the facts are different. In that case

AC (B) = 1 will remain same but

EC (B) = three which is not the same.

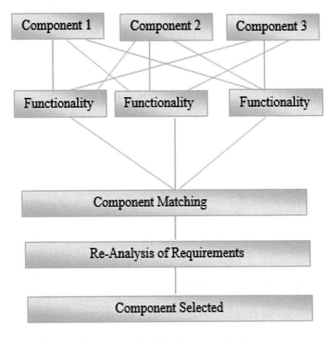

**Fig. 1.** Component identification and selection process

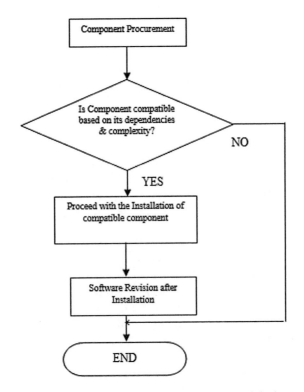

**Fig. 2.** Showing the flowchart of component procurement and deployment process.

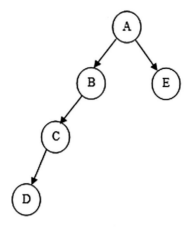

**Fig. 3.** Showing a dependency tree with five components

Considering such cases, AC & EC does not do complete justice to the problem as one essential criterion are worth consideration, i.e., Dependency Depth of tree.

Dependency depth is the number of nodes on the longest path of the dependency tree. Nodes are designated as components. Component dependency path plays a vital role in analyzing the complexity of the component. The complexity of the component is the critical criteria while recognizing the component for reuse or removal or replacement.

## 3 Mathematical Representation of Dependencies in Component-Based System

In order to gain a better understanding of component dependencies, the following representations are proposed, where S is a non-empty set of components

$$S = \{Ci|Ci{\rightarrow}Cj, (1 \leq i, j \leq n), Ci, Cj \in S\}$$

Ci denotes a component
n is the total number of components in system S
Ci → Cj represents the dependency of Ci on Cj
Moreover, this relation may go randomly between any components.

With the help of the dependency graph, we can find out dependency depth for a particular component which may help in calculating complexity. The complexity of a component can be calculated after considering component C as public reusability functionality. The maximum of the public classes is the maximum the entry points for that component. If cyclomatic complexity is used for measuring the complexity, it will be the class within the class that might get invoked by using the public element. If a publicly exposed class for the object-oriented system and publicly exposed functions for procedural programming are invoked, how many maximum numbers of types (i.e., either a class or function) that get executed in response to it.

As mentioned in [7], a framework has been proposed for analyzing the relationship between the components. These relationships play an essential role in the process of configuration/removal of a component. Various tasks responsible for being done by SCDA can be divided into three phases. The input of the SCDA is the source code of the software, and it is compatible with.Net source code only. The responsibility of phase 1 is parsing the source code, and that parsed data is represented with the help of XML.

Further, it shows the parsed data in a tree type structure. Separate components are held responsible for various tasks. E.g., Parser & Resolver is parsing the source code. The report of parsing is generated by report generator whose input is generated in XML format by P&R.

Data is of no use if it is not interpreted or informative. So, the next step is to compare the reports of the multiple versions. Then the variations among various reports are analyzed. The analysis has to be quantified, so some metrics are proposed to measure the performance of the system. A metric suit has been proposed to study the relationship between the components.

## 4   Proposed Metrics for Component Reusability Assessment

In order to assess the reusability of software components, specific metrics are proposed. These metrics are helpful in determining the candidate component for reuse or maintainability. These metrics are based on dependencies among components of a system.

Component Dependency Depth (CDD):

CDD is a paradigm agnostic metric, i.e., which is not confined to one single paradigm of software development. It can be defined as the maximum path length for a component C which is a publicly exposed reusable functionality.

Max (Path (C $\rightarrow$ Xj))

Where $1 \leq j \leq n$

and

n is the number of components in the system. This process has to be repeated for all the publicly exposed reusable functionalities/ types/ components of the system.

$$CDD = \frac{\sum_{i=1}^{n}(\max(path(Ci \rightarrow Xj)))}{n}$$

where $1 \leq j \leq n$ and $1 \leq i \leq n$

This metric can be used for comparing two components considering their evolutions. So the comparison can be performed considering multiple versions of the same component. If the component is less complicated it means, it is easy to remove/ replace that component for software evolution & the component is a better fit for reuse too.

The reason behind calculating complexity is reusability or removal or replacement of the component.

(a) In case we want to evaluate whether the component is fit for reuse, then we have to consider an average of depth of dependency-out. As Fan-out can be considered for just up to the 1st level of dependency hierarchy, Reusability factor is directly proportional to the depth of dependency-out.

(b) If we want to remove/replace a component, the consideration should be the depth of dependency-In. Fan-In may not be sufficient enough to cover the entire component which may be directly or indirectly dependent on the candidate component.

(i) Component Dependency Complexity (CDC):

$$CDC = \sum_{j=1}^{n} path(C \rightarrow Xj)$$

where c is a publicly exposed component, and it will measure direct as well as transitive dependencies of a publicly exposed component.

Reusability Points (RP):

It will calculate the number of components which can be reused by other systems. It can be derived from publicly exposed components of the systems.

Component Dependency Density Metric (CDDM):

It will measure the ratio between a total number of dependencies among the components of the system & a total number of components.

$$CDDM = \frac{Total\ number\ of\ dependencies}{Total\ number\ of\ Components}$$

## 5  Tool for Implementing Mathematical Model and Analyzing Dependencies: SCDA

A tool named Source Code Dependency Analyzer (SCDA) is developed to implement the framework proposed in [7]. To configure the component most effectively, the developed Analyzer, SCDA, first parses the assemblies existing in the system, and then shows the assemblies dependent on each other in two ways:

- Files dependent on other files
- Files used by other files

The developed Analyzer saves the dependency report in XML files. SCDA is developed in C#.Net 4.6.1 of Microsoft Visual Studio with compiler Roselyn. Initially started the project in Visual Studio 2013 with compiler named NRefactory but during the research, with the passage of time, the project itself has also undergone the evolution. As some of the versions of the software ShareX, which has been used for a case study, are developed in Visual Studio 2015. So we have to upgrade our software for Visual Studio 2017 and the compiler named Roselyn has been chosen for compiling the code.

## 6    Case Study: NUNIT

A case study has been prepared with the help of the software NUNIT. NUnit is an open source software used for Unit Testing. In order to study the evolution of the software, 20 versions are selected. The source code has been downloaded from the web portal www.codeproject.com. Since its inclusion on GitHub, i.e., Sept.2014, there are approximately 2549 commits, and 90 contributors worked on it. It was earlier an independent application. It was used for testing the software. Then over a period, it was converted to as add-in of Visual Studio. However, now again, its GUI console has been provided. So they decided to provide this as an application for mobile software testing "Xamarin" Total 20 versions of NUnit are analyzed. With the help of them, the proposed metrics can be validated (Tables 1, 2 and 3).

**Table 1.** Information related to the number of Classes, Methods, dependencies, LOC, P_LOC, P_CLS, P_DEP, and P_MET

| S.No. | Ver No | Classes | Methods | Dependencies | LOC | P_LOC | P_CLS | P_DEP | P_MET |
|-------|--------|---------|---------|--------------|-------|-------|-------|-------|-------|
| 1 | 2.6.2 | 1334 | 6429 | 1396 | 99607 | | | | |
| 2 | 2.6.3 | 1294 | 6363 | 1382 | 98747 | −1 | −3 | −1 | −1 |
| 3 | 3.0.0 | 1227 | 5847 | 1231 | 78373 | −26 | −5 | −12 | −9 |
| 4 | 3.0.a1 | 1003 | 4609 | 1017 | 62078 | −26 | −22 | −21 | −27 |
| 5 | 3.0.a2 | 995 | 5118 | 1010 | 62591 | 1 | −1 | −1 | 10 |
| 6 | 3.0.a3 | 997 | 4770 | 1006 | 62912 | 1 | 0 | 0 | −7 |
| 7 | 3.0.a4 | 1025 | 4818 | 981 | 65945 | 5 | 3 | −3 | 1 |
| 8 | 3.0.b3 | 1280 | 6373 | 1340 | 91415 | 28 | 20 | 27 | 24 |
| 9 | 3.0.b4 | 1318 | 6490 | 1435 | 92941 | 2 | 3 | 7 | 2 |
| 10 | 3.0.b5 | 1215 | 5743 | 1208 | 76285 | −22 | −8 | −19 | −13 |
| 11 | 3.0.r1 | 1229 | 5852 | 1230 | 78372 | 3 | 1 | 2 | 2 |
| 12 | 3.0.r2 | 1229 | 5852 | 1230 | 78428 | 0 | 0 | 0 | 0 |
| 13 | 3.0.r3 | 1227 | 5847 | 1231 | 78374 | 0 | 0 | 0 | 0 |
| 14 | 3.0.1 | 1230 | 5880 | 1236 | 78811 | 1 | 0 | 0 | 1 |
| 15 | 3.2.0 | 1264 | 6103 | 1262 | 81425 | 3 | 3 | 2 | 4 |
| 16 | 3.2.1 | 1276 | 6286 | 1264 | 82146 | 1 | 1 | 0 | 3 |
| 17 | 3.4.1 | 1308 | 6476 | 1299 | 83685 | 2 | 2 | 3 | 3 |
| 18 | 3.5.0 | 1136 | 5510 | 1058 | 68439 | −22 | −15 | −23 | −18 |
| 19 | 3.6.0 | 1172 | 5798 | 1115 | 71403 | 4 | 3 | 5 | 5 |
| 20 | 3.6.1 | 1172 | 5801 | 1116 | 71483 | 0 | 0 | 0 | 0 |

**Table 2.** Variations for all the parameters for NUNIT

| | Classes | Methods | Dependencies | LOC |
|------|---------|---------|--------------|------|
| Overall variation | −162 | −628 | −280 | −28124 |
| P_VAR | −12.1439 | −9.76824 | −20.0573 | −28.235 |
| A_P_VAR | −0.6072 | −0.48841 | −1.00287 | −1.41175 |

**Table 3.** List the data calculated by SCDA version 2

| S.No. | Ver No | RP | CDD | CDC | CDDM |
|---|---|---|---|---|---|
| 1 | 2.6.2 | 1168 | 2.474 | 8.716 | 1.046 |
| 2 | 2.6.3 | 1142 | 2.784 | 9.584 | 1.068 |
| 3 | 3.0.0 | 1081 | 2.537 | 14.437 | 1.003 |
| 4 | 3.0.a1 | 901 | 2.630 | 13.026 | 1.014 |
| 5 | 3.0.a2 | 894 | 2.632 | 13.177 | 1.015 |
| 6 | 3.0.a3 | 895 | 2.556 | 12.649 | 1.009 |
| 7 | 3.0.a4 | 921 | 2.519 | 12.502 | 0.957 |
| 8 | 3.0.b3 | 1081 | 2.697 | 13.390 | 1.047 |
| 9 | 3.0.b4 | 1119 | 2.689 | 13.512 | 1.089 |
| 10 | 3.0.b5 | 1079 | 2.480 | 12.871 | 0.994 |
| 11 | 3.0.r1 | 1083 | 2.532 | 14.406 | 1.001 |
| 12 | 3.0.r2 | 1083 | 2.532 | 14.406 | 1.001 |
| 13 | 3.0.r3 | 1081 | 2.537 | 14.437 | 1.003 |
| 14 | 3.0.1 | 1083 | 2.541 | 14.437 | 1.005 |
| 15 | 3.2.0 | 1107 | 2.558 | 14.544 | 0.998 |
| 16 | 3.2.1 | 1119 | 2.542 | 14.347 | 0.991 |
| 17 | 3.4.1 | 1147 | 2.541 | 14.533 | 0.993 |
| 18 | 3.5.0 | 987 | 2.512 | 15.028 | 0.931 |
| 19 | 3.6.0 | 1014 | 2.650 | 16.380 | 0.951 |
| 20 | 3.6.1 | 1014 | 2.650 | 16.380 | 0.952 |

## 6.1   Observations

Following observations have been drawn from the metrics calculated with the help of SCDA Version 2. Appropriate graphs have been plotted to depict the behavior of various parameters for multiple versions of NUnit (Figs. 4 and 5)

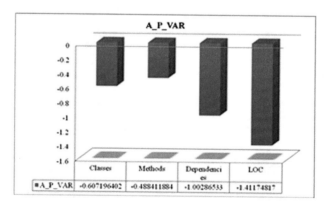

**Fig. 4.** Average percentage variation in all the parameters for NUNIT

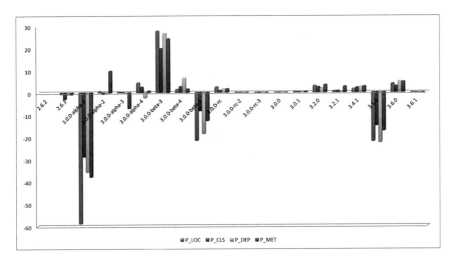

**Fig. 5.** Graphically showing the relationship of P_LOC, P_MET, P_CLS, and P_DEP for NUNIT

- The decrease in both No of Methods and LOC has been observed when existing features have been deprecated. For example, in version 3.0.0-alpha-1 major refactoring of the product was done to make the product future ready for mobile testing. Same did happen in release 3.0.b5 (Figs. 6, 7 and 8).

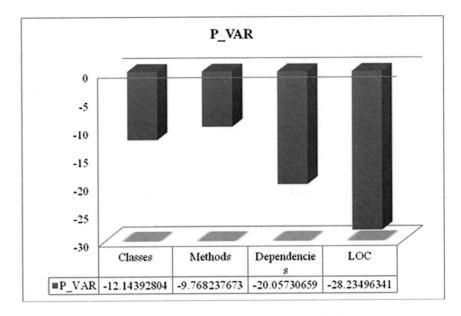

**Fig. 6.** Percentage change in all the parameters for NUNIT

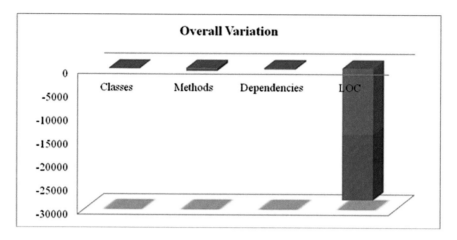

**Fig. 7.** Overall variation in the parameters namely, classes, methods, LOC and dependencies for NUNIT

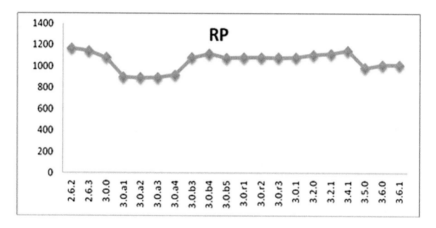

**Fig. 8.** Graph showing the trends in the metric value of Reusability points

- The decrease in CDDM also indicates removal of components which were watchfully related to the other components, e.g., in version 3.5.0 of NUnit code related to console UI was removed. As a result, 172 classes were removed which result in a decrease of 241 dependencies. (Figure 11)
- Any increase in CDC indicates that newly added publicly exposed component has used the functionality of another component more as compared to the existing publicly exposed components. But it also indicates that it is more difficult to introduce any change in its functionality which increases the overall complexity of the system, e.g., in version 5.0.1 11 new features have been added and CDC got increased by 23%. (Figure 10)

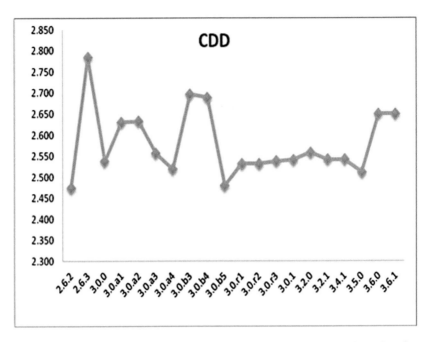

**Fig. 9.** The trends analysis in the metric value of Component Dependency Depth

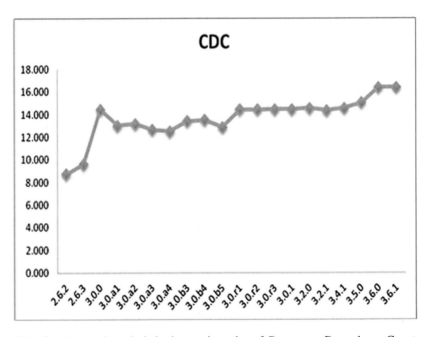

**Fig. 10.** The trends analysis in the metrics value of Component Dependency Count

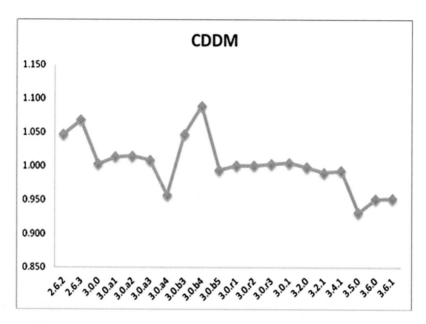

**Fig. 11.** The growth graph for Component Dependency Density Metric

- The CDD indicates the average no of the component that might get executed while invoking any publicly exposed component. That could decrease performance at execution time because of context switches. Increase in CDD means more deeply nested publicly exposed components have been added, e.g., in version 4.5.1 CDD got increased by 5%. (Figure 9)

## 7   Conclusions

The quantitative analysis can be better presented with the help of the software metrics. Some metrics which have been used from past many years are used again. Moreover, some metrics have been proposed to measure the growth of the software components. It can be concluded from the results that even if the software is open source in nature, the evolution pattern may be smooth but impulsive changes. Moreover, this also made evident that the attribute advancements are made at regular intervals. However, few observations may go in the opposite direction too. That is, due to a large number of developers in the development team, it is possible to have sudden oscillations. The size of the code may increase many-fold, but nothing is evident from the functionality point of view. From all these observations, the decision to reuse replace or remove regarding a component can be made more intelligently.

# 8  Future Work

Shortly, the efforts can be made to develop such a framework which can parse the system developed in any language for dependency analysis. This research project has contributed only to the .Net projects. So a comprehensive framework is required which can work in any development environment.

# References

1. Belguidoum, M., Dagnat, F.: Dependency Management in Software Component Deployment. Electron. Notes Theor. Comput. Sci. **182**, 17–32 (2007)
2. Estublier, J., Casallas, R.: The Adele, configuration manager. Config. Manag. (Trends in Software), 99–134 (1995). http://doi.org/10.1.1.92.6064
3. Henry, S., Kafura, D.: Software structure metrics based on information flow. IEEE Trans. Softw. Eng. **SE-7**(5), 510–518 (1981)
4. Hutchinson, J., Kotonya, G.: A Review of Negotiation Techniques in Component-Based Software Engineering (2006)
5. Mahmood, S., Lai, R., Kim, Y.S.: Survey of component-based software development. IET Softw. **2**(1), 57–56 (2008)
6. McIlroy, M.D.: Mass produced software components. In: Proceedings of NATO Conference on Software Engineering, pp. 88–98 (1969)
7. Ratti, N., Kaur, P.: A conceptual framework for analyzing the source code dependencies. In: Bhatia, S.K., et al. (eds.) Advances in Intelligent Systems and Computing. Advances in Computer and Computational Sciences, vol. 554, pp. ISBN 480–483. 978-981-10-3772-6
8. Rhinelander, R.: Components have no Interfaces! Wcop2007. Retrieved from http://research.microsoft.com/en-us/um/people/cszypers/events/wcop2007/initialsubmissions/p10-rhinelander.pdf (2007)
9. Sametinger, J.: On a taxonomy for software components. In: Workshop on Component-Oriented Programming, pp. 1–6 (1996)
10. Sametinger, J.: Software Engineering with Reusable Components. Springer, Berlin (1997). ISBN 3-540-62695-6
11. Jalender, B., Govardhan, A., Premchand, P.: A pragmatic approach to software reuse. J. Theor. Appl. Inf. Technol. (JATIT) **14**(2), 87–96 (2010)

# Improve the Tracking of the Target in Radar System Using the Particular Filter

Zarai Khaireddine[1]([⊠]), BenSlama Sami[2], and Cherif Adnane[1]

[1] Faculty of Sciences of Tunis El-MANAR, Analysis and Signal Processing of
Electrical and Energetic Systems, Tuni, Tunisia
khaireddine.zarai@fst.utm.tn, adnane.cher@fst.rnu.tn
[2] Information System Department, FCIT, King Abdulaziz University, Jeddah,
Saudi Arabia
benslama.sami@gmail.com

**Abstract.** In this paper we are interested to improve the state estimation of a maneuvering target, which improve the tracking of it in real time in radar system. Trajectory estimation is posed as a continuous filtering problem with a simple closed –form solution, which is used a turn to update the label cost. Many of the most successful tracking methods at present perform tracking by detection, the target is presented by an object model that can be detected in every frame independently. The advantage of using an object detector are that it naturally handles re-initialization if a target has been lost, and that it avoids excessive model drift. The detector yields the per-frame evidence for the presence or the absence of the target using the false alarm in discrete time. The aim is to improve the process of the state estimation for too random target at the given instant in order to smooth their true state path for a long time, it simplifies the process of real-time tracking. In this frame, we propose the numerical methods presented by MONTE CARLO (MC) counter-part the methods conventionally used named Extended KALMAN Filter (EKF), in order we showed that the first are more successful.

**Keywords:** Radar · Tracking · EKF · MC

## 1 Introduction

In this paper we are interested to improve the state estimation of a maneuvering target, which improve the tracking of it in real time in radar system.

Trajectory estimation is posed as a continuous filtering problem with a simple closed–form solution, which is used a turn to update the label cost. Many of the most successful tracking methods at present perform tracking by detection, the target is presented by an object model that can be detected in every frame independently.

The advantage of using an object detector are that it naturally handles re-initialization if a target has been lost, and that it avoids excessive model drift. The detector yields the per-frame evidence for the presence or the absence of the target using the false alarm in discrete time.

C. Benavente-Peces et al. (Eds.): SEAHF 2019, SIST 150, pp. 55–62, 2019.
https://doi.org/10.1007/978-3-030-22964-1_6

The aim is to improve the process of the state estimation for too random target at the given instant in order to smooth their true state path for a long time, it simplifies the process of real-time tracking.

In this frame, we propose the numerical methods presented by MONTE CARLO (MC) counter-part the methods conventionally used named Extended KALMAN Filter (EKF), in order we showed that the first are more successful.

### 1.1    Problem Statement

The problems of the tracking in the radar systems are directly connected by the filtering Doppler, the number and the type of targets presented to every scanning. If the latter is important his pose a problem of filtering and in the tracking in real time. Besides the random movement of every target (Maneuvering target) its pose also a problem of stochastic filtering according to the SWERLING models [11].

In this paper we are interest in estimation and tracking problems, new approach more used are exposed such as MONTE CARLO developed to improve the tracking process compared by Extended KALMAN Filter.

## 2    Bloc Diagram of Radar Signal Processing

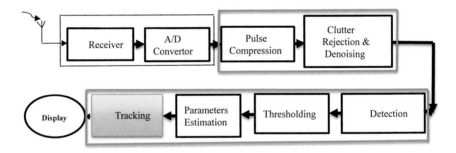

**Fig. 1.**  Radar signals processing blocs.

The signal processor is that part of the system which separates targets from clutter on the basis of Doppler content and amplitude characteristics. In modern radar sets the conversion of radar signals to digital form is typically accomplished after IF amplification and phase sensitive detection. At this stage they are referred to as video signals, and have a typical bandwidth in the range 250 KHz to 5 MHz. The Sampling Theorem therefore indicates sampling rates between about 500 kHz and 10 MHz. Such rates are well within the capabilities of modern analogue-to-digital converters (ADCs).

In this paper we are interest by the part framed in red in synoptic above presented in Fig. 1 which corresponds to the target tracking block.

# 3 Adaptive Algorithm for Tracking of the Target in Radar System

## 3.1 The State Model of the Maneuvering Target

According to the nature of movement of the target we can represent his model mathematical discreet in time. In the reality targets radars move with a time-varying speed so their accelerations varied abruptly, we say that they are (maneuvering) no linear targets, this characteristic raises a problem of the tracking.

We can describe the evolution of state of the target in the discreet time by the following state model:

$$\begin{cases} X_k = F \cdot (X_{k-1}) + W_k \\ \quad Y_k = H \cdot (X_k) + V_k \end{cases}$$

In this section we will present two types of methods of the target tracking in radar system, such as Extended KALMAN filter counterpart the Particular Filer based on MONTE CARLO algorithm.

## 3.2 Extended KALMAN Filter (EKF): Algorithm A

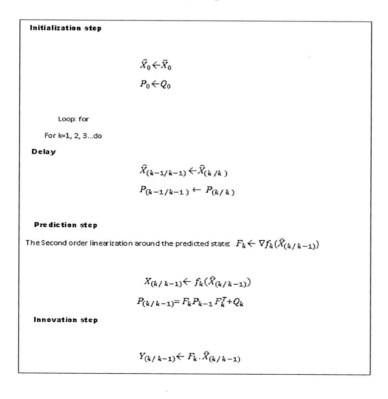

**Initialization step**

$$\hat{X}_0 \leftarrow \bar{X}_0$$

$$P_0 \leftarrow Q_0$$

Loop: for

For k=1, 2, 3...do

**Delay**

$$\hat{X}_{(k-1/k-1)} \leftarrow \hat{X}_{(k/k)}$$

$$P_{(k-1/k-1)} \leftarrow P_{(k/k)}$$

**Prediction step**

The Second order linearization around the predicted state $F_k \leftarrow \nabla f_k(\hat{X}_{(k/k-1)})$

$$X_{(k/k-1)} \leftarrow f_k(\hat{X}_{(k/k-1)})$$

$$P_{(k/k-1)} = F_k P_{k-1} F_k^T + Q_k$$

**Innovation step**

$$Y_{(k/k-1)} \leftarrow F_k \cdot \hat{X}_{(k/k-1)}$$

### 3.3  The Numerical Method: Monte Carlo Method Particular Filter (Algorithm B)

The Monte Carlo method consists in representing the probability of the region observed by the concentration of their sequential samples Np. The goal is to approach a probability distribution that it is generally impossible to calculate analytically.

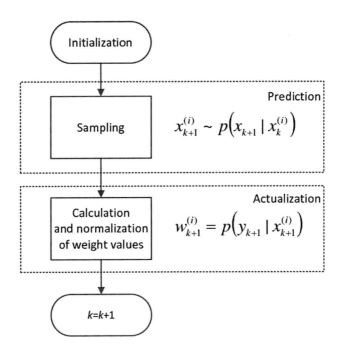

## 4  Simulation Results

In this part, we will simulate the theoretical results was compared in the last section, it nevertheless has a certain practical interest in showing the contribution of the Monte Carlo methods compared by the classical methods such as the EKF which are sometimes questioned and especially by proposing simple solutions to difficult tracking problems. We will compare the performance of EKF on the problem of tracking and we will confirm that the particular Filter based on Monte Carlo (SIR algorithm) is a particular solution of this problem is simple and efficient compared to the solutions of the algorithm conventionally used.

Tracking Model:

X = 0.5*X + [05 *X /[ ⟦ (1 + X ⟧ ^2] +8 cos [1.2*(k1-)] +processing noise
Y = ⟦X⟧ ^2 /20 + measurement noise

Simulation Parameters:

X = 0.1: the first state
Q = 1: the state process covariance matrix
R = 1: the covariance matrix of the measurement equation
t_f = the simulation time interval in seconds = 100 s.
Number of samples (NS) = 500, 4000 and 6000: the number of samples generated
in t_f in first.

## 4.1   Steps Simulation

- Step A: we simulate with N = 500 samples in 100 s

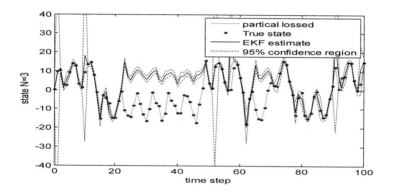

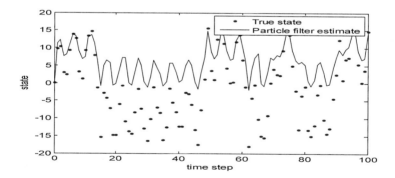

- Step B: we simulate with N = 4000 samples in 100 s

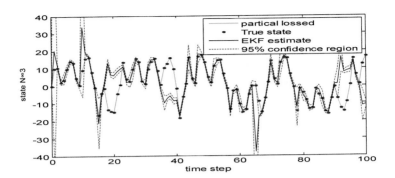

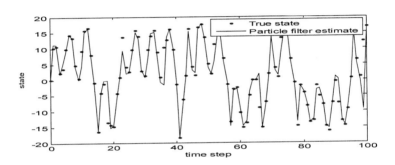

- Step C: we simulate with N = 6000 samples in 100 s

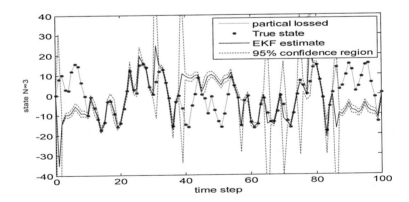

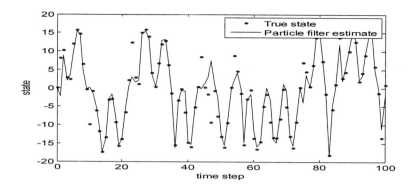

## 4.2   Comparative Study

See Table 1.

**Table 1.** Judgment metrics

| T=100 s  N=500 | EKF ( Algorithm A) | P.F (Algorithm B) |
|---|---|---|
| RMSE | 14.2193 | 15.6059 |
| PSNR | -23.0576 | -23.8658 |
| NRMSE | 1.4864 | 1.6313 |
| SNR | -3.3790 | -4.1872 |
| T=100s  N=6000 | EKF ( Algorithm A) | P.F (Algorithm B) |
| RMSE | 13.8900 | 2.5438 |
| PSNR | -22.7666 | -8.1098 |
| NRMSE | 1.4429 | 0.2669 |
| SNR | -2.9163 | 11.7405 |

The simulation results exhibits that the state estimation accuracy is almost same for all two filters. Yet, the PF shows better state estimate on target with small error. The PF based on SIR algorithm estimates the state of the target more accurately than the EKF about 100 s. Therefore, it is possible to conclude that PF shows similar or better estimation compared to existing EKF. The additional analysis on initial convergence rate of PF of Monte Carlo is required.

## 5   Conclusion and Future Works

The simulation results are verified, we have worked to overcome the constraints related to the nonlinearity of the target, we have pursued a manoeuvring target and we obtain the desired results and we confirm that the particular filtering (PF) based on Monte Carlo approach is more efficient in the complex conditions with respect to the KAL-MAN Extended filter. PF is fast in calculation, so it is capable to perform the real-time tracking in a non-stationary white noise gain, optimal filtering capability and making the state trajectory smoothed clear in a long period.

## References

1. Kulikov, G.Y., Kulikova, M.V.: The accurate continuous-discrete extended KALMAN filter for radar tracking. IEEE Trans. Signal Process. **64**(4) (2016)
2. Lafarge, F.: Introduction to KALMAN Filtering Discrete KALMAN Filter Theory and Applications, Edition (2009)
3. Jishy, K.: Tracking of moving targets in passive radar using Gaussian particle filtering. Doctoral thesis. National Institute of Telecommunications, Paris (2011)
4. Bel, K.L., Baker, C.J., Smith, G.E.: Cognitive radar framework for target detection and tracking. IEEE J. Sel. Top. Signal Process. **9**(8) (2015)
5. Cho1, M.-H., Tahk, M.-J.: Modified gain pseudo-measurement filter design for radar target tracking with range rate measurement. In: 25th Mediterranean Conference. Valletta, Malt, July 3–6 2017

# Robot Intelligent Perception Based on Deep Learning

Sehla Loussaief[1,2(✉)] and Afef Abdelkrim[1,2]

[1] Laboratory of Research in Automatic, National Engineering School of Tunis,
University of Tunis El Manar, Tunis, Tunisia
sehla.loussaief@gmail.com, afef.a.abdelkrim@ieee.org
[2] National Engineering School of Carthage, University of CarthageTunis,
Carthage, Tunisia

**Abstract.** Robotic and automation field is continuously in expansion. Robotic systems are now operating in unknown and dynamic environments. Therefore, they must not only classify sensory pattern but also determine the decision and action to be made. The well making decision of robot will depend on its efficiency when processing raw sensor data. In this work, we propose an innovative approach for robot intelligent perception and decision making process. We investigate the ability of deep learning methods to be brought to bear on robotic system decision making and control. Our challenging researches consist on providing robots the ability to autonomously recognize obstacle without a pre-programming need. For this purpose, we design a deep learning based framework to compute a high-quality convolutional Neural Network (CNN) model for image classification. The designed approach is labeled Enhanced Elite CNN Propagation Method. Simulations demonstrate the effectiveness of robot decision making when exploring its environment based on our approach.

**Keywords:** Robotic system · Deep learning · Convolutional neural network · Robotic intelligent perception · Robotic making decision

## 1 Introduction

For a long time, robotic systems have been used in industrial environments. Nevertheless, these robotic systems are pre-programmed with repetitive assignments and work in a structured way lacking the ability of autonomy. With the emergency of mobile robots, such structured environment can no longer be adopted since it doesn't provide them with autonomy. Mobile robots must be able to make their own decisions regarding navigation paths, whether objects in their surrounding are obstacles, recognize images and audio [1, 2], etc. While operating in dynamic and unknown environments, robots must be capable to perceive obstacles and react accordingly. Therefore, sensing and intelligent perception are becoming vital because they will determine the robotic system performance [3, 4]. In this paper, robot intelligent perception is investigated. Our contribution is to provide a mobile robot with the capability

© Springer Nature Switzerland AG 2019
C. Benavente-Peces et al. (Eds.): SEAHF 2019, SIST 150, pp. 63–70, 2019.
https://doi.org/10.1007/978-3-030-22964-1_7

of recognizing surrounding obstacles and decision making autonomously based on deep learning techniques.

This paper is organized as follows. Section 2 exposes the usage of deep learning for robotic system perception. In Sect. 3, problem statement and related work are presented. In Sect. 4, developed approaches and simulation results are illustrated. The last section includes our concluding remarks.

## 2　Deep Learning for Robotic System Perception Training

Deep learning is defined as the field of science that involves training extensive artificial neural networks using complex functions for data sense extraction [5]. Deep learning models use supervised learning technique to reach their goals [6]. For instance, for image classification task a neural net will be required to be trained with a collection of labeled data [7]. Thanks to deep learning more advances are observed in many areas such as image recognition, natural language processing, etc. It is on this backdrop that deep learning has taken the forefront position in developing innovative methods to improve robotic system perception capabilities. Robotic system perception can be defined as the ability of robots to detect their surroundings [8]. It is reliant on several sensory information sources and requires an immense high-dimensional data processing capabilities [9]. Structured old approaches show their limits when robots operate in an unstructured way to process information about their environment [10, 11]. As such, deep learning introduces new techniques which allow processing surrounding data from robotic system sensors using perception and final decision-making strategies [12].

## 3　Problem Statement and Related Work

Robotic systems are increasingly used to automate routine tasks in many areas. In traditional approach, robots are limited to operate in a structured manner and require a pre-programming task from researchers. As shown in Fig. 1, this operation way includes tedious effort to manually implement methods and policies for raw sensor treatment and robot motor control. This traditional approach shows its limit to address robots which operate in a dynamic and instructed way. In our work, we aim to design an approach allowing robots to perceive intelligently surrounding obstacles and make the corresponding decision. For instance, make a stop decision while recognizing a stop sign obstacle. For robot intelligent perception many researches use deep learning neural network. For each observation robots are able to autonomously make decision about action to be executed (see Fig. 2). Thus, choosing an optimal deep neural network is the most challenging aspect. In related works, some researchers use very deep convolutional neural networks (CNN) [13]. Others use Leaky Rectifiers to get faster training [14]. In this work, we present a competitive framework based on an innovative designed method labeled Enhance CNN Model Propagation.

**Fig. 1.** Robotic system making decision process based on traditional methods

**Fig. 2.** Robotic system making decision based on deep learning

## 4 Proposed Approach and Robotic Experiments

For mobile robotic system, the most challenging issue is moving from sensory input to control output. Our contribution is to offer to the robotic system the ability of making decision while navigating based on intelligent perception. For obstacle recognition and decision making, the robotic system is interfaced with a machine learning framework for image classification. This framework delivers to the robot a high-quality pre-trained CNN model able to classify robot captured images on line and return their labels so that the robot makes the corresponding decision. Many techniques were investigated to design the prediction model such as Bag of Features, AlexNet pre-trained CNN, and Transfer Learning. But, the best reached image classification accuracy not exceeded 93.33% [15]. For this purpose, we design an innovative approach labeled Enhanced Elite CNN Model Propagation (Enhanced E-CNN-MP) which allows as to compute an optimal CNN for robot perceived obstacles classification [16]. To solve such non-deterministic problem, we use genetic algorithms (GA) to search for a best CNN structure offering a high-quality pre-trained CNN suitable for robot decision making. The designed framework is shown in Fig. 3.

The Enhanced E-CNN-MP approach uses specific designed GAs methods for chromosome encoding and recombination. Each chromosome is a candidate solution representing a CNN architecture. The training process error is chosen as the GA fitness function of a chromosome to be optimized. Enhanced E-CNN-MP main algorithm is presented in Algorithm I. Figure 4 represents the Enhanced E-CNN-MP approach simulation which gave as the best fit CNN structure offering a classification accuracy of 98.94% [16]. The optimal saved CNN structure is described in Table 1. The proposed approach for robot making decision is described in Algorithm II. For robot experiments, we use the Indigo Gazebo[1] simulator with the TurtleBot robotic system implementation.[2] The designed scene is described in Fig. 5. As shown in Figs. 6 and 7, simulations demonstrate that the robot is able to label obstacle in its surroundings

---

[1] http://gazebosim.org/.

[2] https://www.turtlebot.com/.

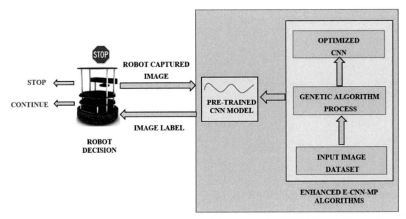

**Fig. 3.**  Robotic system perception based on deep learning framework

during navigation. When getting a StopSign label for captured image, the TurtleBot robot stops. For others obstacle labels it applies the navigation algorithm with obstacle avoidance.

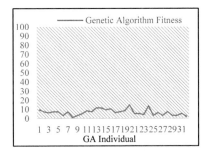

**Fig. 4.**  Enhanced E-CNN-MP approach simulation

**Table 1.** Best fit CNN structure

| Layer | Layer name | Layer properties |
|---|---|---|
| 1 | Image input | 227 x 227 x 3 images with 'zerocenter' normalization |
| 2 | Convolution | 59 19 x 19 x 3 convolutions with stride [1 1] and padding [9 9 9 9] |
| 3 | Batch normalization | Batch normalization with 59 channels |
| 4 | ReLU | ReLU |
| 5 | Max pooling | 2x2 max pooling with stride [2 2] and padding [0 0 0 0] |
| 6 | Convolution | 37 4 x 4 x 59 convolutions with stride [1 1] and padding [2 2 2 2] |
| 7 | Batch normalization | Batch normalization with 37 channels |
| 8 | ReLU | ReLU |
| 9 | Max pooling | 2 x 2 max pooling with stride [2 2] and padding [0 0 0 0] |
| 10 | Convolution | 81 6 x 6 x 37 convolutions with stride [1 1] and padding [3 3 3 3] |
| 11 | Batch normalization | Batch normalization with 81 channels |
| 12 | ReLU | ReLU |
| 13 | Max pooling | 2x2 max pooling with stride [2 2] and padding [0 0 0 0] |
| 14 | Convolution | 79 17 x 17 x 81 convolutions with stride [1 1] and padding [8 8 8 8] |
| 15 | Batch normalization | Batch normalization with 79 channels |
| 16 | ReLU | ReLU |
| 17 | Max pooling | 2 x 2 max pooling with stride [2 2] and padding [0 0 0 0] |
| 18 | Convolution | 41 5 x 5 x 79 convolutions with stride [1 1] and padding [2 2 2 2] |
| 19 | Batch normalization | Batch normalization with 41 channels |
| 20 | ReLU | ReLU |
| 21 | Max pooling | 2 x 2 max pooling with stride [2 2] and padding [0 0 0 0] |
| 22 | Fully connected | 256 fully connected layer |
| 23 | Softmax | Softmax |
| 24 | Classification output | crossentropyex |

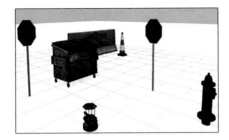

**Fig. 5.** TurtleBot simulation scene

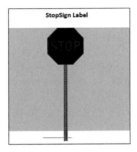

**Fig. 6.** Captured image with a stop made decision

**Fig. 7.** Captured image with continuous navigation decision

---

**Algorithm I: Enhanced E-CNN-MP**

**Input**: Images dataset D, NumConLayers, (MaxG), Generation size N

**Output**: Best individual Fit*I*, Fit*CNN*

Initialization of TrainingDS, TestDS, Best Accuracy, Fraction of elites ($f_e$), Fraction of crossover created children($f_c$)

Randomly (Uniform distribution) create an initial population P of N chromosomes.

**for** g **in** 1 **to** MaxG  **do**     /* Max generations number*/

[$S_1,S_2$ ,..,$S_N$, BestAccuracy, Fit*I*, Fit*CNN*]← **FitnessCNN** (P, NumConLayers, BestAccuracy, TrainingDS, TestDS);

P'← **Recombination** (P , $f_e$, $f_c$);

P←P'; **end for;**

return Fit*I*; save Fit*CNN*;

| Algorithm II: Robot Navigation with Usage of Intelligent Perception |
| --- |
| **Input:** CNN pre-trained Model designed with Enhanced CNN MP method |
| **Output:** Robot decision |
| **/\*Load the CNN pre-trained model\*/** |
| CnnRobot←Load ('PretrainedCNN'); |
| **/\*Robot navigation with Stop Sign obstacle recognition\*/** |
| Start Robot Navigation with obstacle avoidance; |
| Image ← ('camera input'); |
| Label ← **classify**(CnnRobot, Image); |
| **if** Label== 'StopSign' **then**  pause(10); **else**   continue(); **end if;** |

## 5   Conclusion

Robots are now used in complex and dynamic environments. Pre-programming them by researchers is becoming impractical task. Their performances are tightly depending on sensing and intelligent perception.

In this article we investigate deep learning as technique for robot intelligent perception development. Current work contributions demonstrate how we can extend deep learning into robot decision making and control. We present an innovative approach allowing a mobile robot to drive machine learning algorithms for intelligent perception during its environment exploration. The developed intelligent robotic system is able to recognize obstacles based on an optimal designed pre-trained CNN for image classification. For this CNN hyper-parameters computation, we use our innovative designed approach "Enhanced Elite CNN Model Propagation". This approach allows as to interface the robotic system with a high-quality prediction model offering 98.94% of classification accuracy. Robotic simulation results demonstrate the ability of robot to make, for example, a stop decision during its navigation when recognizing a stop sign. Our designed framework can be applied in many field of researches based on robotic system use where machines must not only capture surrounding images but also choose accordingly the best action without requiring tedious job on the part of the researcher.

## References

1. Tai, L., Li, S., Liu, M.: A deep-network solution towards model-less obstacle avoidance. In: IEEE/RSJ International Conference on Intelligent Robots and Systems (IROS), pp. 2759–2764 (2016)
2. Sun, D., Kleiner, A., Nebel, B.: Behavior-based multi-robot collision avoidance. In: Proceedings of IEEE International Conference on Robotics and Automation, pp. 1668–1673 (2104)
3. Tai, L., Liu, M.: Deep-learning in mobile robotics—from perception to control systems: A survey on why and why not. abs/1612.07139 (2016)

4. Shao, J., Loy, C.: Deeply learned attributes for crowded scene understanding. In: IEEE Conference on Computer Vision and Pattern Recognition (CVPR), pp. 4657–4666 (2015)
5. LeCun, Y., Bengio, Y., Hinton, G.: Deep learning. Nature **521**(7553), 436 (2015)
6. Levine, S., Pastor P., Krizhevsky, A., Quillen, D.:Learning hand-eye coordination for robotic grasping with large-scale data collection. In: International Symposium on Experimental Robotics, pp. 173–184. Springer, Berlin (2016)
7. Yang, Y., Fermuller, C., Li, Y., Aloimonos, Y.: Grasp type revisited: A modern perspective on a classical feature for vision. In Proceedings of the IEEE Conference on Computer Vision and Pattern Recognition, pp. 400–408 (2015)
8. Duan, Y., Chen, X., Houthooft, R., Schulman, J., Abbeel, P.: Benchmarking deep reinforcement learning for continuous control. In: International Conference on Machine Learning, pp. 1329–1338 (2016)
9. Gongal, A., Amatya, S., Karkee, M.: Sensors and systems for fruit detection and localization: A review. Comput. Electron. Agric. **116**, 8–19 (2015)
10. Lenz, I., Lee, H., et Saxena, A.: Deep learning for detectingrobotic grasps. Int. J. Robot. Res. **34**(4–5), 705–724 (2015)
11. Ghahramani, Z.: Probabilistic machine learning and artificial intelligence. Nature **521**(7553), 452 (2015)
12. Giusti, A., Guzzi, J.: A machine learning approach to visual perception of forest trails for mobile robots. IEEE Robot. Autom. Lett. **1**(2), 661–667 (2016)
13. Veeriah, V., Zhuang, N., Qi, G.-J.: Differential recurrent neural networks for action recognition. In: 2015 IEEE International Conference on Computer Vision (ICCV), pp. 4041–4049. IEEE (2015)
14. Brighton, H., Selina, H.: Introducing Artificial Intelligence: A Graphic Guide, ser. Introducing…Icon Books Limited (2015)
15. Loussaief, S., Abdelkrim, A.: Deep learning vs. bag of features in machine learning for image classification. In: International Conference on Advanced Systems and Electrical Technologies, IC'ASET (2018)
16. Loussaief, S., Abdelkrim, A.: Convolutional neural network hyper-parameters optimization based on genetic algorithms. (IJACSA). Int. J. Adv. Comput. Sci. Appl. (2018). https://doi.org/10.14569/ijacsa.2018.091031

# Performances Study of Speed and Torque Control of an Asynchronous Machine by Oriented Stator Flux

M. Benmbarek[1(✉)], F. Benzergua[2], and A. Chaker[1,2]

[1] National Polytechnic School, ENP, Oran, Algeria
benmbarek-med@hotmail.com
[2] University Science and Technologies USTO, Oran, Algeria

**Abstract.** In this article, we proceeded to the study and analyzing of the speed and torque control performance of an asynchronous machine. This research presents in particular a new control scheme, whose principle is to control the operation of this machine similarly to a DC machine. Many control methods dealing this subject have been proposed in the publications and studies, the control by orientation of the rotor flux remains the most used given the high dynamic performance it offers for a wide range of applications. In this respect, our strategy of orienting the stator flux has shown by numerical simulation, the robustness of the proposed control against parametric variations as well as the working conditions. The results make it possible to illustrate, both in terms of performance and robustness, the contribution of such a control to impose on the machine dynamic behaviors similar to those of a control with oriented rotor flux. The objective of the regulators (PI) that we used is to regulate the stator flux and the speed as well as the torque.

**Keywords:** Asynchronous machine · Performances · Direct vector control · Flux oriented · Regulators · Speed · Torque

## Nomenclature

| | |
|---|---|
| a, b, c | Indices of stator and rotor phases |
| s, r | Indices relating to stator and rotor |
| $d,q$ | Indices of direct and quadrature orthogonal components |
| $I_{rd}, I_{rq}$ | Rotor current of the axes d-q |
| $V_{sd}, V_{sq}$ | Stator voltage of the axis d-q |
| $V_{rd}, V_{rq}$ | Rotor voltage of the axis d-q |
| $\varnothing_{sd}, \varnothing_{sq}$ | Stator flux of the axis d-q |
| $\varnothing_s$ | Amplitude of the stator flux |
| $R_s, R_r$ | Stator and rotor resistances |
| $L_s$ | Cyclic inductance of the stator |
| $L_r$ | Cyclic inductance of the rotor |
| $M$ | Mutual cyclic inductance stator-rotor |
| $T_s$ | Stator time constant ($L_s/R_s$) |
| $T_r$ | Rotor time constant ($L_r/R_r$) |

© Springer Nature Switzerland AG 2019
C. Benavente-Peces et al. (Eds.): SEAHF 2019, SIST 150, pp. 71–84, 2019.
https://doi.org/10.1007/978-3-030-22964-1_8

| | |
|---|---|
| $\sigma$ | Leakage coefficient of the Blondel |
| $w_s$, $w_r$ | Pulsations of the stator and the rotor respectively |
| $w_m$ | Mechanical pulsation |
| $J$ | Moment of inertia |
| $F$ | Coefficient of viscous friction |
| $C_{em}$, $C_r$ | Electromagnetic and load torques |
| $[P],[P]^{-1}$ | Park matrix, inverse matrix of Park |
| $S$ | Laplace operator |
| $G$ | Gain |
| IM | Induction Machine |
| $\sim$ | Symbol indicating the compensation |
| $\wedge$ | Symbol indicating the estimate |
| $*$ | Symbol Indicating the reference |
| (*) | Other notations and symbols are defined in the article |

## 1 Introduction

Robustness, low cost, ease of maintenance make the interest of the asynchronous machine in many industrial applications. The absence of natural decoupling between the inductor and the armature gives this machine a non-linear multivariable dynamic model which is the opposite of the simplicity of its structure. This poses a theoretical problem in his control, however, its qualities justify the renewed interest of the industry with respect to this type of machine [1]. In the asynchronous machine, the torque is the result of an interaction between the stator and rotor magnitudes. Thus, any torque variation by variation of the rotor current results in a stator flux evolution. To obtain a torque control with good dynamic performance, it is necessary, by a control system external to the machine, realize a decoupling between torque and flux. This decoupling is realized out by applying the oriented flux control, which consists in separating the torque control from that of the flux by orientation of the latter. This technique (vector control) gives the asynchronous machine the same performance as the DC machine with independent excitation whose torque is governed by the rotor current and the flux by the inductor current [2].

Many control methods dealing this subject have been proposed in publications and studies, the control by orientation of the rotor flux remains the most used given the dynamic performance it offers for a wide range of applications. In this respect, our objective is to highlight the orientation of the stator flux by a new control strategy that has been shown by numerical simulation, good dynamic performance and robustness against parametric variations [1]. The stator flux oriented in this work is regulated by a feedback loop requiring a good knowledge of its module and its phase, this one, must be verified whatever the transitional regime. This control mode guaranteed a correct decoupling between flux and torque, and the performances are generally correct, as well as the control proved a good robustness against parametric variations. This technique takes into consideration three integral proportional regulators (PI), whose objectives are to regulate the stator flux and the speed as well as the torque. The outputs of the torque and flux regulators directly select the components of the appropriate voltage vector through vector modulation.

## 2   Model of the Asynchronous Machine

The model of the asynchronous machine destined for vector control by orientation of the stator flux is obtained by considering the components of the stator flux vector and rotor current as state variables, and the components of the voltage vector as control variables [3]. This model is described by the following matrix representation:

$$\frac{dX}{d} = [A]X + [B]U \tag{1}$$

$$[A] = \begin{bmatrix} -\gamma & w_r & a_1 & -a_2 w_m \\ -w_r & -\gamma & a_2 w_m & a_1 \\ \frac{M}{T_s} & 0 & -\frac{1}{T_s} & w_s \\ 0 & \frac{M}{T_s} & -w_s & -\frac{1}{T_s} \end{bmatrix}, X = \begin{bmatrix} I_{rd} \\ I_{rq} \\ \emptyset_{sd} \\ \emptyset_{sq} \end{bmatrix}, [B] = \begin{bmatrix} -a_2 & 0 & a_3 & 0 \\ 0 & -a_2 & 0 & a_3 \\ 1 & 0 & 0 & 0 \\ 0 & 1 & 0 & 0 \end{bmatrix}, U = \begin{bmatrix} V_{sd} \\ V_{sq} \\ V_{rd} \\ V_{rq} \end{bmatrix}$$

$$J\frac{d\Omega}{dt} = C_{em} - C_r - f\Omega$$

$$\gamma = \left[\frac{1}{T_r\sigma} + \frac{(1-\sigma)}{T_s\sigma}\right], a_1 = \frac{1}{T_s M}\frac{(1-\sigma)}{\sigma}, a_2 = \frac{1}{M}\frac{(1-\sigma)}{\sigma}, a_3 = \frac{1}{\sigma l_r}, \sigma = 1 - \frac{M^2}{l_r l_s}, w_r = w_s - w_m \tag{2}$$

The asynchronous machine is a non-linear system where the torque is a crossed product of two magnitudes. In addition, this system is non-stationary because all magnitudes are likely to vary with time.

## 3   Principle of the Vector Control with Oriented Stator Flux

By neglecting parasitic phenomena such as armature reaction or switching, the electrical machine that best meets the decoupling hypotheses is the separate excitation DC machine.

Indeed, in this type of structure, it is simple to imagine independent controls of the flux and the torque respectively by the inductor and armature currents [4]. The objective for a control of the induction machine is to perform the previous operation using two action variables. The expression of the electromagnetic torque as a function of the rotor current and the stator flux is given by:

$$C_{em} = p\frac{M}{L_s}(\emptyset_{sq}I_{rd} - \emptyset_{sd}I_{rq}) \tag{3}$$

To simplify the control it is necessary to make a judicious choice of referential. For it, we place ourselves in a referential $d$–$q$ linked to the rotating field with an orientation of the stator flux (the axis 'd' aligned with the direction of the stator flux), we obtain:

$$\begin{cases} \varnothing_{sd} = \varnothing_s \\ \varnothing_{sq} = 0 \end{cases} \tag{4}$$

The expression of couple becomes:

$$C_{em} = -p\frac{M}{L_s}(\varnothing_s I_{rq}) \tag{5}$$

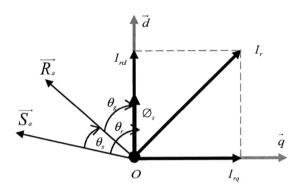

**Fig. 1.** Oriented stator flux illustration

Taking into account the choice of the referential ($\varnothing_{sq} = 0$) and the third equation of the system (1), the evolution of the flux is given by:

$$\frac{d}{dt}\varnothing_{sd} = -\frac{1}{T_s}\varnothing_{sd} + \frac{M}{T_s}I_{rd} + V_{sd} \tag{6}$$

We can notice from relations (5) and (6) that only the direct component $I_{rd}$ determines the amplitude of the stator flux while the torque only depends on the quadrature component $I_{rq}$ if the stator flux is kept constant. As a result, a structure similar to that of a DC machine is obtained.

By applying the condition (4) only the stator voltages, rotor currents, speed and/or rotor position are accessible. The model of the asynchronous machine established in the oriented stator flux domain given by (7) serves, on the one hand, to constitute simple observers that function naturally in the open loop (estimators). On the other hand, to the synthesis of control loops:

$$\begin{cases} V_{rd} = \frac{1}{a_4}\left[\frac{d}{dt}I_{rd} + \gamma I_{rd} - w_r I_{rq} - a_1\varnothing_s + a_3 V_{sd}\right] \\ V_{rq} = \frac{1}{a_4}\left[\frac{d}{dt}I_{rq} + \gamma I_{rq} + w_r I_{rd} - a_2 w_m\varnothing_s + a_3 V_{sq}\right] \\ \frac{d}{dt}\varnothing_s = -\frac{1}{T_s}\varnothing_s + \frac{M}{T_s}I_{rd} + V_{sd} \\ 0 = -w_s\varnothing_s + \frac{M}{T_s}I_{rq} + V_{sq} \\ C_{em} = -p\frac{M}{L_s}(\varnothing_s I_{rq}) \end{cases} \tag{7}$$

## 4 Estimators and the Self-piloting Used for the Control

Since the current $I_{rd}$ is measurable, the estimation of the stator flux becomes simpler according to Eq. (7):

$$\hat{Ø}_s = \frac{MI_{rd} + T_s V_{sd}}{[T_s S + 1]} \tag{8}$$

Figure 2 illustrates the estimation scheme of the stator flux vector with the determination of its position. The vector control methods differ essentially in the calculation of the Park angle $\theta_r$, (essential magnitude for the control). In direct control, the calculation of this parameter is done directly from the measured or estimated magnitudes. About that, we used a self-piloting said implicit [5, 6], hence the setting of the referential does not require the direct knowledge of $w_r$, the latter is deduced from the estimate of the speed $\hat{w}_s$:

$$\hat{w}_r = \hat{w}_s - p\Omega \tag{9}$$

The speed $w_s$ of the axis reference d, q is directly outcome by using the rotor component $I_{rq}$, which is accessible from the measurement of the real rotor currents after the transformation of Park:

$$\hat{w}_s = \left[\frac{MI_{rq}}{T_s} + V_{sq}\right] \frac{1}{\hat{Ø}_s} \tag{10}$$

In order to make Eq. (10) usable, we add an initial value $1 \times 10^{-3}$ wb to the stator flux $(\hat{Ø}_s)$ which is zero during machine start-up. The estimated electromagnetic torque is given by the following equation:

$$\hat{C}_{em} = -p\frac{M}{L_s}(\hat{Ø}_s I_{rq}) \tag{11}$$

The articulation between the estimation and self-piloting blocks is indicated by Fig. 2.

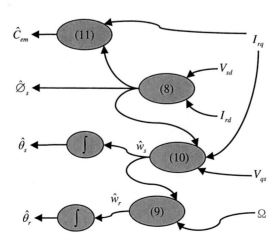

**Fig. 2.** Informational graph on estimation and self-piloting (implicit) of the machine

## 5  Direct Control

The schematic diagram of the direct control by oriented stator flux is illustrated in Fig. 3.

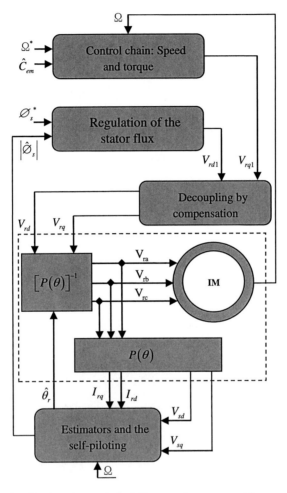

**Fig. 3.** Diagram of a direct vector control of the asynchronous machine powered with voltage by a balanced three-phase network

This method requires a good knowledge of the flux module and its phase [3, 7]. The flux vector is rarely measured directly by sensors. The practical realization of the sensors is delicate (poor accuracy, filtering of the measured signal, high cost, etc.). That's why, the techniques of estimation (or observation) of the flux from the measurable magnitudes are used. This control mode guarantees a correct decoupling between the flux and the torque whatever the point of operation, because it depends less on machine parameter variations [8, 9].

We can define according to the model (7) two control chains in the axes d,q (Fig. 3):

- Chain 1: Regulation of the stator flux by the rotor-current of the axis d;
- Chain 2: The regulation of the torque by the rotor-current of the axis q.

## 5.1   Decoupling by Compensation

The objective is, as far as possible, to limit the effect of an entry to a single output. We can then model the process as a set of monovariable systems evolving in parallel. The controls are then non-interactive. The purpose of this method is to define two new control variables $V_{rd1}$ and $V_{rq1}$ from Eq. (7). Such that $V_{rd1}$ only acts on $I_{rd}$ and $V_{rq1}$ on $I_{rq}$. The two new control variables are defined as follows:

$$\begin{cases} V_{rd} = V_{rd1} - e_{rd} \\ V_{rq} = V_{rq1} - e_{rq} \end{cases} \tag{12}$$

With:

$$\begin{bmatrix} V_{rd1} \\ V_{rq1} \end{bmatrix} = \sigma l_r \left( \begin{bmatrix} \frac{d}{dt} I_{rd} \\ \frac{d}{dt} I_{rq} \end{bmatrix} + \gamma \begin{bmatrix} I_{rd} \\ I_{rq} \end{bmatrix} \right) \tag{13}$$

$$\begin{bmatrix} e_{rd} \\ e_{rq} \end{bmatrix} = \sigma l_r w_r \begin{bmatrix} I_{rq} \\ -I_{rd} \end{bmatrix} + \frac{M}{l_s} \emptyset_s \begin{bmatrix} \frac{R_s}{l_s} \\ w_m \end{bmatrix} - \frac{M}{l_s} \begin{bmatrix} V_{sd} \\ V_{sq} \end{bmatrix} \tag{14}$$

The electrical part Eq. (12) appears as two monovariable processes coupled by the disturbance magnitudes $e_{rd}$ and $e_{rq}$ The voltages $V_{rd}$ and $V_{rq}$ are then reconstituted from the voltages $V_{rd1}$ and $V_{rq1}$. These voltages respectively allow the regulation of flux and torque, Fig. 4.

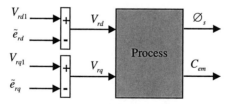

**Fig. 4.**  Reconstitution of voltages $V_{rd}$ and $V_{rq}$

By explicitly showing the flux and the torque from Eqs. (7) and (13), we obtain the reconstruction of the decoupling given in Fig. 5.

Figure 6 represents a functional diagram of the control which, a priori, can be envisaged following two strategies, depending on whether the non-linear disturbances ($e_{rd}$ and $e_{rq}$) are compensated or not. The compensation has the effect of decoupling the

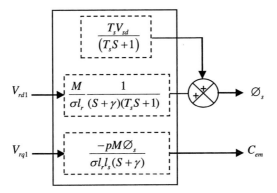

**Fig. 5.** Representation of the decoupling

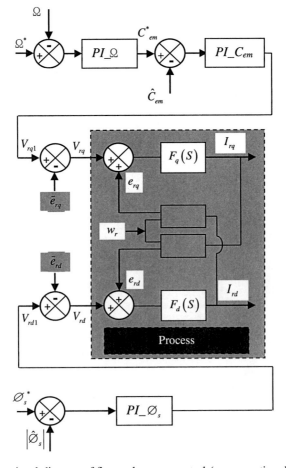

**Fig. 6.** Functional diagram of flux and torque control (compensation decoupling)

two processes through a real-time reconstruction of these reciprocal disturbances which, in this case, are measurable.

The disturbances reconstituted by the control device ($e_{rd}$, $e_{rq}$) ensure a decoupling only if:

$$\begin{cases} \tilde{e}_{rd} = e_{rd} \\ \tilde{e}_{rq} = e_{rq} \end{cases} \tag{15}$$

In such conditions, the system becomes linear, it comes:

$$I(S) = F(S)[V(S)] \tag{16}$$

With:

$$I(S) = \begin{bmatrix} I_{rd}(S) \\ I_{rq}(S) \end{bmatrix}, \ F(S) = \begin{bmatrix} F_d(S) \\ F_q(S) \end{bmatrix}, \ V(S)$$
$$= \begin{bmatrix} (V_{rd1}(S) + e_{rd}(S)) - \tilde{e}_{rd}(S) & 0 \\ 0 & (V_{rq1}(S) + e_{rq}(S)) - \tilde{e}_{rq}(S) \end{bmatrix}$$

Expressions (12) and (13) result in the transfer functions $F_d$ (S) and $F_q$ (S) noted in (16):

$$\begin{cases} F_d(S) = \frac{I_{rd}(S)}{V_{rd}(S) + e_{rd}(S)} \\ F_q(S) = \frac{I_{rq}(S)}{V_{rq}(S) + e_{rq}(S)} \end{cases} \tag{17}$$

## 5.2   Synthesis of Control Loops

The control system uses three regulators, namely:

– The stator flux regulator is determined by the use of the transfer equation linking $Ø_s$ to the component $V_{rd1}$:

$$\frac{Ø_s}{V_{rd1}} = \frac{M}{\sigma l_r} \frac{1}{(S + \gamma)(T_s S + 1)} + \frac{T_s V_{sd}}{(T_s S + 1)} \tag{18}$$

– The speed regulator is determined from the following equation:

$$\Omega = \frac{(K_{p\Omega}S + K_{i\Omega})}{JS^2 + (f + K_{p\Omega})S + K_{i\Omega}} \Omega^* - \frac{S}{JS^2 + (f + K_{p\Omega})S + K_{i\Omega}} C_r \tag{19}$$

The characteristic equation has second-order dynamics. In pursuit ($C_r = 0$), taking into account the coefficient of friction, we will write:

$$\Omega = \frac{\left(\frac{K_{p\Omega}}{K_{i\Omega}} S + 1\right)}{\frac{1}{K_{i\Omega}} S^2 + \frac{(f + K_{p\Omega})}{K_{i\Omega}} S + 1} \Omega^* \qquad (20)$$

- The torque regulator is determined by the transfer equation linking $C_{em}$ to the component $V_{rq1}$:

$$\frac{C_{em}}{V_{rq1}} = \frac{-pM\varnothing_s}{\sigma l_r l_s} \frac{1}{(S + \gamma)} \qquad (21)$$

The regulators used in this work are of integral proportional type (PI) [10]:

$$G_{pi}(S) = K_p + \frac{K_i}{S} \qquad (22)$$

$K_p$: coefficient of proportionality
With:
$K_i$: coefficient of integration.

## 6   Résultats de Simulation

The results exposed in Fig. 7 are relative to two controls of the asynchronous machine, to $\varnothing_s$ oriented as well as $\varnothing_r$, these two controls have been designed with the same strategy and test against the different disturbances that can affect the optimal functioning, such as, the variation of speed and load. The control by $\varnothing_s$ oriented was also tested for parametric variations due mainly to temperature during operation.

The simulations were carried out on an asynchronous machine of power equal to 4 kW, the parameters of this machine are given in appendix:

The results obtained for a $\varnothing_s$ oriented control, calculated with linearization compensation, show that the expected dynamics have been respected, from the outset, we note the effectiveness of the proposed control and self-piloting strategy, which effectively maintained the orientation of the axis 'd' on the stator flux, and this by observing the pace of the currents $I_{rd}$ and $I_{rq}$, these last are respectively in perfect agreement with the evolution of the flux $\varnothing_{sd}$ and the couple. The existing negative sign in the torque expression of Eq. (7) justifies the reflected image of its pace with respect to the current $I_{rq}$. In the presence of the compensation, $\varnothing_{sq}$ equal to a zero value and $\varnothing_{sd}$ remains constant once the transitional regime has been extinguished and independently of the torque evolution. During the speed-up phase, the torque takes the value of 41 Nm since the speed reaches 150.79 rad/s in 0.75 s with a moment of inertia equal to 0.07 kg m².

The speed response is without overshoot, low static error, with a rapid rejection of disturbances, the characteristic obtained follows its reference model with a slight transient error. The sudden change of speed between 2.5 and 3.5 s shows good prosecution of references, so the control is robust at low speeds.

The observation of the characteristics of the torque, speed and even the currents $(I_{sd}, I_{sq})$ for the case of the control by $\emptyset_r$ oriented does not raise of particular comment, previous analysis ($\emptyset_s$ orienté) is substantially recovered, on the other hand, the flux curves ($\emptyset_{rd}$ and $\emptyset_{rq}$) are a little disturbed.

we can say from the results obtained (Fig. 7) that the control strategy proposed in this article is efficient, it allowed to impose on the machine <u>dynamic</u> behaviors similar to those of a control with à $\emptyset_r$ oriented. The model of the asynchronous machine destined for vector control by orientation of the rotor flux is given in the appendix. In order to test the robustness of the proposed control ($\emptyset_s$ orienté), we studied the influence of the variations of the parameters on the performances of setting. Two cases are considered:

1. variation of $R_s$, $R_s = m\ R_s$, $0.5 \leq m \leq 1.5$, with $J = J_n$;
2. variation of $R_s$, $R_s = m\ R_s$, $0.5 \leq m \leq 1.5$, with $J = 2J_n$.

We did not present the results obtained during rotor resistance variation ($R_r = mR_r$, $0.5 \leq m \leq 1.5$), since there is practically no impact on the control which remains largely insensitive, this is understandable, since the time constant $T_r$ does not intervene in the calculation of estimators.

On the other hand, the observation of the Eqs. (8) and (10) of Fig. 2, show that the stator resistance intervenes in the computation of the estimators of the magnitudes to control, which causes during its variation (in particular when $R_s$ decreases by 50%) a true justification of the somewhat oscillatory behavior, and of short duration, of $\emptyset_{sq}$ and $C_{em}$ curves precisely during the transient phase, (Fig. 8). This is because of the influence of the inaccuracy of the measurement of $R_s$ on the evaluation of the observation error, therefore a less precise correction. Despite this, the error remains acceptable since it does not lead to a static error in steady regime, but the dynamic regimes are slightly disturbed.

Figure 9 presents the impact of the moment of inertia variation $J$ and the resistance $R_s$ on the control, we noticed that there is practically no influence of these disturbances on the static performances. On the other hand, transitional regimes (starting and changing the direction of rotation) have undergone an increase in the amplitude of the flux and torque curves. It is also shown that, speed is obtained, without overshoot, largely insensitive to load couples with a fast disturbance rejection and a negligible static error. The errors recorded in the dynamic regimes remain acceptable for the control which guaranteed the desired performances.

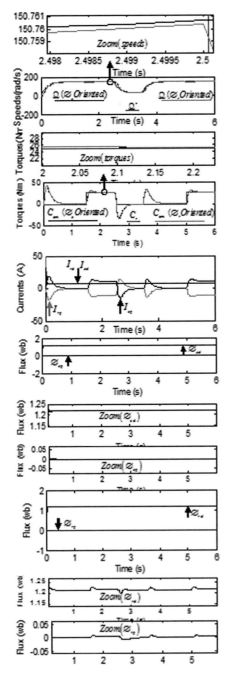

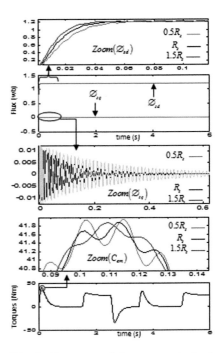

**Fig. 8.** Response of the system to load and resistance variations $R_s$ with $J = J_n$

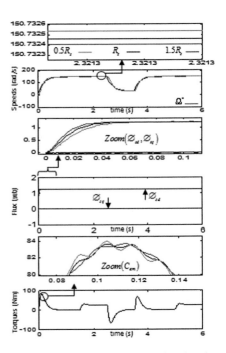

**Fig. 7.** Answers of the machine for $\varnothing_s$ and $\varnothing_r$ oriented along the axis'd'

**Fig. 9.** System response to load and resistance variations $R_s$ with $J = 2 J_n$

# 7 Conclusion

The objective of this work is to highlight the direct control of the asynchronous machine by the orientation of the stator flux, and this by:

(1) To ensure persevering decoupling for the different types of disturbances that the machine may undergo during its operation, It should be noted in this context that, the main problem of the control resides essentially in the robustness of the estimates against the variations of the parameters and speed.
(2) Give control by $\varnothing_s$ oriented of the asynchronous machine good dynamics performances similar to a control by $\varnothing_r$ oriented.

It is shown that the corrective feedback loop used based on classical regulator (PI) helps to accelerate the convergence of flux estimates, this reduces the sensitivity of the estimates to parameter variations. Consequently,
we had good dynamics, and perseverance in the orientation of the $\varnothing_s$ on the axis 'd' for any type of disturbance.

# Appendix

(1) The matrix representation of the model of the asynchronous machine established in the oriented rotor flux domain:

$$\frac{dX'}{d} = [A']X' + [B']U \tag{23}$$

$$[A'] = \begin{bmatrix} -\gamma' & w_s & a'_1 & a_2 w_m \\ -w_s & -\gamma' & -a_2 w_m & a'_1 \\ \frac{M}{T_r} & 0 & -\frac{1}{T_r} & w_r \\ 0 & \frac{M}{T_r} & -w_r & -\frac{1}{T_r} \end{bmatrix}, \quad X' = \begin{bmatrix} I_{sd} \\ I_{sq} \\ \varnothing_{rd} \\ \varnothing_{rq} \end{bmatrix} \quad [B']$$

$$= \begin{bmatrix} a'_3 & 0 & -a_2 & 0 \\ 0 & a'_3 & 0 & -a_2 \\ 0 & 0 & 1 & 0 \\ 0 & 0 & 0 & 1 \end{bmatrix}, \quad U = \begin{bmatrix} V_{sd} \\ V_{sq} \\ V_{rd} \\ V_{rq} \end{bmatrix}$$

With: $\gamma' = \left[\frac{1}{T_s\sigma} + \frac{(1-\sigma)}{T_r\sigma}\right], a'_1 = \frac{1}{T_r M}\frac{(1-\sigma)}{\sigma}, a'_3 = \frac{1}{\sigma l_s}$

(2) The data of the asynchronous machine with two pole pairs: 4 kW, 220/380 V – 50 Hz, 15/8.6 A, 1440 tr/min, $C_{em} = 25$ Nm.

Parameters: $R_s = 1.2\ \Omega$, $R_r = 1.8\ \Omega$, $L_s = 0.1554$ H, $L_r = 0.1568$ H, $M = 0.15$ H, $J = 0.07$ kg kg m², $f = 0.0001$ N.m.s/rd.

# References

1. Morand, F.: *Techniques d'observation sans capteur de vitesse en vue de la commande des machines asynchrones.* Thèse de doctorat École doctorale de Lyon, 07 janvier 2005
2. Baghli, L.: Contribution to induction machine control, using fuzzy logic, neural networks and genetic algorithms. Thèse de doctorat de University de Henri Poincare, France (2009)
3. Youbi, L., Craciunescu, A.: Commande directe du couple et commande vectorielle de la machine asynchrone. Rev. Roum. Sci. Techn, Électrotechn. et Énerg 2(53), 87–98 (2008)
4. Grellet, G., Clerc, G.: Actionneurs Electriques, Principe, Modèles. Commande. Collection Electrotechnique, Edition Eyrolles (1997)
5. Caron, J.P., Hautier, J.P.: Modélisation et commande de la machine asynchrone. Edition Technip, Paris (1995)
6. Poitiers, F.: Etude et Commande de Génératrices Asynchrones pour L'utilisation de l'énergie Éolienne, Thèse de Doctorat en Electronique et Génie Electrique. Ecole Polytechnique de l'Université de Nantes (2003)
7. Youbi, L.: A. craciunescu, Etude comparative entre la commande vectorielle à flux orienté et la commande directe du couple de la machine asynchrone, U.P.B. Sci. Bull, Series C **69**(2) (2007)
8. Comnac, V.: Sensorless direct torque and stator flux control on induction machine using an extended KALMAN filter. In: Proceedings of IEEE International Conference on Control Application, pp. 674–679. Maxico (2001)
9. Jelassi, K.: Positionnement d'une Machine Asynchrone par la Méthode du flux Orienté. Thèse de Doctorat, INPT, Toulouse (1991)
10. Rivoir, M., Ferrier, J.L.: Asservissement, régulation, Commande analogique, Tome 2. Edition. Eyrolles, Paris (1996)

# If You Want to Well Understand Today, You Have to Search History, Babylon is a Good Example

Emad Kamil Hussein[✉]

Al-Mussaib Technical College, TCM, Al-Furat al-Awsat Technical University, ATU, Babylon, Iraq
emad_kamil72@tcm.edu.iq

**Abstract.** Based on the general scope of this conference, main aim of this paper is to extend bridges between different people from different countries and cultures depending on the common points that are connecting them by employing the historical ancient civilizations including the well-known historical city named "Babylonia" as a live example of such very old nations. A detailed explanation about this city has been stated in this paper in order to stimulate readers, cultured people, archeologists, and others to get a clear idea about how was the life at the past time and how such people were managing their life, daily interaction, financial issue, and other essential aspects of ordinary life. Babylonia located on the two banks of the river Euphrates in middle southern region in Iraq is mentioned many times at the late third millennium B. C.E. and firstly came to importance as the royal kingdom of the famous king Hammurabi of about 1790–1750 B.C.E.

## 1 Introduction

### Babylonia at a Glance

The word Babylon has been mentioned and spoken in an ancient Akkadian civilization (2335–2154 BC) at the center of the so called zone Mesopotamia and present day Iraq, where this city was representing the center of administrative city at that time. After many years, this small city has been widely expanded during the famous historical figure King Hammurabi at the early years of the eighteenth century BC, then transformed into the formal and biggest capital of the Mesopotamia, and in many cultures this city was commonly named as the country of Akkad [1]. Babylon has many different names in diverse languages, for example, the first name was Babilu, in old Babylonian Babilim, in Arabic language Atlal Babil, and in Hebrew Bavel. In present days, it locates exactly on the Euphrates river of about 5 km to the north of Hillah city and about 90 km to the south of Baghdad. See Fig. 1.

At that ancient time there were a lot of famous scientist that tried to create a general map of the Babylon kingdom as indicated in the following one in Fig. 2.

Almost of the kings that were governing Babylon were trying many times to unify the remain region of the south of Iraq at that time into one strong kingdom, so that

© Springer Nature Switzerland AG 2019
C. Benavente-Peces et al. (Eds.): SEAHF 2019, SIST 150, pp. 85–90, 2019.
https://doi.org/10.1007/978-3-030-22964-1_9

**Fig. 1.** Location of Babylon City in present days [2].

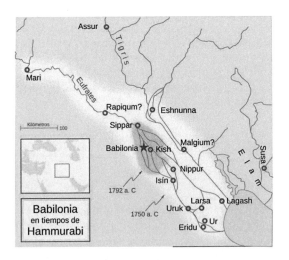

**Fig. 2.** Ancient map of the Babylon city [2].

makes Babylon as a center of all branches of science including medicine, astronomy, mathematics, establishing learning centers and libraries including some kinds of printed pieces of clay and sheets of breed in addition it becomes very active in commercial trade plus exchanging knowledge and goods that imported and exported from different near and far countries like Egypt, Iran (Persian Empire), and other very ancient kingdoms.

## Babylonian Kings

As a result of the nature of Babylon city including its very important and strategic location near the Euphrates river and representing the center of roads, to the east towards the Persian Empire, and the Egyptian Kingdom at the west direction, a famous and multiple strong kings have been governed Babylon kingdom, so in the coming paragraph, there will be a brief explanation about the most famous kings:

Starting from the 612 BCE, the two Babylonian kings were **Nabopolassar** and **Nabuchadnezzar (II),** see Fig. 3. They were reassembling Babylon city based on unifying the Assyrian empire with the other small cities and local administrative zones into one very civilized, commercially and military strong kingdom, so that leads to consider Babylonian Kingdom as the biggest and tougher power in the Mesopotamia. One of the most important events that have been done under ruling of those two great kings were, building of high classical building named Ziggurat Tower, where it was used for worship and communication with the god at that time, plus constructing one of the most popular gardens around the world named the Hanging Gardens and there are some its remaining structures until our days in the city of Babil in Iraq [3].

**Fig. 3.** King Nabuchadnezzar (II) [3].

Nabuchadnezzar ordered at his brilliant historical era to build the famous entrance of Babylon, called the Ishtar Gate, see Fig. 4.

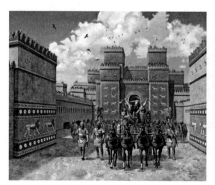

**Fig. 4.** Ishtar Gate [4, 5]

Another achievement of this king is the so called Hanging Gardens where they were a very strange thing at that time comparing with other historical things, see Fig. 5.

**Fig. 5.** Hanging Gardens [5].

At (555–539 B.C.E.) another famed king called Nabonidus, he was the last Babylonian king but he was crushed by the Persian Emperor Cyrus (II) so that representing the ending point of this great historical city [3].

One the most legendary kings of Babylon is the King Hammurabi (1792–1750 BCE), where he was the brilliant king at that time not only in Babylon but in the whole region of the middle east in present days, and he have important achievements in most aspects of life including all branches of ordinary life and also he did arrange the civilized life of the ancient people and nation of Babylon via establishing the so called the Hammurabi Code, see Fig. 6 below, and renewing the center of Babylon city, in addition there was a very famous historical lion of Babylon, see Fig. 7.

**Fig. 6.** Hammurabi Code [6]

**Fig. 7.** Lion of Babylon [6, 7]

It is possible to summarize all of these attainments existed in the Babylon city as follows:

1- Hammurabi Code
2- Hanging Gardens
3- Ishtar Gate
4- Babylon Lion

## 2  Conclusions

From the above mentioned cultural information about the historical city, Babylon, these conclusions may be drawn:

- Inspite of now advance technologies similar to the existing today at that time, Babylon city civilization abled to construct a unique pieces of historical things like the Hanging Gardens, where no electricity or water pumps, so it is concluded that this nation was very advanced in sciences and engineering.
- There was a political and commercial thinking that control and governing public life in Babylon city via creating the fame code of the king Hammurabi.
- This nation was very civilized because of the presence of many societies of knowledge within the high building named the Ziggurat Tower, which was used for education and worship.
- Creating a political system that control the relationship with the other important empires and kingdoms, exactly as the same of our today life in many civilized countries.

**Recommendations**

- Our today culture is representing an extension to such historical cities including the Babylon city, so it is important to complete the chain of progress towered spreading piece around the world.

- Many landmarks have been constructed at this city, so let the coming generations be informed about the ancient civilizations by visiting such unique historical locations especially the Babylon city.
- This is kind invitation to visit the Babylon city to see the on earth events and historical facts and fame figures.
- Supporting Babylon city to be mentioned on the international historical list.
- Doing more and deep studying about this city under license of the United Nation with cooperation if Iraqi academicians.

**Acknowledgements.** I'd like to thank governorate of Babylon in Iraq and the authority of historical cities in Babylon for their priceless support and very useful information and photos about Babylon plus some of specialist academician from Babylon where they provided me with rich historical information about our beloved city Babylon.

# References

1. https://en.wikipedia.org/wiki/Babylonia
2. https://www.google.com/maps/search/Babylon+location/@32.4619995,44.7768481,7z
3. https://www.khanacademy.org/humanities/ancient-art-civilizations/ancient-near-east1/babylonian/a/babylonia-an-introduction
4. https://www.britannica.com/place/Babylon-ancient-city-Mesopotamia-Asia
5. https://en.wikipedia.org/wiki/Hanging_Gardens_of_Babylon#/media/File:Ancient_seven_wonders_timeline.svg
6. https://www.youtube.com/watch?v=xLk1NP_nwwE
7. Governorate of Babylon
8. Authority of Historical Cities in Babylon

# WhatsApp for Defamiliarizing Foundation Year English Learners: A Collaborative Action Research Evaluation

Najat Alsowayegh and Ibrahim Garba(✉)

King Abdulaziz University, Jeddah, Saudi Arabia
n.alsowayegh@hotmail.com, igarba90@gmail.com

**Abstract.** We undertook two cycles within a collaborative action research by asking how using WhatsApp supports second language learners understanding of instruction, promoting, task achievement, enjoyment and creativity of learning English in class and online in Blackboard. We gathered data using a survey which indicated the advantages, challenges and learning opportunities of using WhatsApp in the first cycle. In the second cycle, the focus group, data from Blackboard and observations indicate using WhatsApp was '*enjoyable*' and '*refreshing*'. The collaborative action research prompted our conclusion for using provided a positive evaluation for using WhatsApp to encourage learning across the university for providing language learning instructions either in class or online in Blackboard.

**Keywords:** Second language learning · M-learning · Collaborative action research · Transformational learning

## 1 Introduction

We begin our research with the challenges that teachers face when using students' mobile phone social media for educational purposes. The lack of experience teachers possess indicates a trial and error to harness mobile learning (m-learning) (AlTameemy 2017; Awada 2016; Stockwell and Hubbard 2013).

Second language teachers (SLT) using m-learning on traditional settings to second language learners (SLLs) have a monumental task of understanding how the technology achieves meaningful learning (O'Brien 2004). We find the challenge narrows towards using m-learning to inform practice by practitioners evaluating their roles as SLT. Teaching on traditional settings brings pressures for learners of English who need to build their abilities for learning within a new learning environment that SLT incorporate through the m-learning.

A new context is created within the environment where SLT and SLLs become inexperienced and instruction turns the familiar upside down through the need to communicate (Valdes and Jhones 1991). In our language teaching context of foundation year program at a university in the Middle East, we piece the m-learning puzzle in our classroom environment (Carr and Kemmis 2003). As language teachers with experience of teaching English we to discover how SLLs will enjoy our incorporation

© Springer Nature Switzerland AG 2019
C. Benavente-Peces et al. (Eds.): SEAHF 2019, SIST 150, pp. 91–97, 2019.
https://doi.org/10.1007/978-3-030-22964-1_10

of m-learning through a social media platform called WhatsApp (Burns 2005, p. 7). Our education includes the educational values that support our experiences to relax the process around providing our students instruction on WhatsApp to support formal learning to take place online captured at the University's learning management system, Blackboard.

Our collaborative action research (CAR) evaluated how WhatsApp supports SLLs understanding of instruction, promoting SLLs task achievement, enjoyment and creativity of learning English.

### 1.1 Research Questions

We asked the following questions to guide our choice of research approach and the review of literature to support the data and analysis:

- What is enjoyment and creativity within the use of technology for learning?
- How is enjoyment and creativity defined and linked to communication?
- Why is enjoyment and creativity necessary for creating and sustaining the learning?
- How do we improve our teaching when using WhatsApp for supporting learners of English language?

## 2 Methodology

Our collaboration began with the decision for using CAR method for planning how to incorporate the use of WhatsApp in our teaching practices in two cycles of CAR. The female and male sections of the foundation year programs have separate buildings. The learners joining the foundation year courses select their respective programs and continue studying in English. The first researcher used WhatsApp for teaching the female section and used a survey to find out the SLLs opinions about using WhatsApp for learning English, observations of the SLLs' responses in WhatsApp and recorded the findings. The second researcher used WhatsApp for learning English online in Blackboard and recorded the views of the SLLs through a focus group an interview with the SLLs and observations of the SLLs' responses to the instruction given in WhatsApp to conduct learning activities on Blackboard and the findings recorded.

## 3 Findings

### 3.1 The Female Group

The female group were administered a questionnaire (Burns 2005; Cohen et al. 2005; Hamad 2017) to find out the perception of the learners towards the use of WhatsApp as a supplementary tool to improve their English language proficiency. Amongst the 24 learners at the language institute, we asked and received a 100% response rate questions on the following the advantages, challenges and learning opportunities using WhatsApp:

1. 14 itemed question on advantages for using WhatsApp N: 14 M: 4.23 SD: 0.24. The 14 itemed questionnaire responses of SLLs' sample responses see Fig. 1 shows positive response for learning using WhatsApp.

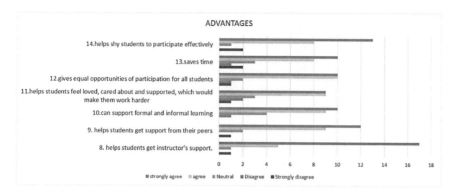

**Fig. 1.** The SLLs views of advantages for using WhatsApp for learning English

2. The 3 itemed question on challenges for using WhatsApp N: 3 M: 3.71 SD: 0.20. The 3 item questionnaire responses of SLLs' sample responses see Fig. 2 shows positive response for the challenges represented from using WhatsApp for learning English.

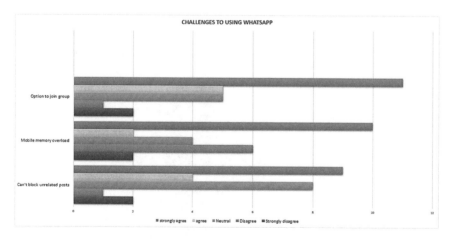

**Fig. 2.** The SLLs views of challenges to using WhatsApp for learning English

3. 13 item question on learning English through WhatsApp N: 13 M: 4.04 SD: 0.19.

The 13 item questionnaire responses of SLLs' sample responses see Fig. 3 shows positive response for the learning represented from using WhatsApp for learning English.

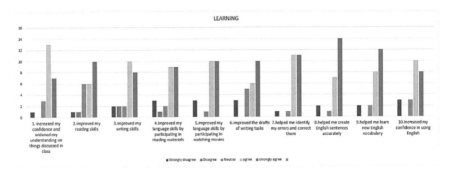

**Fig. 3.** The SLLs views of learning using WhatsApp for learning English

We coded and cleaned the WhatsApp text for personal references and replaced the learners and their SLT as STU and AR.t The tallied posts shows the SLT making 70% less posts that the learners during the period 2 September 2018 to 10 November 2018 as see Fig. 4.

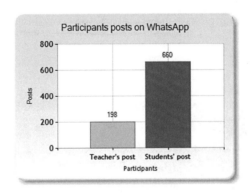

**Fig. 4.** The number of posts made in WhatsApp between the teacher and the students

Using NVivo 11, we transcribed the focus group conversation recorded at the end of the first part of the semester, 2018. We identified themes related to activities, content and setting. Finally, observing the classroom reflection of the teaching session (Burns 2005; Lightbown and Spada 2013; Littlejohn 2011), we identified the instruction, feedback and learners' enjoyment as reoccurring themes from the focus group.

## 3.2   The Male Group

The male group consisting of 18 SLLs were observed (Burns 2005; Cohen et al. 2005; Sonia and Rawekar 2017), using a focus group, and the SLLs' learning activities on Blackboard (Burns 2005; Cohen et al. 2005; Hiemstra 2001).

### Focus group

We found through the focus group that using WhatsApp for learning understanding instructions for online activities on Blackboard was '*convenient*' according to the SLLs. We also found that the feedback the SSL received from the instructor was useful for learning in Blackboard because the messages for online activities were also followed like *news items* on WhatsApp. Finally, we also learnt that the SLLs enjoyed the process of learning because it was '*enjoyable*' and '*refreshing*'. However, some of the SLLs preferred messages to be sent at specific times of the day so they can get online and carry out the activities.

### Blackboard

Activities on Blackboard included posting on a forum, doing quizzes, and participating on a discussing. The instruction on WhatsApp was links to where the forum was and what to do with a sample of the language and pieces missing such as the subject pronouns and nouns learners can replace with their information. 5 out of 18 learners were active in asking questions on WhatsApp and the rest situated around them. After clarifying with the instructor what the expectations were all learners followed the link given in WhatsApp to make their final post on bb so they could be graded. Faced with the task to perform and the sense of responsibility 16 learners work were marked and feedback given. **Also due to the high participation rate in** WhatsApp **the use of** WhatsApp **contributed to learners knowing where to go and where to post for the forum activities**.

The use of the quizzes provided learners an ability to practice the language where feedback was built in the quiz as part of the feature in Blackboard. In class, learners were given. The common exam with their peers however a comparison could not be made with the peers and the learners being taught for this research. **The exams in Blackboard gave learners an opportunity to check their work from the class activity and their understanding by having the answers available once the quiz was finished**.

The key functions for creating the discussion was for learners to work together. Groups were created and learners watched a video the teacher created and uploaded. Students were asked to create videos about the target language where a colleague asked them questions and the video uploaded on Blackboard. The initial uploading of the video could have been taught in class or the lab with learners following the teacher in a teacher led activity. However, having the video on Blackboard and learners using the discussion to clarify expectations allowed transforming videos on Blackboard.

# 4 Implications

We evaluated how WhatsApp supports SLLs understanding of instruction, promoting SLLs task achievement, enjoyment and creativity of using English. Our evaluation was used in the context of the classroom we taught English, the instructions we provided our SLLs on WhatsApp and the online activities carried out on Blackboard. As the first researcher incorporated a survey of the classroom and WhatsApp, a survey was used to find out the views of the learners. We reflected on the challenge of using WhatsApp in the context and hence asked questions related to measuring the advantages, challenges and learning opportunities using WhatsApp. The questionnaire represented what has been included as instruments for measuring the satisfaction of learners in the work of Awada (2016).

We also used a second cycle that viewed the process of learning from the SLLs' and SLT's perspective. We considered the meaning that SLLs attached to their learning using WhatsApp, Blackboard through observing and recording their views during a focus group and their work in Blackboard. We reflected on how SLLs viewed their activities as enjoyable, refreshing which contributed to our considerations of what we had done as language teachers using WhatsApp to influence the enjoyability of the learning outcome. Our observation of our teaching sets our work along the sides of Lightbown and Spada (2013) who observe the importance of interaction between SLL, the environment and the SLT. Though we could not have more SLT use WhatsApp similar to our CAR, we recognize the need for more instruction in this area because our SLLs indicate an interaction with content that highlights positive changes to learning English. When the learning reaches a level of change that can be described as permanent, our colleagues and administrators in the traditional settings can be enticed to consider how to provide access to a wider body of SLLs' learning. As we incorporated an CAR, the lessons we have learnt characterized for sustainment by Burns (2005, p. 210) can provide further support for using WhatsApp for instructing SLT/SLLs when using online learning tools like Blackboard.

We have evaluated our providing second language learners instructions on WhatsApp to support the learning of English in the classroom and on Blackboard. As inexperienced in the combination of m-learning and pedagogy we have incorporated the use of experts in the field to validate the source of data we can generate for evaluating our research. Our data gathering technique borrowed from research method experts like Burns (2005), Cohen et al. (2005), Hamad (2017).

In *Some emerging principles for mobile-assisted language learning* Stockwell and Hubbard (2013) publish their focus paper on mobile assisted language learning (MALL). The paper extensively contextualizes the development and generalizations of MALL as issues relating to physical, pedagogical, psycho-social issues. To ensure our work can be sustained for language teachers to use in our practice our evaluation uses the work of Lightbown and Spada (2013). Categorizing reflection within the observation of learning and teaching which provides the means to examine our teaching with WhatsApp through observing the interaction that occurs. Our CAR has provided a positive evaluation for supporting SLLs through WhatsApp for understanding instruction, promoting SLLs task achievement, enjoyment and creativity when learning English.

# References

AlTameemy, F.: Mobile phones for teaching and learning. J. Educ. Technol. Syst. **45**(3), 436–451 (2017). https://doi.org/10.1177/0047239516659754

Awada, G.: Effect of WhatsApp on critique writing proficiency and perceptions toward learning. Cogent Educ. **3**(1) (2016). https://doi.org/10.1080/2331186X.2016.1264173

Burns, A.: Collaborative Action Research for English Language Teachers. Cambridge University Press, Cambridge (2005)

Carr, W., Kemmis, S.: Becoming Critical: Education Knowledge and Action Research. Taylor & Francis e-Library. Routledge, London (2003)

Cohen, L., Manion, L., Morrison, K.: Research Methods in Education, 5th edn. Routledge, London (2005)

Hamad, M.M.: Using WhatsApp to enhance students' learning of English language "experience to share". High. Educ. Stud. **7**(4), 74–87 (2017). https://doi.org/10.5539/hes.v7n4p74

Hiemstra, R.: Uses and benefits of journal writing. New Dir. Adult Contin. Educ. (Summer) (90), 19–26 (2001). https://doi.org/10.1002/ace.17

Lightbown, P.M., Spada, N.: How languages are learned. In: Oxford Handbooks for Language Teachers, 5th edn. Oxford University Press, Oxford (2013). https://doi.org/10.1017/CBO9781107415324.004

Littlejohn, A.: The analysis of language teaching material: inside the Trojan Horse. In: Tomlinson, B. (ed.) Materials Development in Language Teaching. Cambridge UP, Cambridge, pp. 179–211 (2011). https://books.google.co.uk/books?id=TmhyTQji2UEC

O'Brien, T.: Writing in a foreign language: teaching and learning. Lang. Teach. **37**(1), 1–28 (2004). https://doi.org/10.1017/S0261444804002113

Sonia, G., Rawekar, A.: Effectivity of E-learning through Whatsapp as a teaching learning tool. J. Med. Sci. **4**(1), 19–25 (2017). https://doi.org/10.18311/mvpjms/2017/v4i1/8454

Stockwell, G., Hubbard, P.: Some Emerging Principles for Mobile-Assisted Language Learning. TIRF. Monterey, CA (2013). http://www.tirfonline.org/english-in-the-workforce/mobile-assisted-language-learning

Valdes, A.I., Jhones, A.C.: Introduction of communicative language teaching in tourism in Cuba. TESL Can. J. **8**(2), 57 (1991). https://doi.org/10.18806/tesl.v8i2.588

# Exploratory Study on SDI-ICMM Implementation: Case Study Tunisia

Abdelghaffar Khorchani, Fatima Kies$^{(\boxtimes)}$, and Cesare Corselli

Department of Earth and Environmental Sciences, Università degli Studi di
Milano-Bicocca, Milan, Italy
f.kies@campus.unimib.it

**Abstract.** Management and planning of coastal and marine areas are complex
processes that are more and more required to effectively support a coordinated
development of socio-economic activities. In managing coastal and marine
activities, it is important to have good governance with assistant from spatial
data infrastructure (SDI), to ensure sustainable planning and development of
marine area. GIS and related technologies are becoming increasingly used to aid
in the administration and management of the marine environment. This article
describes the implementation of SDI, demonstrating how an interoperable system can provide strong support in implementing the integrated coastal and
marine management. It also focuses on the degree of GIS implementation, the
issues, the limitation and engagement in establishing SDI-ICMM in Tunisia.

**Keywords:** Integrated coastal and marine management · Spatial data
infrastructure · Land–marine interface · GIS

## 1 Introduction

Management and planning of coastal and marine areas are complex processes that are
more and more required to effectively support a coordinated development of socio-
economic activities while preserving the environment using ecosystem-based approaches [1, 2]. Spatial data lies at the heart of current conservation and management
efforts. Currently, the need for available digital spatial data is growing rapidly, as such
data are required for all sorts of applications [3]. Digital spatial data arose due to the
worldwide development of spatial data infrastructure (SDI). SDI describes a framework
of technologies, policies, and institutional arrangements that together facilitate the
creation, exchange, and use of geospatial data and related information resources across
information sharing communities [4].

In managing coastal and marine activities, it is important to have good governance
with assistant from SDI, to ensure sustainable planning and development of marine
area. GIS and related technologies are becoming increasingly used to aid in the
administration and management of the marine environment, and the development of an
SDI will provide the foundation upon which such decision support tools can be based
[5].

The developments of GIS for marine and coastal areas were autonomous, heterogeneous and distributed between organisations, and need for spatial data sharing

© Springer Nature Switzerland AG 2019
C. Benavente-Peces et al. (Eds.): SEAHF 2019, SIST 150, pp. 98–113, 2019.
https://doi.org/10.1007/978-3-030-22964-1_11

between marine organisations for sustainable marine management [6, 7]. To improve marine GIS development, marine spatial data infrastructure (SDI) was introduced to facilitate and coordinate the spatial data exchange and sharing between stakeholders in the marine spatial community [8–10]. GIS and SDI assist in overcoming issues related to the availability, accessibility, and interoperability of spatial datasets [3].

A Spatial Data Infrastructure for Integrated Costal and Marine Management (SDI-ICMM) covering the land and marine environments on a holistic platform would facilitate greater access to more interoperable spatial data and information across the land-marine interface enabling a more integrated to the management of the coastal zone. SDI-ICMM is a framework that comprising a system of information, products and enabling technologies that are critical to sustainable development and management of coastal, marine and freshwater areas, and will improve spatial data quality and availability, reduce spatial data duplication and redundancies, and make the spatial data interoperability [11]. SDI-ICMM consist of five components; spatial data, policies, access network and people [12]. Besides that, marine SDI involves cooperation between marine organisations, and required partnership engagement which included engagement management, information management and capacity building [13].

The shoreline and marine area contain marine resources and producing economic growth in Tunisia, especially in Gulf of Gabes. The Gulf of Gabes or "*Small Syrte* (*petite Syrte*)" located in the eastern basin of the Mediterranean and in the south of Tunisia (Fig. 1), is probably among the most affected Mediterranean ecosystems by global change. It extends over a length exceeding 260 m, about 20% of the Tunisian coastline. Indeed, a recent analysis of the intensity and distribution of the cumulative impacts of 22 anthropogenic pressures on the whole Mediterranean basin (including fishing activities, climate change, biological invasions, coastal erosion, pollution, etc.) has shown that the Tunisian continental shelf and more particularly the Gulf of Gabes is an area with multiple influences [14].

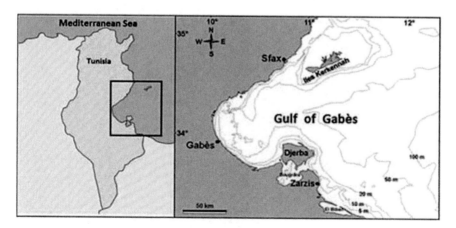

**Fig. 1.** The study area: the Gulf of Gabes (Central Mediterranean Area)

Gulf of Gabes is managed by national and local governments. Local governments have jurisdiction above low water mark. The National government is responsible for the area off-shore: the waters and seabed.

## 2   Research Methodology

This study used survey method, using questionnaire as an instrument for data collection and analysis. The aim of the questionnaire is to examine the different components and sub-components of SDI-ICMM, to assess the level of implementation, and to identify current use, management and sharing of spatial data about Gulf of Gabes from the perspective of the people involved in managing this area.

For this study, a multi-view SDI assessment framework as proposed by Grus et al. [15] was adopted. To recognise the multi-faceted character of SDI and to assess the SDI from different viewpoints we need of multi-view SDI assessment framework. Five viewpoints were established and these are: (i) Policy and Legal issues; (ii) Technical issues; (iii) Funding; (iv) People; (v) Data.

The main points of reference for the questionnaire are the viewpoints identified in the adopted methodology (Policy and Legal issues; Technical issues; Funding; People; Data) and broken down into a set of seventeen indicators. The Information was compiled to establish scores against the indicators (Table 1). For all possible indicators, there were six possible responses namely: Not Sure = 0; Absolutely False = 1; Slightly False = 2; Slightly True = 3; Fairly True = 4; Absolutely True = 5.

**Table 1.**   Scores by the indicators

| Resp. | Policy issues | | | | | Technical | | | | | | Funding | | | People | | | Score |
|---|---|---|---|---|---|---|---|---|---|---|---|---|---|---|---|---|---|---|
| | A | B | C | D | E | A | B | C | D | E | F | A | B | C | A | B | C | |
| R1 | 2 | 3 | 5 | 3 | 2 | 5 | 3 | 5 | 5 | 1 | 1 | 3 | 1 | 1 | 3 | 3 | 4 | 58.8% |
| R2 | 5 | 3 | 5 | 4 | 5 | 3 | 5 | 4 | 5 | 3 | 2 | 3 | 4 | 4 | 1 | 4 | 5 | 75.3% |
| R3 | 5 | 5 | 5 | 2 | 5 | 1 | 3 | 2 | 3 | 2 | 1 | 4 | 2 | 1 | 2 | 2 | 4 | 52.9% |
| R4 | 5 | 5 | 5 | 3 | 5 | 0 | 1 | 3 | 0 | 1 | 2 | 2 | 4 | 2 | 1 | 2 | 4 | 48.2% |
| R5 | 5 | 5 | 1 | 3 | 5 | 1 | 3 | 3 | 3 | 0 | 1 | 3 | 4 | 1 | 0 | 3 | 5 | 49.4% |
| R6 | 3 | 5 | 1 | 4 | 3 | 1 | 2 | 2 | 1 | 2 | 1 | 4 | 4 | 4 | 2 | 2 | 4 | 54.1% |
| R7 | 5 | 5 | 4 | 4 | 5 | 4 | 4 | 4 | 4 | 4 | 5 | 3 | 4 | 4 | 2 | 3 | 5 | 80.0% |
| R8 | 5 | 5 | 5 | 0 | 5 | 1 | 2 | 5 | 1 | 0 | 1 | 5 | 4 | 1 | 1 | 5 | 5 | 54.1% |
| R9 | 5 | 0 | 3 | 0 | 5 | 5 | 1 | 1 | 1 | 1 | 0 | 1 | 0 | 0 | 0 | 1 | 1 | 23.5% |
| R10 | 0 | 0 | 0 | 0 | 0 | 4 | 1 | 1 | 1 | 0 | 0 | 1 | 0 | 1 | 0 | 1 | 0 | 12.9% |
| R11 | 1 | 1 | 2 | 1 | 1 | 5 | 1 | 0 | 1 | 0 | 0 | 1 | 1 | 1 | 2 | 2 | 0 | 22.4% |
| R12 | 5 | 3 | 5 | 3 | 5 | 4 | 4 | 2 | 1 | 4 | 1 | 1 | 1 | 1 | 2 | 3 | 1 | 52.9% |
| R13 | 4 | 4 | 2 | 3 | 4 | 5 | 4 | 4 | 3 | 3 | 2 | 3 | 2 | 3 | 3 | 2 | 3 | 64.7% |

A total of 20 questionnaires were sent out for this survey. The questionnaires were sent to the stakeholders and users of geo-information in Tunisia, both in government and private sectors; producers and users; within the central and local administration; NGOs and academia. There were significant limitations observed during the data collection process between December 2016, and January 2017.

## 3 Experimental

The results of the questionnaire collected from the respondents are presented in Table 1. The questionnaire was sent to 20 people in Tunisia by email. Out of these total, 9 questionnaires were returned which is 45% of the questionnaire sent out, while the other 4 questionnaires are done with a face-to-face interview (20%). Moreover, the respondents are from relevant people and are here considered as a true representative of the population.

The raw result is presented in Table 1 after which the data are analysed from different perspectives. In Table 1, the scores from the respondents are presented against the indicator classes. The respondents which are thirteen in number are represented by numbers 1–13. Each indicator class is divided into specific indicators represented by alphabets. Each of these alphabets represents and corresponds to a question in the questionnaire. The response from each respondent for each specific indicator is scored on a scale of 0–5. The scores of each respondent for all the specific indicators in all the indicator classes are summed and converted to percentage. This percentage now represents the total score given to the SDI-ICMM by the respondent.

The aim of the second part of the questionnaire is to identify critical factors in implementing spatial data sharing in Tunisia's land, coastal and marine organisations. The objectives of the questionnaire are identified the level of GIS implementation in the organisations, identify critical factors in implementing GIS, and finally identify the relationship between these critical factors. The questionnaire consists of three parts: (i) Information on respondent's background; (ii) Information on GIS implementation in the organisations; (iii) Level of spatial data sharing implementation. The questions were used a Likert Scale to measure the extent of agreement describe by each item. The scale ranged from 1 to 5, where: 1 = Strongly disagree; 2 = Disagree; 3 = Fair; 4 = Agree; 5 = Strongly agree.

Cronbach's Alpha was conducted to measure the internal consistency of the research instrument. Suppose that we measure a quantity which is a sum of $K$ components (K-items or testlets): $X = Y_1 + Y_2 + \cdots + Y_K$ Cronbach's $\alpha$ is defined as [16]

$$\sigma = \frac{K}{K-1}\left(1 - \frac{\sum_{i=1}^{K}\sigma_{Yi}^2}{\sigma_X^2}\right)$$

Where

$\sigma_X^2$    the variance of the observed total test scores

$\sigma_{Yi}^2$    the variance of component $i$

To analyse correlation from the questionnaire, the inferential analysis (Spearman's Rho) was selected to analyse the correlation from the Likert scale question and Pearson Chi-Square to analyse the correlation of nominal data.

$$r = \frac{\sum (X - \overline{X}) * (Y - \overline{Y})}{\sqrt{\sum (X - \overline{X})^2} * \sqrt{\sum (Y - \overline{Y})^2}}$$

The Spearman correlation coefficient is defined as the Pearson correlation coefficient between the ranked variables [17]. For $n$ raw scores $Xi, Yi$ are converted to ranks $rgXi, rgYi$ is computed from:

$$r_s = \rho_{r_{gX},r_{gY}} = \frac{\mathrm{cov}(r_{gX}, r_{gY})}{\sigma_{rgX}, \sigma_{rgY}}$$

where
$\rho$ denotes the usual Pearson correlation coefficient but applied to the rank variables.
cov $(rgX, rgY)$ is the covariance of the rank variables.
$\sigma rgX$ and $\sigma rgY$ are the standard deviations of the rank variables.

The generally agreed value of the lower limit for Cronbach's alpha is 0.70. The analysis was performed separately for the items of each factor, the summaries of the reliability analysis given in Table 2. All items show the results higher than 0.70 therefore it is reliable.

**Table 2.** Scores by the indicators

| Component | alpha |
|---|---|
| GIS | |
| • Geospatial data | 0.8507 |
| • Technologies | 0.8844 |
| • Human resources | 0.9248 |
| SDI | |
| • Data user | 0.9793 |
| • Data provider | 0.8619 |
| • Data exchange | 0.9532 |
| Collaboration | |
| • Within organisation | 0.9234 |
| • Between organisation | 0.7877 |

# 4    Results and Discussion

## 4.1    Analysis by Respondents

The responses to each specific indicator vary greatly across respondents, across position rank, across sectors of the economy and across geographical location (Fig. 2).

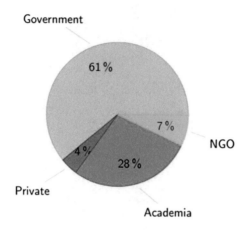

**Fig. 2.**  Analysis by sector

61% of the respondents are from the government sector, 4% from the private sector,
28% from the academia and 7% from NGOs. SDI-ICMM mainly concerns the government stakeholders; therefore, the participants are mostly people working in public
sector. The respondents are mostly drawn from government establishments. Even
though the government policy makes room for public-private participation, the reality
is that the people that constitutes the geospatial data creators' disseminators and users
fall within government sector.

Figure 3 shows that 77% of respondents are working in ministry and central
administration, while 23% of the respondents are outside (local). The administrative
data of Tunisia are strongly tinted of centralization. This state of affairs means that
existing competences appear only in terms of the competencies traditionally reserved
for the centre, that it competences exercised by the State or by national public institutions. So structurally and functionally, the central public administrative institutions
play a role preponderant. All the ministries are located in the capital city Tunis (Fig. 4).

Most government decisions are taken in the headquarters of the ministries. Though
the questionnaire is sent nationwide, the subjects at Tunis seems to be more informed
of SDI-ICMM, as most people from local administration did not respond.

The respondents consist of directors from government (39%), university professor
(21%), senior civil servants (32%), and field professionals (8%). This is more or less an
equitable distribution of respondents.

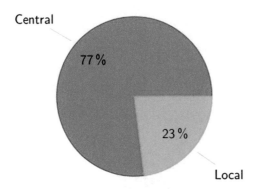

**Fig. 3.** Analysis by location

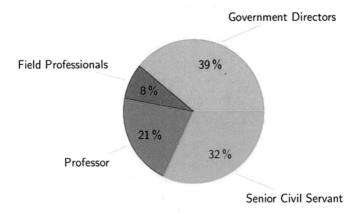

**Fig. 4.** Analysis by position rank

## 4.2    Analysis by Indicators

Here an analysis of the results based on responses to each specific indicator is made. Table 3 summarizes how research subjects responded to each specific indicator. Each alphabet on the left column of the table represents a specific indicator (question in the questionnaire), while the figures inside the table represent the number of respondents that scored the SDI-ICMM a particular ranking. For instance, in specific indicator, "A" in Policy and Legal Issues component class, nine respondents answered 'Absolutely True' in the questionnaire (69%), while two respondents answered 'Fairly True' and two respondents each answered 'Slightly False' and 'Not Sure' respectively. This means that there is certainly the presence of the variable which specific indicator "A" is assessing. Table 3 is represented and analysed in the following charts and paragraphs respectively.

**Table 3.** Summary of respondents to each specific indicator

| | Absolutely True | Fairly True | Slightly True | Slightly False | Absolutely False | Not Sure |
|---|---|---|---|---|---|---|
| *Policy and legal issues* | | | | | | |
| A | 69% | 15% | | 8% | | 8% |
| B | 54% | 8% | 23% | 8% | | 8% |
| C | 23% | 15% | 31% | | 15% | 15% |
| D | 15% | 15% | 15% | 23% | 8% | 23% |
| E | | 38% | | 23% | 23% | 15% |
| *Technical* | | | | | | |
| A | 15% | 15% | 8% | 8% | 46% | 8% |
| B | 8% | 15% | 31% | 8% | 38% | |
| C | 23% | 23% | 23% | | 15% | 15% |
| D | 8% | 8% | 31% | 15% | 23% | 15% |
| E | | 31% | | 23% | 23% | 23% |
| F | 8% | 23% | 8% | 15% | 38% | 8% |
| *Funding* | | | | | | |
| A | 8% | 15% | 23% | 8% | 38% | 8% |
| B | | 15% | 38% | 15% | 23% | 8% |
| C | | 38% | 15% | 8% | 31% | 8% |
| *People* | | | | | | |
| A | 8% | 8% | 8% | 23% | 23% | 31% |
| B | 15% | 23% | 31% | 8% | 23% | |
| C | 31% | 31% | 31% | | | 8% |

### 4.2.1 Policy and Legal Issues

Analysis of the result of questionnaire on the Policy and Legal Issues component class indicate that SDI-ICMM can started well with this component. There is almost unanimous agreement on the necessity of creation a national coastal and marine SDI coordinating body. The response to the specific question on the SDI at highest political level was scored well. Here we mean a politician in the National Assembly pioneering and pushing for SDI awareness, funding and law. On the legal framework for spatial data creation and pricing, the respondents scored it poorly. Actually, there is policy framework guiding these activities but they are not signed into law yet (Fig. 5).

### 4.2.2 Technical

The technical aspect of any coastal and marine SDI system is the pivot on which its data sharing rotates. With respect the SDI access network, the intention is to put in place a high-speed and high bandwidth backbone carrier as the main gateway and master server and implement a database server at each mode. This is not available in reality yet.

The bad shape of access network facilities notwithstanding at public administration level, the analysis from the questionnaire responses indicates weak accessibility to

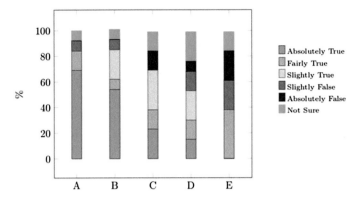

**Fig. 5.** Policy and legal issues indicator class

geospatial data through geo-portal. There is an equal good effort towards interagency coordination of spatial data creation. Metadata capturing is also scored highly by few respondents.

The responses (Fig. 6) however show lack of standardization in spatial data creation and absence of clearinghouse. Data is acquired and stored for own use and applications, with the difficulties of unnecessary overlaps and duplication, lack of accessibility, and varying standards and formats.

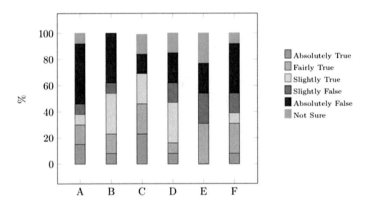

**Fig. 6.** Technical indicator class

### 4.2.3   Funding

Figure 7 highlighted the policy statements on coastal and marine SDI funding. But that have not been fulfilled probably due to lack of SDI Directive. And funding is earmarked as major problem in the SDI-ICMM implementation. The responses of the subjects to this component class are not very encouraging. The major source of income for SDI implementation is from national budget.

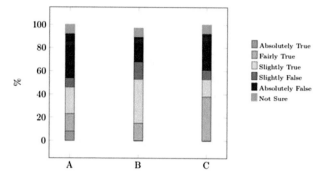

**Fig. 7.** Funding indicator class

There is an effort towards fund generation from access charges and data sales, but this is not viable yet. In addition, Tunisia has received several international grants through WWF, Global Environment Facility, UE, etc. but in the context of specific project (ICZM, climate change, etc.). Even there is no agreement on the existence of policy for spatial data pricing.

### 4.2.4   People

There is sound organizational framework for the SDI-ICMM implementation. Responses from the questionnaire however indicates that there is not enough public-private participation. The major stakeholders, predominantly government however participate in the implementation. On the specific component of skilled personnel, there is reasonable number of skilled personnel to man the coastal and marine SDI implementation. Though availability of skilled personnel especially in technical areas is still a problem (Fig. 8).

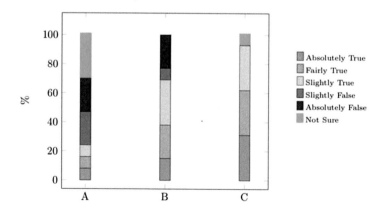

**Fig. 8.** People indicator class

#### 4.2.5    Data

Spatial data plays an important role in aiding planning and management decisions in both the terrestrial and marine environments. The issues of access to and requirements of such data are well documented for land, but less so for the marine environment.

However different activities are involved in the management and administration of the marine and coastal environments which will require access to spatial information for better decision-making. Therefore, a common theme from many of the initiatives that aim to improve coastal and oceans management is the desire for access to appropriate and reliable spatial information to support these initiatives. Often the various spatial datasets are collected and stored by different organisations which can make them difficult to determine their existence and access. The questions in the survey were concerned with spatial data use, availability, accessibility, sharing, collection, standards and policies.

### 4.3    Respondent's Background

Respondent's background focusing on respondent's experience in using GIS in the organisations. Table 4 shows the general background of the respondents such as user type, number of years using GIS and GIS function being used by respondents.

**Table 4.** Respondents background in using GIS

|  | n | % |
| --- | --- | --- |
| *Background of respondents in using GIS* | | |
| Less than one year | 1 | 7.7 |
| One to two years | 1 | 7.7 |
| Two to five years | 2 | 15.4 |
| More than five years | 9 | 69.2 |
| *Respondent's GIS user type* | | |
| Data User | 4 | 30.8 |
| Data Provider | 1 | 7.7 |
| Both (Data user and provider) | 8 | 61.5 |
| *Respondent's GIS functionality* | | |
| View information | 1 | 7.7 |
| Collect data | 2 | 15.4 |
| Analyse information | 1 | 7.7 |
| Integrated with other system | 1 | 7.7 |
| View, collect and analyse | 8 | 61.5 |

Results from Table 4 show that, the respondents were mostly using GIS for more than five years, most of the respondents were both data user and data provider and using most of GIS functions. From these results, it can conclude that the respondents have knowledge of GIS background. But according to their specialties (university diploma) they have other training (e.g. geologist, hydrologist, agronomist) that is to say they are not real GIS specialties.

## 4.4    GIS Implementation in the Organisation

GIS implementation discusses about the respondent's knowledge and experience in handling GIS, and the important aspect of GIS that need to have in the organisations. Analysis was based on three main components of GIS; data, personnel and software, hardware and network. Figure 9 shows the mean for GIS personnel understanding of the importance of spatial data components.

Figure 10 shows the mean of organisations or people factors in succeeding spatial information system implementation in the organisations. From the results, it shows most of the respondents are aware of the GIS components. The mean of the answer was mostly above 4, which indicated the respondents agreed with the importance of each component.

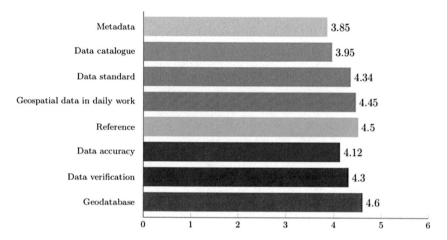

**Fig. 9.** Importance of spatial data components by GIS personnel

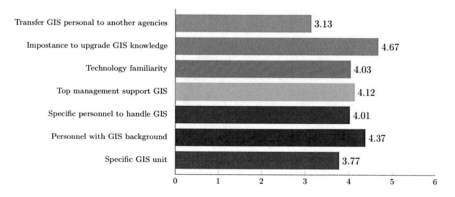

**Fig. 10.** Organisations or people factors in succeeding SDI-ICMM implementation

## 4.5   Spatial Data Sharing Implementation

Implementation of spatial data sharing discussed on the knowledge of respondents in spatial data sharing, limitation in implementing spatial data sharing correlation between knowledge of GIS and SDI in the organisational (Land, Coastal and Marine) implementation of spatial data sharing. For descriptive statistical analysis, three analyses were done: understand the knowledge on cooperation on spatial data exchange in the organisation; understand the cooperation for GIS implementation in the organisation; understand the opinion on collaboration in enabling spatial data sharing between all organisations (Land, Coastal and Marine) implemented in SDI-ICMM.

Table 5 shows the summary of the descriptive analysis. The hypothesis was constructed based on three main criteria; the knowledge and implementation of GIS in the organisation, the knowledge and implementation of spatial data sharing in the organisations, and the collaborative process in enabling spatial data sharing between organisations. To understand the correlation between the three main criteria, these hypotheses have been listed:

**Table 5.**  Descriptive analysis

| Factors | Mean score |
|---|---|
| *Knowledge on cooperation on spatial data exchange* | |
| • Get data from other unit/division to assist processing | 4.37 |
| • Give data to other divisions to assist other divisions | 4.31 |
| • Get spatial data from other agencies | 4.13 |
| • Give spatial data with other agencies | 4.15 |
| • Integrate system with other divisions | 4.03 |
| • Integrate system with other agencies | 2.19 |
| *Cooperation for GIS implementation* | |
| • Geospatial data collection | 4.53 |
| • Geospatial data upgrading | 4.23 |
| • Cooperation on developing GIS | 4.11 |
| • Cooperation on upgrading GIS | 4.04 |
| *Opinion on collaboration in enabling spatial data sharing between organisations* | |
| • Frequent meeting facilitating geospatial data updating | 4.18 |
| • Frequent meeting facilitating geospatial data sharing | 3.28 |
| • Formal collaboration for spatial data sharing | 4.27 |
| • Collaboration to improve GIS | 4.49 |

For the **first hypothesis**, there is correlation between duration using GIS in the organisation with personnel knowledge in GIS, Spearman's rho analysis was used to determine the relationship between duration using GIS and knowledge on GIS. There was a positive correlation, which was statistically significant, $r = 0.339$, $p < 0.05$.

For the **second hypothesis**, there is a correlation between GIS user in the organisation with the personnel level of knowledge in GIS, A Pearson Chi-Square test was used to determine whether there was significant correlation between types of GIS user in the organisation with the personnel level of knowledge in GIS. There was no significant correlation between types of GIS user with a level of knowledge on GIS, $X^2 = 23.76$, DF = 13, p > 0.05.

The **third hypothesis**, there is a correlation between GIS knowledge on spatial data with GIS technologies, using Spearman's Rho analysis, to determine the relationship between respondent's knowledge on GIS data with knowledge of GIS technologies and GIS management. There was a positive correlation, which was statistically significant between respondent's knowledge on GIS data with knowledge of GIS technologies, r = 0.728, p < 0.05. There was also a positive correlation between respondent's knowledge of GIS technologies with GIS institutional management, r = 0.362, p < 0.05.

The **fourth hypothesis**, there is a correlation between spatial data sharing implementation with knowledge about GIS, was used Spearman's Rho analysis to determine the relationship between respondent's knowledge on GIS with the implementation of spatial data sharing. There was a positive correlation with statistically significant, r = 0.335, p < 0.05, which indicate that to successfully implement spatial data sharing, the knowledge on GIS in important.

The **fifth hypothesis**, there is correlation between knowledge about spatial data sharing with knowledge, spatial data sharing implementation, was analysed using Pearson Chi-square analysis, to determine the correlation between respondent's knowledge of spatial data sharing with the spatial data sharing implementation, where there was also a significant correlation between the two components, $X^2 = 60.31$, DF = 13, p > 0.05. From the inferential statistical analysis, the findings show that:

- There is a correlation between personal knowledge of SDI-ICMM component in the SDI implementation in the organisations.
- There is a correlation between personal knowledge of SDI-ICMM with the implementation of spatial data sharing in the organisations.
- There is a correlation between spatial data sharing with SDI-ICMM implementation.
- There is a correlation between cooperation in the organisations with spatial data sharing implementation.
- There is a correlation between collaboration with other organisation with spatial data sharing implementation.

## 5  Conclusion

The analysis of results has shown some of the limitations and problems for the development of a SDI that can contain data from terrestrial as well as the marine and coastal environments. The assessment of management and planning framework, demonstrated the complexity of the management framework. The stakeholders of land, coastal and marine environments have different rights, interests, or responsibilities of this area. The task of efficiently and effectively managing all stakeholders is complicated by the fact that their rights can often overlap which gives rise to the need for cooperation between agencies. However, these problems can be overcome through coordination for collaborative planning. There should be proper regulation to enforce that all spatial data providers should be involved in and contribute to the development of a SDI-ICMM.

The biggest impediment to interoperability was that not all organisations used the same data format, and so their data could not be integrated with other data. The lack of interoperability of different dataset from custodians is the most significant problems found during the integration of land and marine spatial data. The other problem It would like to mention is the differences in scales, quality and coverage of spatial data and the lack of or poor quality of metadata. An issue that was brought up in this part was the need for interoperability across the land—marine interface. The stakeholders in Gulf of Gabes are responsible for managing not only marine and coastal areas, but also terrestrial areas, and activities (i.e. tourism, etc.) that may cover all of these environments.

The responses to each specific indicator vary greatly across respondents, across position rank, across economy sectors, and across geographical location. This is expected as SDI is a complex and dynamic concept, with each respondent approaching it from where it matters to him most. This study analysed the current use, access and sharing of spatial data from the perspective of the selected stakeholders responsible for managing this area. It highlighted the fact that marine and coastal spatial data is used by many different organisations and sectors and comes from different environments land and marine. The lack of a formalised approach to data collection, maintenance and sharing in the marine and coastal environments showed a lack of interoperability from different data formats. Determining what data is available is difficult because there is no one organisation or authority that holds all spatial data and this data is usually collected for a particular project and is rarely made available for other organisations to use.

The survey highlighted that there is need for a common and holistic platform which leads to the promotion of data sharing and communication between organisations thus facilitating better decision-making involving marine and coastal spatial information.

# References

1. European Union: Directive 2014/89: establishing a framework for maritime spatial planning. The European Parliament and of the Council (2014)
2. Douvere, F.: The importance of marine spatial planning in advancing ecosystem-based sea use management. Mar. Policy **32**(5), 762–771 (2008)
3. Sutrisno, D., Gill, S.N., Suseno, S.: The development of spatial decision support system tool for marine spatial planning. Int. J. Digit. Earth, 1–17 (2017)
4. ESRI: Spatial Data Infrastructure A Collaborative Network. In *Esri*, p. 8 (2010)
5. Rajabifard, A., Binns, A., Williamson, I.: Administering the marine environment—the spatial dimension. J. Spat. Sci. **50**(2), 69–78 (2005)
6. Masser, I., Rajabifard, A., Williamson, I.: Spatially enabling governments through SDI implementation. Int. J. Geogr. Inf. Sci. **22**(1), 5–20 (2008)

7. Tarmidi, Z., Rashid, A., Shariff, M., Mahmud, A.R., Ibrahim, Z.Z.: The important of information integration in marine management: a review. World Appl. Sci. J. **22**(6), 870–876 (2013)
8. Strain, L., Williamson, I., Rajabifard, A.: An SDI model to include the marine environment. Dep. Geomatics, Sch. Eng. Cent. SDI L. Adm., vol. M.Sc., p. 134 (2006)
9. Rajabifard, A.: Diffusion for regional spatial data infrastructures: particular reference to Asia and the Pacific. Dep. Geomatics, Sch. Eng. Cent. SDI L. Adm., vol. Doctor Phi, no. March, 229 (2002)
10. Crompvoets, J.: Geoportals. Int. Encycl. Geogr. People Earth, Environ. Technol., 1–6 (2017, March)
11. Meiner, A.: Spatial data management priorities for assessment of Europe's coasts and seas. J. Coast. Conserv. **17**(2), 271–277 (2013)
12. Strain, L., Rajabifard, A., Williamson, I.: Marine administration and spatial data infrastructure. Mar. Policy **30**(4), 431–441 (2006)
13. Tarmidi, Z.M., Rashid, A., Shariff, M., Mahmud, A.R., Ibrahim, Z., Hamzah, A.H.: Issues and challenges in managing Malaysia's marine spatial information sharing. Fed. Surv. no. Boateng 2006, 1–8 (2014)
14. Micheli, F., et al.: Cumulative human impacts on Mediterranean and Black Sea marine ecosystems: assessing current pressures and opportunities. PLoS One **8**(12) (2013)
15. Grus, L., Crompvoets, J., Bregt, A.K.: Multi-view SDI assessment framework. Int. J. Spat. Data Infrastruct. Res. **2**(2), 33–53 (2007)
16. DeVellis, R.: Scale Development. Sage, Newbery Park, CA (1991)
17. Myers, J., Arnold, D.: Well. 2003. Research Design and Statistical Analysis (ed.). Lawrence Erlbaum

# Studying the Corrosion Behavior of Pd–Cu Alloy Used in Hydrogen Purification Systems

Hatem M. Alsyouri[1,2], Farqad F. Al-Hadeethi[3(✉)],
and Sohaib A. Dwairi[4]

[1] Chemical Engineering Department, The University of Jordan, Amman, Jordan
[2] Department of Chemical Engineering, College of Engineering and
Technology, American University of the Middle East, Egaila, Kuwait
[3] Science Center, Royal Scientific Society, Amman, Jordan
farqad.hadeethi@rss.jo
[4] Water and Environment Center, Applied Science Sector, Royal Scientific
Society, Amman, Jordan

**Abstract.** The use of hydrogen as an energy carrier is considered a global goal within the coming years. Therefore, using various alloys is unavoidable in purification processes; for example, hydrogen produced by any suitable upstream process is purified using a hydrogen-selective metallic membrane made of Pd or Pd-based alloys. During this work the corrosion behavior of Pd and Cu in acidic media (0.25 M HCl) at various operating conditions [temperatures (30 and 50 °C) and speeds of agitation (0, 300, 600, and 900 RPM)] was studied. Multiple regression analysis with respect to ANOVA was utilized to generate mathematical correlations. The derived correlations and three-dimensional mapping (i.e. surface response) revealed that increasing temperature and speed of agitation affected the corrosion rate of Pd and Cu. The corrosion rate of copper was much higher than palladium for the same conditions. This result indicates that alloying copper with palladium in the hydrogen purification systems is not acceptable due to the degradation that might occur in copper as a direct result of the galvanic corrosion phenomenon.

**Keywords:** Corrosion · Pd–Cu · Hydrogen purification systems

## 1  Introduction

Corrosion is defined as the dissolution process of a metal or alloy by electrochemical reaction with the surrounding environment. Basically, corrosion has three main reasons of concern and study; safety, economics, and conservation [1, 2]. Therefore, several factors should be taken into consideration when selecting material for construction used in a particular application; physical, mechanical, corrosion resistant and cost [2].

Due to the excellent thermal conductivity and good mechanical workability, copper is used in several industrial activities. Copper dissolution in acidic solutions has been studied by several researchers [3–5]. On the other hand, Palladium and palladium alloys have attracted considerable attention owing to their perfect permeability and selectivity towards hydrogen [6–11].

© Springer Nature Switzerland AG 2019
C. Benavente-Peces et al. (Eds.): SEAHF 2019, SIST 150, pp. 114–118, 2019.
https://doi.org/10.1007/978-3-030-22964-1_12

The purpose of this work is to investigate the corrosion behavior of Pd and Cu because of their importance in fabricating hydrogen purification membranes. The work was focused on accelerated conditions using acidic medium (i.e. 0.25 N HCl) at various operating temperatures (30 and 50 °C) and speeds of agitation (0, 300, 600, and 900 RPM).

## 2  Experimental Work

### 2.1  Materials

Specimens of copper (99.93% purity) with an area 30 × 15 mm, and palladium (99.97% purity) with an area of 30 × 10 mm were utilized during the experimental work.

### 2.2  Tests

Corrosion tests were performed using the Potentiostat and Galvanic corrosion monitoring and analyzing system at 0.25 N HCl.

## 3  Results and Discussion

### 3.1  Corrosion of Copper

It is clear from the following generated mathematical correlation based on multiple regression and the surface response given in Fig. 1 the interaction among the independent variables [temperature (T) and speed of agitation (S)]. Increasing speed of agitation from 0 to 900 RPM at temperatures (30 and 50 °C) resulted in an increase in the corrosion rate of copper in 0.25 N HCl which caused a shift to greater values of corrosion rate, except at 50 °C and 600 RPM, where the rate was shifted to lower value (i.e. 0.288 mm/year). The reason behind the decrease in the corrosion rate is due to the formation of a protective film made of the corrosion products on the copper surface, which slowed down the rate of mass transfer of hydrogen and oxygen ions to the copper surface. While, increasing the speed from (0 to 900 RPM) caused a removal of the mentioned protective film and as a result it increased the diffusion or migration of hydrogen and oxygen ions to copper surface. This trend is in agreement with Stupnišek-Lisac et al. [4], Abdelhadi et al. [5], Modestov et al. [12], Smyrl et al. [13], Habbache et al. [14], and Nielsen et al. [15].

$$C.R. = a + b \times \frac{T}{(S+1000)} + c \times T \times S + d \times S^3 + f \times T + g \times \frac{(S-200)^3}{T}$$

where:

T: Temperature in (°C)
S: Speed of agitation in (RPM).

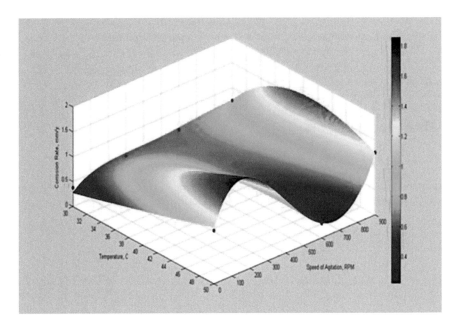

**Fig. 1.** Surface response of the generated mathematical correlation for the corrosion of copper.

### 3.2 Corrosion of Palladium

It is clear from the following generated mathematical correlation based on multiple regression and the surface response given in Fig. 2 the interaction among the independent variables [temperature (T) and speed of agitation (S)]. Corrosion rate of palladium at 30 °C was increasing with increasing speed of agitation from (0–300) RPM, then decreasing at (300–600 RPM), followed by an increase at (600–900 RPM). Whereas, at 50 °C corrosion rate of palladium was continuously increasing at (0–600 RPM) followed by a decrease at (600–900 RPM). In other words this condition caused palladium to lose its passivity as a result of the interactions among the operating conditions (i.e. temperature and speed of agitation). Maximum corrosion rate was observed at 30 °C and 300 RPM with 0.107 mm/year, while the minimum rate was observed at 50 °C and 0 RPM with 0.018 mm/year. It is important to report that corrosion rates of palladium were significantly lower than that of copper under all studied operating conditions.

$$C.R. = a + b \times T \times (S - 200) + c \times T^2 \times (S - 400) + d \times T^3 \times (S - 400)^3 + f \times \frac{S \times (S + 500)}{T^2} + g \times \frac{(S - 100)^2}{T^3 \times (S + 100)}$$

where:

T: Temperature in (°C)
S: Speed of agitation in (RPM).

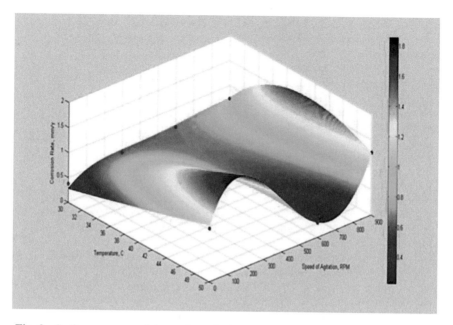

**Fig. 2.** Surface response of the mathematical correlation for copper corrosion behavior.

## 4 Conclusion

The results revealed that increasing temperature and/or speed of agitation will affect the corrosion rate of Pd and Cu couple. The derived correlations for Pd and Cu and three-dimensional mapping (i.e. surface response) showed clearly the interactions among different variables (temperature and speed of agitation) on the corrosion rate of Pd and Cu. Copper which is considered more anodic to Pd in the galvanic series, so alloying Pd with Cu causes it to corrode rapidly as a direct result of galvanic corrosion generated between Pd/Cu. This result indicates that it is not acceptable to use membranes made of Pd/Cu for hydrogen purification.

## References

1. Schweitzer, P.A.: Fundamentals of Corrosion, 1st edn. CRC Press, New York (2010)
2. Revie, R.W., Uhlig, H.H.: Corrosion and Corrosion Control, 3rd edn. Wiley, New York (2008)
3. Khaled, K., Hackerman, N.: Ortho-substituted anilines to inhibit copper corrosion in aerated 0.5 M hydrochloric acid. Electrochim. Acta **49**, 485–495 (2004)
4. Stupnišek-Lisac, E., Galić, N., Găsparac, R.: Corrosion inhibition of copper in hydrochloric acid under flow conditions. Corrosion J. **56**(11), 1105–1111 (2000)
5. Abdelhadi, A., Abbasi, G., Saeed, F.: Estimating the Life Time of Industrial Mechanical Structures Facing Metallic Corrosion Phenomena in Acidic Environment. M.S. Thesis, University of Jordan, Jordan (2010)

6. Hu, X., Huang, Y., Shu, S., Fan, Y., Xu, N.: Toward effective membranes for hydrogen separation: multichannel composite palladium membranes. J. Power Sources **181**, 135–139 (2008)
7. Lin, Y.M., Rei, M.H.: Separation of hydrogen from the gas mixture out of catalytic reformer by using supported palladium membrane. Sep. Purif. Technol. **25**, 87–95 (2001)
8. Ryi, S.K., Park, J.S., Kim, S.H., Cho, S.H., Hwang, K.R., Kim, D.W., Kim, H.G.: A new membrane module design with disc geometry for the separation of hydrogen using Pd alloy membranes. J. Membr. Sci. **297**, 217–225 (2007)
9. Decaux, C., Ngameni, R., Solas, D., Grigoriev, S., Millet, P.: Time and frequency domain analysis of hydrogen permeation across PdCu metallic membranes for hydrogen purification. Int. J. Hydrog. Energy **35**, 4883–4892 (2010)
10. Didenko, L.P., Savchenko, V.I., Sementsova, L.A., Bikov, L.A.: Hydrogen flux through the membrane based on the Pd-In-Ru foil. Int. J. Hydrog. Energy **41**, 307–315 (2016)
11. Shahrouz, N., John, S., David, B.: Effects of low Ag additions on the hydrogen permeability of Pd–Cu–Ag hydrogen separation membranes. J. Membr. Sci. **451**, 216–225 (2014)
12. Modestov, A.D., Zhou, G.D., Ge, H.H., Loo, B.H.: A study by voltammetry and the photocurrent response method of copper electrode behavior in acidic and alkaline solutions containing chloride ions. J. Electroanal. Chem. **380**, 63–68 (1995)
13. Smyrl, W.H., Stephenson, L.L.: Digital impedance for faradaic analysis: III. Copper corrosion in oxygenated 0.1 N HCl. J. Electrochem. Soc. 1563–1567 (1985)
14. Habbache, N., Alane, N., Djerad, S., Tifouti, L.: Leaching of copper oxide with different acid solutions. Chem. Eng. J. **152**, 503–508 (2009)
15. Nielsen, B., Dogan, O., Howard, B.: Effect of temperature on the corrosion of Cu-Pd hydrogen separation membrane alloys in simulated syngas containing H2S. Int. J. Corros. Sci. **96**, 74–86 (2015)

# New Parametrization of Automatic Speech Recognition System Using Robust PCA

Sonia Moussa[1(✉)], Zied Hajaiej[1], and Ali Garsallah[2]

[1] Laboratory of Signal Image and Information Technology (LSITI), National Engineering School of Tunis (ENIT), BP 37, Belvedere, 1002 Tunis, Tunisia
{Sonia.Moussa,Zied.Hajaiej}@enit.rnu.tn
[2] Laboratory of High Frequency Electronic Circuits and Systems, Faculty of Mathematical, Physical and Natural Sciences of Tunis (FST), University of Tunis El Manar Tunis University, Campus PB 94, Rommana 1068, Tunis, Tunisia
ali.gharsallah@fst.rnu.tn

**Abstract.** The speech signals can be severely distorted by additive noise and reverberation, therefore speech recognition performance degrades significantly in different environments. The usefulness of perceptual features is emphasized in this work, which is derived from input speech and adaptive filtering algorithms for developing a robust automatic speech recognition system is proposed also. Thus, MFCC is one of the most commonly used features for automatic speech recognition systems. However, it has been observed that the performance of MFCC-based systems is more influenced by changing in different noise types and their levels. In this paper, we develop a new approach such as MFCC-RPCA in order to obtain a higher recognition rate. In the experimental part, we took isolated words from TIMIT database with SNR (signal to-noise-ratio) ranges from −3 to more than 9 db. Experimental results have shown that the proposed method has enhanced the quality of signals by reducing the noise level.

**Keywords:** Speech recognition · MFCC · MFCC-RPCA · TIMIT database

## 1 Introduction

Recently, several techniques of signal analysis have been applied in audio and speech denoising with relatively good results in controlled conditions. However, the signal may be corrupted by a wide variety of sources in such environments including: additive noise, linear and non-linear distortion, transmission and coding effects, and other phenomena. Thus, a number of works have explored the use of auditory models for building robust speech recognition system. However, a commonly approach to recover speech signals from noisy observations is a speech enhancement technique which estimates and removes the noise from the spectrum of the input speech signal [2].

In this paper, we consider the possibility of improving the performance of a noise robust automatic speech recognition (ASR) system by integrating of MFCC-Robust Principal Component Analysis (RPCA) algorithm for noise suppression. Furthermore, RPCA has the potential to recover clean speech from distorted speech under various types of noises conditions.

C. Benavente-Peces et al. (Eds.): SEAHF 2019, SIST 150, pp. 119–124, 2019.
https://doi.org/10.1007/978-3-030-22964-1_13

We have taken advantage of the correlation of noise spectrum, such as low rank and sparseness of speech spectrum in short-time Fourier transform (STFT) domain to develop a RPCA based denoising scheme for recovering clean speech under various types of noise conditions.

In the following sections, we describe how a speech signal is converted into sequences of MFCC feature vectors and how these vectors are then processed by an HMM-based recognition system in order to estimate the sequences of spoken words.

This paper is organized as follows: In Sect. 2, we explain the implementation of our proposed method. In Sect. 3, the architecture of our provided model, the experimental results and their improvements in different scenarios are presented. Finally, we provide some concluding remarks with suggestions for future directions of this work.

## 2  The Proposed Model (MFCC-RPCA)

### 2.1  MEL Frequency Cepstral Coefficients (MFCC)

Mel-frequency cepstral coefficients (MFCC) are used as the primary audio features. They have been widely used in audio signal processing problems, for example, speech recognition, audio retrieval, and emotion recognition in speech.

The process of traditional MFCC is shown in Fig. 1 [5].

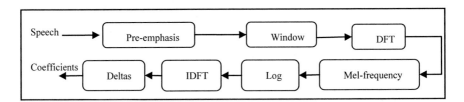

**Fig. 1.**  The process of traditional MFCC

In the MFCC extracting stage, the log operation makes convolutional noise in the spectral domain to be additive and simple in the log-Mel-filter-bank and cepstral domains, but leads the additive noise in the spectral domain to be very complex in the cepstral domain.

The measurement sequence $y(m)$ is preprocessed to obtain $y_i(m)$, where $i$ is the number of frame, then each frame speech signal transition to the frequency domain transforms from time domain using FFT or DCT, and it can be expressed as [3]:

$$Y_i(\omega) = FFT[y_i(m)] \tag{1}$$

And the energy of each frame can be expressed as:

$$E_i(\omega) = [Y_i(\omega)]^2 \tag{2}$$

Thus the actual frequency f, for each tone, is measured in Hz. An individual pitch is measured on a scale called the "Mel-Scale". Below 1000 Hz the mel-frequency scale is linear whereas above 1000 Hz is logarithmic spacing. 1 kHz tone is a pitch as reference point and the perceptual hearing threshold is above 40db, which is well-defined as 1000 Mels. Hence, the following estimated formula to calculate the Mels for a given frequency f in Hz [10]:

$$mel(f) = 2595 * \ln\left(1 + \frac{f}{700}\right) \tag{3}$$

During the calculation of MFCC, this intuition is implemented by creating a bank of filters which collect energy from each frequency band, with 10 filters spaced linearly below 1000 Hz, and the remaining filters spread logarithmically above 1000 Hz.

The figure below gives an example of weighted triangular filters of bandwidth proportional to the Mel scale (Fig. 2):

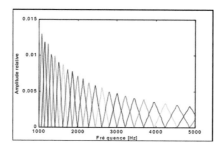

**Fig. 2.** Example of triangular weighted mel band proportional bandwidth filters

## 2.2 Robust Principal Component Analysis (RPCA)

Recently, several authors have proposed a new theory called RPCA, which can remedy the deficiency of PCA, which have talked about this approach in voice/music.

This method assumes that the background music have low-rank structure and vocal components have sparse structures, in the time-frequency domain. RPCA proves to be a very effective tool for extraction of the vocal section from a sample containing mixture of vocal and music [1, 11].

In this paper, we apply the RPCA approach to speech and noise separation problem and we proposed a speech enhancement method based on this approach.

Robust Principal Component Analysis is one of the recent methods used in vocal separation from a mixture of speech and noise. This method assumes that the background speech has low-rank structure and noise components have sparse structures, in the time-frequency domain.

Denote by $Y \in \mathbb{R}^{m \times n}$ the original data matrix, by $L \in \mathbb{R}^{m \times n}$ the low-rank component and by $E \in \mathbb{R}^{m \times n}$ the sparse component, RPCA can be mathematically described as the following convex optimization problem [7–9]:

$$min_{L,E}\|L\|_* + \lambda\|E_1\|s.t\ Y = L + E \tag{4}$$

where $\|L\|_* = \sum_r \sigma_r(L)$ denotes the nuclear norm of $L, \sigma_r(L)(r = 1, 2, \ldots, min(m, n))$ is the $r$th singular value of $L, \|E\|_1 = \sum_{i,j}|e_{i,j}|$ denotes the $L_1$-norm of $E$ and $e_{i,j}$ is the element in the $i$th row and $j$th column of $E$.

Therefore, The RPCA algorithm assumes that the noise signal should be converted in time-frequency domain and the noise is treated as a low-rank component whiles the human speech is analyzed as a sparse component [6].

The figure illustrated above shows our proposed method and its different stages.

Compared to the majority of traditional noise robust speech recognition approaches dedicate to compensate for the noise impact after MFCC or i-vector extracting, not deal

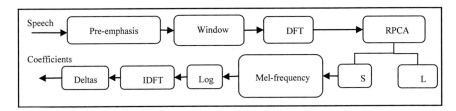

**Fig. 3.** Bloc diagram of our proposed method MFCC-RPCA

with additive noise directly, we embed RPCA based denoising algorithm into MFCC extraction phase, as illustrated in Fig. 3. For given noisy speech, the spectrum of which is firstly generated by preprocessing and STFT. Then RPCA decomposes noisy speech spectrum into two matrices: low-rank matrix and sparse matrix [7, 9].

If we denote, as before, the clean speech signal with $s(t)$ and the noise signal with $n$ $(t)$, we can say that the speech signal $y(t)$ is [10]:

$$y(t) = s(t) + n(t) \tag{5}$$

Using short-time Fourier Analysis (STFT), the speech signal can be written as below:

$$Y(m, k) = \sum_{i=-\infty}^{+\infty} y(i)w(m - i)e^{\frac{-j2\pi ki}{L}} \tag{6}$$

where $k \in \{1, \ldots, L\}$, denotes the index of the discrete acoustic frequency, L is the length of the frequency analysis and m is the index of time-frame, w(m) is an analysis window function [10].

We need to normalize the magnitude $|Y(m, k)|$ in each frame by averaging it with the magnitude values from adjacent three frames, therefore we obtain the enhanced speech signal Fourier transform detailed below [4]:

$$\hat{S}(m, k) = |S(m, k)|^{1/2} e^{j/Y(m,k)} \tag{7}$$

The enhanced speech signal that will result $\hat{s}(t)$ will be the inverse Fourier transform of the $\hat{S}(m, k)$ function.

## 3  Results and Analysis

The isolated words of TIMIT database were used to evaluate the performance of the proposed estimators. These words were corrupted by F16, explosion, door and glass at −3, 0, 3, 6, 9 dB. The sampling frequency used is 16 kHz. The signal to noise ratio SNR is defined as follows:

In order to quantify performance of speech enhancement method, we adopt segmental SNR measures, which are defined as [5]:

$$SNR = 10 \log_{10}(\frac{P_{signal}}{P_{noise}}) \tag{8}$$

**Table 1.** The recognition rate on TIMIT corpus by various types of noises at SNR = −3 and 9 dB (%)

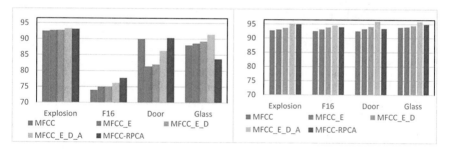

where $P_{signal}$ and $P_{noise}$ represent respectively the power signal and the noise.

Table 1 shows the performance of our proposed method MFCC-RPCA. Comparing MFCC and MFCC-RPCA, we note that the recognition rate has been improved by reducing the noise level at SNR = −3 dB and SNR = 9 dB.

## 4  Conclusion

In previous years, the research on robust principal component analysis (RPCA) has been attracting much attention. In this paper, we presented a RPCA based speech enhancement approach. The advantage of this method is that it can directly estimate enhanced the quality of signals by reducing the noise level. Recently developed robust principal component analysis (RPCA) has been shown effective in separating the speech components from background noise.

## References

1. Tejus, R., et al.: Role of source separation using combined RPCA and block thresholding for effective speaker identification in multi source environment. In: 2017 2nd IEEE International Conference on Recent Trends in Electronics, Information & Communication Technology (RTEICT), IEEE (2017)
2. Martínez, C.E., Goddard, J., Di Persia, L.E., et al.: Denoising sound signals in a bioinspired non-negative spectro-temporal domain. Digit. Signal Process. **38**, 22–31 (2015)
3. Zheng, G., et al.: Speech classification based on compressive sensing measurement sequence. In: 2017 IEEE International Conference on Robotics and Biomimetics (ROBIO), IEEE (2017)
4. Gavrilescu, M.: Noise robust automatic speech recognition system by integrating robust principal component analysis (RPCA) and exemplar-based sparse representation. In: 2015 7th International Conference on Electronics, Computers and Artificial Intelligence (ECAI), IEEE (2015)
5. Sun, C., et al.: Noise reduction based on robust principal component analysis. J. Comput. Inf. Syst. **10.10**, 4403–4410 (2014)
6. Wang, M., Zhang, E., Tang, Z.: Robust principal component analysis based speaker verification under additive noise conditions. In: Chinese Conference on Pattern Recognition. Springer, Singapore (2016)
7. Zhao, Q., et al.: Robust principal component analysis with complex noise. In: International Conference on Machine Learning (2014)
8. Candès, E.J., et al.: Robust principal component analysis? J. ACM (JACM) **58.3**, 11 (2011)
9. Huang, J., et al.: Speech denoising via low-rank and sparse matrix decomposition. ETRI J. **36.1**, 167–170 (2014)
10. Chauhan, P.M., Desai, N.P.: Mel frequency cepstral coefficients (MFCC) based speaker identification in noisy environment using wiener filter. In: 2014 International Conference on Green Computing Communication and Electrical Engineering (ICGCCEE), IEEE (2014)
11. Huang, P.-S., et al.: Singing-voice separation from monaural recordings using robust principal component analysis. In: 2012 IEEE International Conference on Acoustics, Speech and Signal Processing (ICASSP), IEEE (2012)

# Voltage and Reactive Power Enhancement in a Large AC Power Grid Using Shunt FACTS Devices

Mankour Mohamed[1(✉)] and Ben Slama Sami[2]

[1] Department of Electrical Engineering, National Polytechnic School Maurice Audin of Oran, Oran, Algeria
med_mank@yahoo.com
[2] Information System Department, King Abdulaziz University, Jeddah, Saudi Arabia
benslama.sami@gmail.com

**Abstract.** This article presents the study of the impact of the shunt FACTS (alternative AC transmission systems) devices on the AC power grids behavior, In order to show the importance of the control of nodal voltages and the compensation of the reactive power. Two types of shunt FACTS devices are studied in this paper, the first one is the SVC (Static Var Compensators) which is based on thyristor valves, and the second one is the STATCOM (Static Synchronous Compensator) which is based on IGBT/GTO valves. In order to study the reaction of the FACTS devices toward the voltage profile and the reactive power, we perform three cases of test (with and without FACTS devices in the Load flow calculation). Where we correct the voltage violation by ratios of the transformers or by varying the reactive power generated by the capacitor banks, and secondly by including the FACTS devices in a specified location in order to show and compare the influences of these devices against the static equipment (synchronous compensator, capacitor banks). The first test was on the IEEE-05 bus and the second test was on the IEEE-57 bus, using the Newton-Raphson method, the models were implemented in MATLAB/Simulink environment. The results carried out show the robustness of the method with fast number of iterations, and this analysis investigates the shunt FACTS devices influence on the AC system behavior.

**Keywords:** Power distribution · Voltage control · Reactive power compensation · FACTS devices (SVC, STATCOM)

## 1 Introduction

The demand for electrical energy is in increasing continuously and constantly. To satisfy the needs of consumers and distribute this energy to points of dispersed consumption. The electrical networks with thermal limits of lines and transits capacity not to exceed, it was necessary to expand and add new lines of extension. Since it was difficult to provide a reliable energy control in highly interconnected networks using a conventional control device [1]. The development of the power electronics has a

© Springer Nature Switzerland AG 2019
C. Benavente-Peces et al. (Eds.): SEAHF 2019, SIST 150, pp. 125–143, 2019.
https://doi.org/10.1007/978-3-030-22964-1_14

considerable effect to improve the management of the AC power networks on their maximum limit of functionality by the controlling of their parameters (control the voltage, active and reactive power in the AC power grid) by maintaining theses parameters at specified values [2, 3], all that can be achieved by the introduction of the control devices based on the power electronic known by FACTS (alternative AC transmission systems) [4]. FACTS devices have an important role in the AC systems by controlling power transits and maintaining safe operating in the different conditions [5].

The American company EPRI (Electric Power Research Institute) launched, in 1988, a project of study of FACTS systems to better control the transit of power in power lines, to increase their transfer capacities and to put a new possibility of the power electronics in the control [6].

The large AC electrical networks are faced by various operating problems because of the traditional control, using electromechanical control systems with slow response time compared to these new FACTS systems based on static switches and power electronic components which they ensure short response time in the different states (static, transient and dynamic condition) [7, 8]. In references [1, 3, 9, 10] two types of FACTS devices can be found in the real power systems, which are classified according their composition of power electronic components (based on thyristors or on IGBT/GTO (Insulate gate bipolar Transistor)/(Gate turn off thyristors) semiconductors).

The analysis of power distribution in an electrical network composed of a number of generators, transmission lines and loads is very important for the study, planning and operation of an electrical network [11]. The power flow calculation consists of determining all the power transits and the voltages in the network for a given load. Four variables are associated with each node of the system: the active and reactive powers, the module and the voltage phase. Only two of its variables are known in one bus, the two others being determined during the calculation [12].

To solve this problem, as known in the literature and in reference [6], it is necessary to determine the conditions of the steady-state operation of a power system, which are: (i) The formulation of an appropriate mathematical model, (ii) The specification of a number of variables and constraints in the system, (iii) The numerical resolution of the system [3].

In literature, different comparison between FACTS devices towards their configuration and reaction against the AC system behaviour, in this paper we are interested by the shunt FACTS devices (SVC and STATCOM). In reference [13, 18] the authors describes the comparison to decrease the voltage flicker in a high voltage station, In reference [14] the author show the comparison between the two devices in voltage enhancement against wind farms. We present in this paper, the role of these shunt FACTS devices to improve the performance of the AC system by maintaining the voltage and reactive power at the acceptable values against a large AC power network.

Detailed power flow, voltage profile, active, reactive and losses power are investigated in this paper for two different networks (IEEE-05 bus and IEEE-57 bus), with and without incorporation of the shunt FACTS devices. The method of calculation used is Newton Raphson, given its robustness and speed of convergence. The different systems are modelled in the *MTALAB* and *MATALB/Simulink* environment.

## 2  Mathematical Representation of the Power System

The equations connecting the nodal voltages and injected currents for a network with n bus are:

$$I = Y.V$$

$$I_i = \sum_{j=1}^{n} Y_{ij}.V_j \; i = 1,\ldots,n \tag{1}$$

In practice, the injected. The (n) complex equations are decomposed into (2n) real equations

$$S_i = P_i + Q_i = V_i.I_i^* \tag{2}$$

$$S_i^* = P_i - Q_i = V_i^*.\sum_{j=1}^{n} Y_{ij}.V_j \tag{3}$$

(a)  *Newton-Raphson Method* [1]

This method is based on the series development of Taylor. This last one is obtained successively from the approximations of the first order:

$$f(x) \approx f(x^k) + f'(x^k).(x^{k+1} - x^k) = 0 \tag{4}$$

where; $f' = \frac{\partial f}{\partial x}$ is the Jacobin of the $f(x)$.
From an initial value $x^0$, corrections $\Delta x^k$ are obtained by solving the linear system:

$$-f'(x^k).\Delta x^k = f(x^k) \tag{5}$$

And the new values of $x^{k+1}$ is:

$$x^{k+1} = x^k + \Delta x^k \tag{6}$$

In the network, the corrected the angle and magnitude of the voltage given by the Eqs. 7 and 8:

$$\Delta P_i = P_i^{spe} - P_i^{cal} = V_i \sum_{j=1}^{n} V_j(G_{ij} \cos \theta_{ij} + B_{ij} \sin \theta_{ij}) \tag{7}$$

$$\Delta Q_i = Q_i^{spe} - Q_i^{cal} = V_i \sum_{j=1}^{n} V_j(G_{ij} \sin \theta_{ij} - B_{ij} \cos \theta_{ij}) \tag{8}$$

with this notation and dividing the Jacobian into sub-matrices, the expression (5) applied to the problem load flow calculation, it can be converted into a following matrix system given by the Eqs. 9 and 10:

$$\begin{bmatrix} \Delta P \\ \Delta Q \end{bmatrix}^k = \begin{bmatrix} H & N \\ M & L \end{bmatrix}^k \cdot \begin{bmatrix} \Delta V \\ \Delta \theta \end{bmatrix}^k \tag{9}$$

The variable $\Delta V$ can be divided by $V$:

$$\begin{bmatrix} \Delta P \\ \Delta Q \end{bmatrix}^k = \begin{bmatrix} H & N \\ M & L \end{bmatrix}^k \cdot \begin{bmatrix} \Delta \theta \\ \frac{\Delta V}{V} \end{bmatrix}^k \tag{10}$$

With:

$$H_{ij} = \frac{dP_i}{d\theta_j}, M_{ij} = \frac{dQ_i}{d\theta_j}, N_{ij} = \frac{dP_i}{dV_j} \cdot V_j, L_{ij} = \frac{dQ_i}{dV_j} \cdot V_j$$

And the expression (6) into a matrix as fellow:

$$\begin{bmatrix} \theta \\ V \end{bmatrix}^{k+1} = \begin{bmatrix} \theta \\ V \end{bmatrix}^k + \begin{bmatrix} \Delta \theta \\ \Delta V \end{bmatrix}^k \tag{11}$$

The matrix of the Jacobian is given by:
For $i = j$;

$$H_{ii} = -Q_i - B_{ii}.V_i^2 \tag{12}$$

$$M_{ii} = P_i - G_{ii}.V_i^2 \tag{13}$$

$$N_{ii} = P_i + G_{ii}.V_i^2 \tag{14}$$

$$L_{ii} = Q_i - B_{ii}.V_i^2 \tag{15}$$

When $i \neq j$ :

$$H_{ij} = V_i.V_j(G_{ij} \sin \theta_{ij} - B_{ij} \cos \theta_{ij}) \tag{16}$$

$$N_{ij} = V_i.V_j(G_{ij} \cos \theta_{ij} + B_{ij} \sin \theta_{ij}) \tag{17}$$

$$L_{ij} = H_{ij} \tag{18}$$

$$M_{ij} = -N_{ij} \qquad (19)$$

In power flow and stability studies, the passive compensating devices (shunt capacitors, shunt reactors and series capacitors) are modeled as admittance elements or fixed values, the synchronous condensers are modeled as synchronous generators but with no steady-state active power output [3, 7].

## 3  Facts Devices

(a) *General concept and definition*:

According to the IEEE, the abbreviation FACTS (Flexible AC Transmission Systems) can be defined as the transmission system of the AC with incorporation of power electronic devices or another static controller in order to improve the controllability, stability of the AC system and the line transfer capacity. Where these characteristics results by the ability of the FACTS devices to control the parameters given by the Eqs. (1 and 2) and shown in the Fig. 1.

$$P = \frac{E_1 E_2}{X} \sin(\delta_1 - \delta_2) \qquad (20)$$

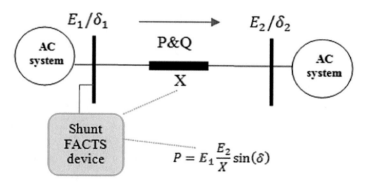

Fig. 1.  Shunt FACTS device regulation.

$$Q = \frac{E_1(E_1 - E_2 \sin(\delta_1 - \delta_2))}{X} \qquad (21)$$

If we consider the voltage regulation and the reactive power compensation, then we talk about the shunt device as explained in the Fig. 1. Regulation of the bus voltage affects indirectly and dynamically the active power flux.

Any device is structured according his connection in the AC network. In general, The FACTS devices are classified into three main categories: Series compensation, shunt compensation and Hybrid (series-shunt) compensation [15].

Another classification of FACTS devices can be given which is according the type or the components of each device. In fact, they are classified into two categories, where in the first category [4], the devices are known by classic HVDC devices and they are composed by the passives elements and thyristors semiconductors such as SVC (Static Var Compensators), TCR (Thyristor Controlled Reactor) or TSR (Thyristor Switched Reactor), TCBR (Thyristor Control Breaking Resistor), TCSC (thyristor controlled series capacitor) [19], TSSC (Thyristor Switched Series Capacitor), TCSR (Thyristor Controlled Series Reactor), TSSR (Thyristor Switched Serie Reactor), TCPAR (Thyristor Controlled Phase Angle Regulator) or SPS (Static Phase Shifter).

The second category [10]: the FACTS devices are mainly composed by IGBT/GTO semiconductors. The most common devices are the STATCOM (Static Synchronous Compensator), SSSC (Static Synchronous Series Compensator), IPFC (Interline Power Flow Controller) and UPFC (Unified Power Flow Controller) [23]. In this paper, two type of shunt FACTS devices are studied (with thyristor and with IGBT/GTO) in order to control both the voltage profile and the reactive power of the AC power network.

(b) *Modeling of the FACTS devices*:

  (b.1) *SVC (Static Var Compensator)*

Static var compensators (SVCs) are shunt-connected static generators and/or absorbers whose outputs are varied so as to control specific parameters of the AC power system [16, 22]. SVC is the association of the TCR and TSC devices (Fig. 2), which the hybrid compensator is the combination of the capacitor banks and harmonic filter (for filtering TCR generated harmonics).

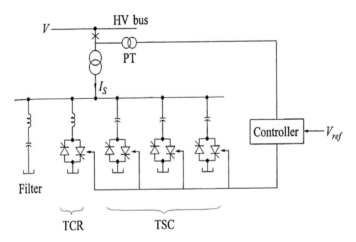

**Fig. 2.** Schematic of typical SVC

The first installation was in south of Africa in 1979, and until now SVCs have been widely used for such application around the world.

The SVC device is based on thyristor valves which they are naturally commutated in the AC system frequency. In the real installations there is two types of SVC, the first one is industry SVC (almost connected with unbalanced loads and the second one is for transmission lines in the AC systems.

SVCs provide continuous and rapid control of reactive power and voltage can provides also an effective and economical means of eliminating the voltage flicker problems [13].

The study-state V/I and V/Q characteristics of the SVC are shown in the Fig. 3. The linear control range lies within the limits determined by the reactor maximum susceptance ($B_{Lmax}$). The total capacitive susceptance ($B_c$) is determined by the filter capacitance and the capacitor banks) [15].

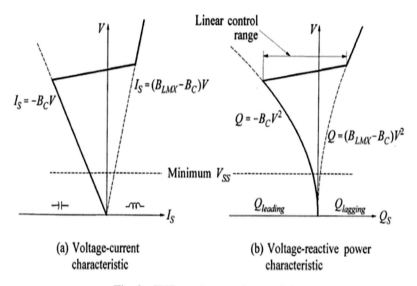

(a) Voltage-current characteristic

(b) Voltage-reactive power characteristic

**Fig. 3.** SVC steady-state characteristic.

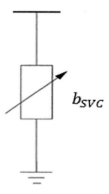

**Fig. 4.** SVC equivalent circuit.

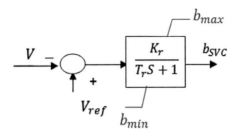

**Fig. 5.** SVC regulator (type 1)

In this paper, as shown in Fig. 4, The SVC device consists that controlled variable is the total susceptance ($b_{SVC}$) and presumes a time regulator (Fig. 5). In this model, the differential equation is as fellow:

$$\dot{b}_{SVC} = \frac{K_r(V_{ref} - V) - b_{SVC}}{T_r} \tag{22}$$

where Kr is the regulator gain, $T_r$ is the regulator time constant, $V_{ref}$ is the voltage reference, $b_{max}$ and $b_{min}$ are the maximum and the minimum susceptance respectively.

This model is completed by the expression of the algebraic equation of the reactive power injected at the bus where the SVC is connected:

$$Q = -b_{SVC}V^2 \tag{23}$$

(b.2) ***STATCOM (STATic COMpensator)***

The STATCOM is also a shunt connected device based on IGBT or GTO semiconductor valves, known by the voltage source converter VSC-FACTS device [17, 20].

The functionality of the STATCOM is closer to the SVC device, but the internal model is little different. Because the STATCOM is composed in reality of three main parts as shown in Fig. 6: (i) DC network, (ii) the VSC (voltage source converter) and (iii) the controllers.

In the reference [20, 21], the STATCOM is proposed as a simplified current injection model. The STATCOM current is always kept in quadrature in relation to the bus voltage so only the reactive power is exchanged between the AC power grid and the STATCOM device [7]. The equivalent circuit of the model is shown in Fig. 7.

The voltage regulator is shown in the Fig. 7 and the differential equation of the current and the reactive power injected by the STATCOM are given respectively by the following equations:

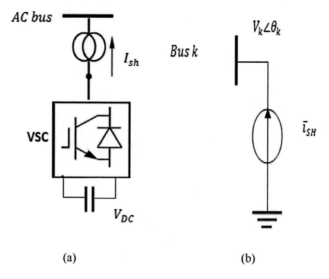

(a)                                    (b)

**Fig. 6.** STATCOM **(a)** model, **(b)** equivalent circuit.

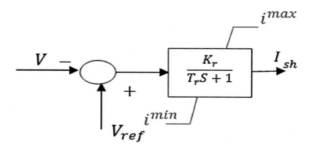

**Fig. 7.** STATCOM voltage regulator.

$$\dot{i}_{SH} = \frac{K_r\left(V_{ref} - V\right) - i_{SH}}{T_r} \tag{24}$$

$$Q = -i_{SH}V \tag{25}$$

where $I_{sh}$ is the current injected, $V_{ref}$ is the voltage reference, Kr is the regulator gain, $T_r$ is the regulator time constant, $I_{min}$ and $I_{max}$ are the minimum and the maximum currents respectively.

**Table 1.** Networks description

|  | 05 bus | 57 bus |
|---|---|---|
| Load bus | 4 | 48 |
| Generation bus | 2 bus (1, 2) | 4 bus (1, 3, 8 and 12) |
| Compensation bus | – | 3 (2, 6 and 9) |
| Transformer | – | 18 |
| Lines | 7 | 60 |
| Condenser batteries | – | 3 batteries (15, 25 and 53) |

## 4   Implementation and Interpretation of Results

The Newton raphson method is implemented firstly for the IEEE-05 bus network then for the IEEE-57 bus. The conception, modelisation and simulation of the AC networks in this paper are performed in the *MATLAB* and *MATLAB/Simulink*. Table 1 gives the description of the two AC systems.

We present the different results, the analyzes the distribution of the powers and the control of the voltages of the network without and with incorporation of FACTS

**Table 2.** Active, reactive power generation and active power losses

| Case number | | | Bus number | | $P_{Losses}$ (MW) |
|---|---|---|---|---|---|
| | | | 1 | 2 | |
| Case 1 | | $P_G$(MW) | 131.12 | 40 | 6.122 |
| | | $Q_G$ (MVAR) | 90.82 | −61.59 | |
| Case 2 | | $P_G$(MW) | 131.06 | 40 | 6.056 |
| | | $Q_G$ (MVAR) | 85.33 | −77.11 | |
| Case 3 | (a) | $P_G$(MW) | 131.06 | 40 | 6.056 |
| | | $Q_G$ (MVAR) | 85.34 | −77.07 | |
| | (b) | $P_G$(MW) | 131.06 | 40 | 6.056 |
| | | $Q_G$ (MVAR) | 85.34 | −77.07 | |

devices (SVC and STATCOM). Both shunt FACTS devices studied in this paper (SVC and STATCOM) are placed for the objective to maintain the voltage at 1 pu where these devices are connected without considering the reactive power generated/absorbed in this bus.

- *First network*: The IEEE-05 bus network is described in the reference [1], Table 2 gives the active and reactive power generated after the three cases of tests (with and without incorporation of the FACTS devices). The results obtained in the first case (base-load flow case) are confounded with the results obtained in [1]. The simulation results are carried out in three cases as fellow:

1. Case N° 1: Initial results (base case -load flow).
2. Case 2: Voltage corrections by placing a capacitor bank in the bus number 3.

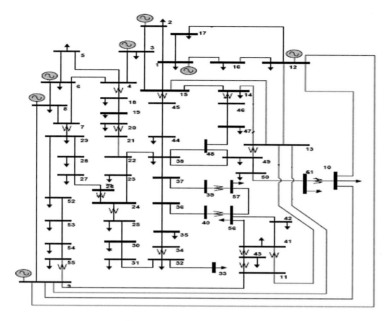

**Fig. 8.** IEEE 57-bus network

3. Case 3: Results after incorporation of FACTS devices.
   (a) Replacing the capacitor bank by SVC.
   (b) Replacing the capacitor bank by STATCOM.

- *Second network*: The load flow method is applied to the IEEE-57 bus networks (Fig. 8) in order to show, envistigate and study the behavior of the AC network (voltage profile and reactive power compensation) in the case of including the shunt FACTS devices into the AC network. Table 3 presents the information of the limits of all the system (voltages, ratio of the transformers, reactive powers and capacitors). The simulation results are carried out in three cases as fellow:

1. Case N° 1: Initial results (base case -load flow).
2. Case 2: Results after voltage corrections.
3. Case 3: Results after incorporation of FACTS devices.
   (a) Replacing the capacitors (18, 25 and 53) by SVCs.
   (b) Replacing the compensators (3, 6 and 9) by STATCOMs.

   *Case N° 1*: Initial results (base case -load flow). In this case, we present the results of the initial state as well as the voltage profile (Fig. 9) (Table 4).

**Table 3.** Voltage, reactive power and capacitors limits.

|  | Min | Max |
|---|---|---|
| V$_1$...........V$_{57}$ | 0.9 | 1.1 |
| a$_1$...........a$_{18}$ | 0.9 | 1.1 |
| Q2 | −17.00 | 50.00 |
| Q3 | −10.00 | 60.00 |
| Q6 | −8.00 | 25.00 |
| Q8 | −140.00 | 200.00 |
| Q9 | −15.00 | 90.00 |
| Q12 | −50.00 | 155.00 |
| C18 | 0 | 20 |
| C25 | 0 | 25 |
| C53 | 0 | 10 |

**Table 4.** Case N° 1 power results.

|  | Active power [MW] | Reactive power [MVAR] |
|---|---|---|
| *Generation* | 1280.09 | 395.66 |
| *Load* | 1250.8 | 319.12 |
| *Losses* | 29.29 | 76.5 |

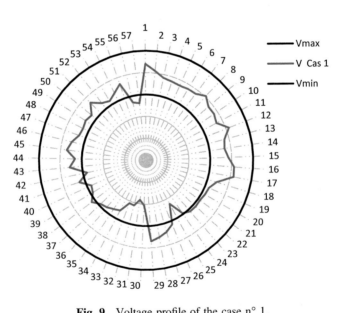

**Fig. 9.** Voltage profile of the case n° 1.

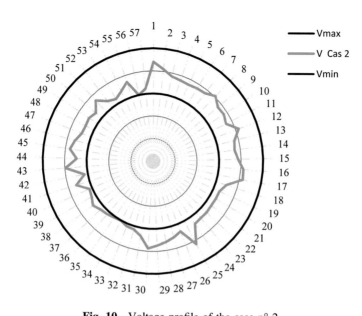

**Fig. 10.** Voltage profile of the case n° 2.

**Table 5.** Case N° 2 power results.

|  | Active power [MW] | Reactive power [MVAR] |
|---|---|---|
| *Generation* | 1279.63 | 375.29 |
| *Load* | 1250.8 | 301.85 |
| *Losses* | 28.827 | 73.447 |

In this case, as shown in the Fig. 9, we notice that the bus 25, 30, 31, 32, 34, 35, 36, 37, 39, 40, 42, 56 and 57 have exceeded their imposed lower limit which is 0.9 pu. The bus 31 is the bus of the greatest violation.

*Case N° 2*: In this case, we present the powers generated after the voltage correction as well as the voltage profile (Fig. 10).

We have tried to control the voltage profile of the network (correction of voltages) by the devices (capacitors banks and the transformers), and that was by the action on the reactive powers generated by the capacitors banks and the transformer ratios. The first action was on the capacitor placed at the bus 25 to produce 20.5 MVAR of reactive power (the action was by a reactive power injection of a step of 1 MVAR). The second action was on the transformers between the bus (11-41), (11-3) and (56-57) by the

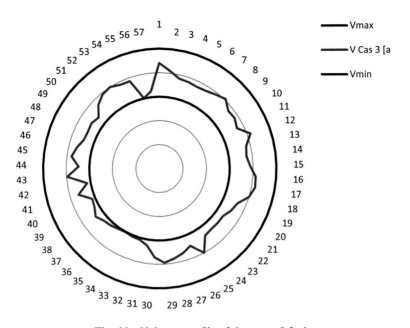

**Fig. 11.** Voltages profile of the case n° 3 -1.

**Table 6.** Case N° 3 -1 power results

|            | Active power [MW] | Reactive power [MVAR] |
|------------|-------------------|-----------------------|
| *Generation* | 1280.32         | 410.91                |
| *Load*     | 1250.8            | 337.4                 |
| *Losses*   | 29.51             | 73.51                 |

decrease of the transformation ratio of 0.04. (The transformation ratios have been changed by a step of 0.01).The results obtained are shown in Fig. 10 and Table 5.

*Case 3*: *Incorporation of FACTS devices.*

1. **Incorporation of SVCs**: we present in this case, the replacement of the capacitors (18, 25 and 53) by SVCs.

Figure 11 shows the voltage profile, and the Table 6 gives the results obtained of the active, reactive powers generated and losses (after the correction of the nodal voltage violations). The values of the power active and reactive are changed if

**Table 7.** Case N° 3 -2 power results.

|            | Active power [MW] | Reactive power [MVAR] |
|------------|-------------------|-----------------------|
| *Generation* | 1280.05         | 374.82                |
| *Load*     | 1250.8            | 301.09                |
| *Losses*   | 29.245            | 73.727                |

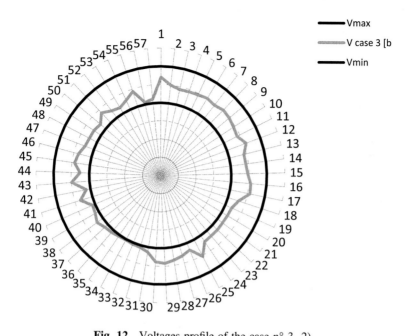

**Fig. 12.** Voltages profile of the case n° 3 -2).

**Table 8.** Total active power losses.

| Case | Case 1 | Case 2 | Case 3 | |
|---|---|---|---|---|
| Bus | | | (1) | (2) |
| $P_{Gen,tot}$ [MW] | 1280.09 | 1279.6 | 1280.3 | 1280.0 |
| $P_{Losses}$ [MW] | 29.29 | 28.827 | 29.51 | 29.245 |

compared to the second case, also we noted an increasing in the reactive power generated around 36 MVAR, this increasing is compensated by the SVCs implanted in place of the capacitors banks, since the reactive powers generated by the SVCs at the bus (15, 25 and 53) are 19.7, 18.52 and 38.54 MVAR respectively.

2. **Incorporation of STATCOMs**: compensators (3, 6 and 9) are replaced by STATCOMs:

Synchronous compensators (2, 6 and 9) have been replaced by STATCOMs, the bus voltage are maintained to 1 pu where these devices are intercalated, as is shown in Fig. 12. The results obtained in this step merge with the previous step, because the characteristics of the two devices (SVC and STATCOM) are based on the same principle. The goal if this replacement is also to study the impact of both the capacitor banks and synchronous compensator against the AC system voltage profile if compared to the shunt FACTS devices (Table 7).

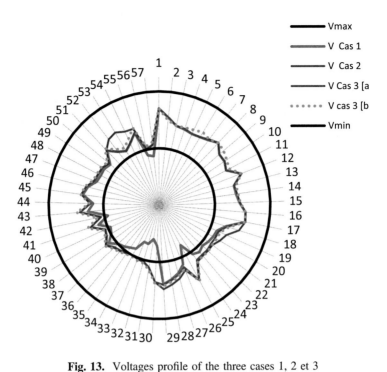

**Fig. 13.** Voltages profile of the three cases 1, 2 et 3

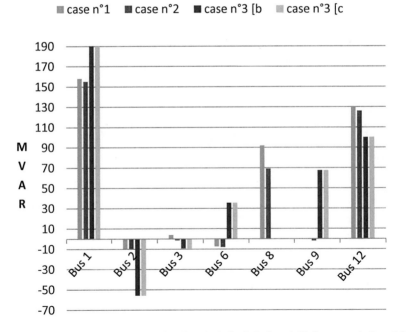

**Fig. 14.** Reactive powers generated at bus 1, 2, 3, 6, 8, 9 and 12 for cases 1, 2 and 3

**Table 9.** Active and reactive power generated.

| Bus Nbr | Case 1 | | Case 2 | | Case 3 | | | |
|---|---|---|---|---|---|---|---|---|
| | | | | | (a) | | (b) | |
| | $P_G$ [MW] | $Q_G$ [Mvar] | $P_G$ [MW] | $Q_G$ [Mvar] | $P_G$ [MW] | $Q_G$ [Mvar] | $P_G$ [MW] | $Q_G$ [Mvar] |
| 1 | 480.09 | 142.9 | 479.62 | 140.12 | 480.31 | 139.63 | 480.05 | 173.83 |
| 2 | 0.0 | 5.72 | 0.0 | 5.72 | 0.0 | 5.74 | 0.0 | −40.75 |
| 3 | 40 | 15.74 | 40 | 10.43 | 40 | 4.15 | 40 | 0.683 |
| 6 | 0.0 | 8.83 | 0.0 | 7.15 | 0.0 | −2.62 | 0.0 | 45.99 |
| 8 | 450 | 68.55 | 450 | 66.40 | 450 | 55.78 | 450 | 3.43 |
| 9 | 0.0 | 12.77 | 0.0 | 9.64 | 0.0 | −3.69 | 0.0 | 80.46 |
| 12 | 310 | 141.12 | 310 | 135.81 | 310 | 135.13 | 310 | 111.19 |

- **Analysis of active, reactive and losses powers**

The Table 8 gives the generated and losses power active of the AC system after the three cases where a slight difference of a 1 MW is found, that mean the results are closer to each other. And, as shown in Fig. 14, there is a big difference in the values of the reactive powers generated, the cause is the shunt characteristic (parallel compensation) of these FACTS devices.

This significant change in reactive powers for the different cases is due to the reaction of the FACTS devices (SVC, STATCOM) against the reactive power. We also find that there is a slight decrease in active losses after correction of the voltage profile (case2) (see Fig. 13). Also, an increase of 1 MW after the insertion of the devices SVCs and STATCOMs (cases 3-a, 3-b), it is explained by the power consumed by those devices due to their own components (static switches, thyristors, IGBT/GTO and the controllers).

From the obtained results shown in the Table 9, we can say that as the reactive power compensation affects the steady state as well as the dynamic performance of the AC power system. The replacement of the shunt FACTS devices by the other static equipment (capacitor banks and synchronous compensator) is done in order to show the impact of these devices in the behavior of the AC system without consideration of the cost of the equipment and the other influences as well as the stability dynamic during either the fault condition or the voltage collapse.

## 5   Conclusion

In this article, three cases of studies were carried out for two different AC networks an (IEEE-05 bus and IEEE-57 bus) to perform the load flow using the *Newthon Raphson* method in the *MATLAB* environment and the *MTALB/Simulink* to model the AC networks. The first case was for the base-case load flow, the second one was for correction of the voltage profile by changing the reactive power generated by capacitor banks and the modifications of the transformers ratios, and the third one was by replacing one time the capacitors banks by SVCs devices and the other time, the synchronous compensators by STATCOMs devices. The results show that the FACTS devices can improve the stability of the AC power networks by keeping the bus voltage at a specified value, by improving the reactive power generated/transmitted in power grid, also by decreasing the power losses of the system especially when these devices are located in an optimal placement.

In the real AC power network, the FACTS devices have a dynamic reaction against the changes of system behavior (active load, power electronics, harmonic generation, voltage collapse, flickers, instability of the system, AC faults...etc.), that's why the FACTS devices are implemented in the AC networks, to resist and react with a certain flexibility to ensure the stability of the system against theses kind of circumstances. Our future work will be focalized in two different area: (i) study of the behavior of the AC in dynamic condition (in presence of the different kind of faults) and (ii) by including costs of the FACTS equipment and the fuel of the generators in order to evaluate the networks studied in this paper in the optimal power flow calculation.

# References

1. Enrique, A., Claudio, R.F.E, Hugo, A.P., César, A.: FACTS Modeling and Simulation in Power Networks. Wiley (2004)
2. Kothari, D.P., Nagrath, I.J.: Modern Power System Analysis, 4th edn. Tata McGraw-Hill Education Pvt. Ltd (2011)
3. Grainger, J.J., Stevensen, W.D.: Power System Analysis. McGraw-Hill, Inc., New York (1994)
4. Mathur, R.M., Varma, R.K.: Thyristor-Based FACTS Controllers for Electrical Transmission Systems. IEEE Press, Wiley, New York (2002)
5. Hingorani, N.G., Gyugyi, L.: Understanding FACTS Concepts and Technology of Flexible AC Transmission Systems. IEEE Press, Los Alimitos (1999)
6. Song, Y.H., Johns, A.T.: Flexible AC transmission Systems (FACTS). The Institute of Electrical Engineers, London (1999)
7. Milano, F.: Power System Modeling and Scripting. British Library Cataloguing in Publication Data. Springer, London, Dordrecht, Heidelberg, New York (2010)
8. Arrillaga, J., Arnold, C.P.: Computer Analysis of Power Systems, pp. 53–102. Wiley (1994)
9. Ammar, S.: Interaction des dispositifs FACTS avec les charges dynamiques dans les réseaux de transport et d'interconnexion. PhD thesis, National Institute Polytechnic of Grenoble (November 2010)
10. Amirnaser, Y., Reza, I.: Voltage-Sourced Converters in Power Systems: Modeling, Control, and Applications. IEEE Press, Wiley (2010)
11. Arrillaga, J., Watson, N.R.: Computer Modelling of Electrical Power Systems, 2nd edn. Wiley, New York (2001)
12. Haque, M.H.: A general load flow method for distribution systems. Electr. Power Syst. Res. **54**(1), 47–54 (2000)
13. Samet, H., Jarrahi, M.A.: A Comparison between SVC and STATCOM in flicker mitigation of electric arc furnace using practical recorded data. In: 30th International Power System Conference, Tehran, Iran (November 2015)
14. Nedzmija, D.: Impact of STATCOM and SVC to voltage control in systems with wind farms using induction generators (IG). In: IET, Mediterranean Conference on Power Generation,

Transmission, Distribution and Energy Conversion (MedPower), Belgrade, Serbia (November 2016)
15. Sood, V.: HVDC and FACTS Controllers: Applications of Static Converters in Power Systems. Kluwer Academic Publishers (2004)
16. Zhang, X.-P., Rehtanz, C., Pal, B.: Flexible AC Transmission Systems: Modeling and Control. Springer (2005)
17. Chun, L., Qirong, J., Xiaorong, X., Zhonghong, W.: Rule-based control for STATCOM to increase power system stability. In: Power System Technology, Proceedings in International Conference on POWERCON, 372–376 (1998)
18. Garcia-Cerrada, A., Garcia-Gonzalez, P., Collantes, R., Gomez, T.: Comparison of thyristor-controlled reactors and voltage–source inverters for compensation of flicker caused by arc furnaces. IEEE Trans. Power Deliv. **15**(4), 1225–1231 (2000)
19. Noroozian, M., Angquist, L., Ghandhari, M., Andersson, G.: Improving power system dynamics by series-connected FACTS devices. IEEE Trans. Power Deliv. **12**, 1635–1641 (1997)
20. Rahim, A.H.M.A., Al-Baiyat, S.A., Al-Maghrabi, H.M.: Robust damping controller design for a static compensator. IEEE Proc. Gener. Transm. Distrib. **149**(4), 491–496 (2002)
21. Haque, M.H.: Improvement of first swing stability limits by utilizing full benefit of shunt FACTS devices. IEEE Trans. Power Syst. **19**(4), 1894–1902 (2004)
22. Nor Adni Binti Mat, L., W Mohd, N.W.M, Nurlida, b.I., Nurul Huda, b.I., Nur Ashida, b.S.: The modeling of SVC for the voltage control in power system. Indones. J. Electr. Eng. Comput. Sci. **6**(3), 513–519 (2017)
23. Sauvik, B., Paresh, K.N.: State-of-the-art on the protection of FACTS compensated high-voltage transmission lines: a review, IET high voltage. High Volt. **3**(1), 21–30 (2018)

# Concept of Internal Momentum Absorber Structure to Reduce Impact During Accidents

Mandip Kumar Nar, Prabhjeet Singh$^{(\boxtimes)}$, and Paprinder Singh

Faculty of Engineering, Design & Automation, GNA University, Phagwara,
Punjab, India
prabhjeet.singh88@gmail.com

**Abstract.** Human life is precious and every technique must be explored to save it. Momentum absorber structure is one of the concept if implemented in vehicle can able to prevent injuries to passengers and damage to vehicle and surrounding as compared to conventional vehicle structure. In this present study an effort is made to design a conceptual momentum absorber structure that has the ability to absorb momentum during collision of vehicle and added as one of the safety features. The working of momentum absorber structure is also explained in the present work.

**Keywords:** Momentum · Vehicle structure · Collision · Car accidents · Safety

## 1 Introduction

The problem of car accident is a big issue around the world as many people lost their precious life with the dangerous car accident. As per stats around 3,000 car accidents occur every day around the globe. Country wise accidents death rate are shown in a Table 1[1].

According to WHO(World Health Organization) there were 1.25 million road traffic deaths globally in 2013. Alcohol and other drugs are found to be the most contributing cause in up to 22% of vehicular accidents on the world's highways and byways [1]. Honestly it is very difficult to change habits of the people so as an engineer we can make that type of car structure which can reduce the damage during the accident which helps in saving the life of people. Internal mass (momentum absorber structure) is capable to reduce the damage during the accident of cars.

According to WHO (World Health Organization) 80% of cars sold in the world are not compliant with main safety standards [2]. The real problem is that around only 40 countries have adopted the full set of most important regulations of car safety [2].

Now a days, the vehicle have number of safety equipment's which could work during accidents in efficient way. But that's increases the cost of the vehicle too high like air bags, traction control etc. The main motive that kept in mind is to save human from any injury but not meant for whole car and surrounding damage.

Internal momentum absorber structure aimed to reduce the injury for humans, damage to whole car and also for surrounding point of view. This can be achieved

© Springer Nature Switzerland AG 2019
C. Benavente-Peces et al. (Eds.): SEAHF 2019, SIST 150, pp. 144–149, 2019.
https://doi.org/10.1007/978-3-030-22964-1_15

**Table 1.** Country wise car accident death rate

| Rank | Country | Car accident deaths (Per 100,000) |
|------|---------|-----------------------------------|
| 1. | Libya | 73.4 |
| 2. | Thailand | 36.2 |
| 3. | Malawi | 35 |
| 4. | Liberia | 33.7 |
| 5. | Democratic Republic of the Congo | 33.2 |
| 6. | Tanzania | 32.9 |
| 7. | Central African Republic | 32.4 |
| 8. | Iran | 32.1 |
| 9. | Rwanda | 32.1 |
| 10 | Mozambique | 31.6 |

through shifting the momentum by internal skeleton of mass (momentum absorber structure).

## 2 Method to Reduce Damage

Firstly we have to understand actually what happen during front to front collision of vehicle. If we analyze the accident of two cars (front to front) there is a transfer of momentum and sharing of momentum between the two cars.

Let us assume there are two cars A and B having different momentum, and the car A is having more momentum than car B. The total momentum is equal to sharing momentum between the two cars.

But the difference of momentum of car A and car B give some value which is the extra momentum and this extra momentum will produce more damage on car with lower momentum as compare to car with higher momentum.

Let us explain the problem with an example:

Suppose we have two cars A and B.

For Car A:
Mass, $m1 = 50$ kg
Velocity, $v1 = 30$ m/s
For Car B:
Mass, $m2 = 40$ kg
Velocity, $v2 = 20$ m/s

Momentum of car A, P1

$$P1 = m1 \times v1$$
$$P1 = 50 \times 30$$

So, Momentum of Car A, P1 is 1500 kg m/s
And momentum of car B, P2

$$P2 = m2 \times v2$$
$$P1 = 40 \times 20$$

So, Momentum of Car B, P2 is 800 kg m/s
Thus the extra or damageable momentum, PD (transfer momentum) is equal to,

$$PD = P1 - P2$$
$$1500 - 800 = 700\text{kg m/s}$$

The final or sharing momentum between two cars is equal to, PS

$$PS = P1 + P2$$
$$1500 + 800 = 2300\text{kg m/s}$$

So 700 kg m/s is extra momentum on car B and after collision total momentum on car B, PB

$$PB = PD + P2$$
$$700 + 800 = 1500\text{kg m/s}$$

It concludes that the vehicle with low momentum is susceptible for high damage.

This is explained with the help of graph in Fig. 1. Shaded area shows the sharing momentum between two cars and extra momentum of 700 kg m/s is displayed on car B which results in more damage to car B.

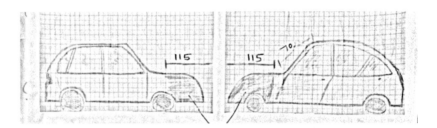

**Fig. 1.** Momentum transfer during collision

## 3   Momentum Absorber Structure

As the transfer of momentum is responsible for the damage to the vehicle, so this study focuses on internal mass adjustment approach to overcome the effect of momentum transformation. Figures 2 and 3 shows the concept of momentum transfer structure.

In this concept shocks cylinders are installed which work as a momentum force absorber. The cylinder contains a helical spring and opposite magnets installed at the ends of spring. The cylinders are covered by a rubber as shown in give Figs. 4 and 5.

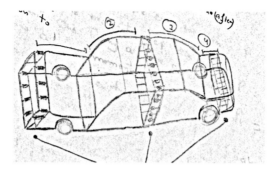

**Fig. 2.** Momentum absorber structure

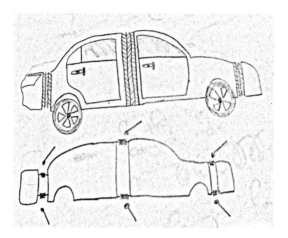

**Fig. 3.** Outer body of momentum absorber structure

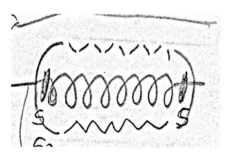

**Fig. 4.** Internal spring structure

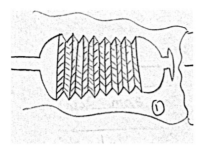

**Fig. 5.** External appearance of cylinder

## 4    Working of Momentum Absorber Structure

If accident occurs momentum force of collision try to contract the springs of the cylinder, as the helical spring installed inside cylinders need energy for contracting this will balanced the extra momentum applied on the vehicle and thus preventing damage. Also there are some polarity magnets are installed at the end of the spring, as equal poles repel each other it prevents the spring from damage and also help in absorbing the extra momentum. So during the accident the car will compress due to action of momentum absorber structure of vehicle. The different views of the momentum absorber structure is shown in Fig. 6.

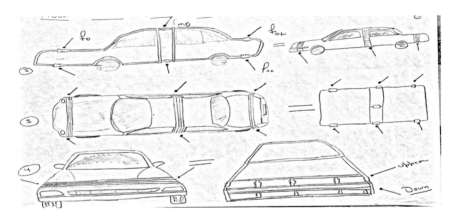

**Fig. 6.** Different views of momentum absorber structure

## 5    Conclusion

If somehow we can absorb the damageable momentum between the two vehicles, the damage can be kept to the minimum. This paper represented the concept for the same with the help of shock absorbing cylinders. As human safety comes first, so the concept of momentum absorber structure if applied to the vehicles helps in saving life, vehicle and surrounding during accident.

# References

1. Car accidents deaths data at worldatlas.com, seen on July 2018
2. Safety standards data from WHO (world health organization) available on internet
3. Rajput, R.K.: Applied Mechanics, 3rd edn. Laxmi Publication (1988)
4. Khurmi, R.S.: Applied Mechanics. S. Chand Publication (2016)

# Denoising Magnetic Resonance Imaging Using Fuzzy Similarity Based Filter

Bhanu Pratap Singh[1]([✉]), Sunil Kumar[1], and Jayant Shekhar[2]

[1] AURO University, Surat, India
{bhanupratap.singh, sunil.kumar}@aurouniversity.edu.in
[2] Adama Science and Technology University, Adama, Ethiopia

**Abstract.** Images are very useful source of information which is often degraded due to presence of noise. Noise present in the image especially in MRI images hides the important information which is very important to diagnose the disease. So to retain the quality of image we need to remove noise. Hence denoising is very essential to obtain precise images to facilitate the accurate observations. Fuzzy Similarity based Non-Local Means (FSNLM) filter is used to select homogeneous pixels for the estimation of noise-free pixels. Rician noise introduces bias which corrupts MRI images. The bias correction has been proposed for the removal of bias from MRI images which increases contrast and PSNR. The proposed scheme has been tested on simulated data sets and compared with existing method.

**Keywords:** dB · FSNLM · IEF · MAE · MRI · NLM · PD · PSNR · SBD · SSIM

## 1 Introduction

Magnetic resonance imaging (MRI), a powerful diagnostic technique, used in radiology to visualize detailed internal structures of the human body. MRI is based on the principles of nuclear magnetic resonance. MRI images contain some degree of noise. Noise is a random variation of image Intensity and visible as grains in the image. It may produce at the time of capturing or image transmission. It is the pixels in the image which show different intensity values rather than true pixel values.

There is uncertainty in several aspects of image processing. The approach of Fuzzy Logic emulates the way of decision making in humans that includes all intermediary possibilities between digital values yes and no. The logic block that a computer can perceive takes precise input and produces a definite output as TRUE or FALSE, which is similar to human's yes or no. The fuzzy logic works on the ranges of possibilities of input to attain the definite output. The range of possibilities like certainly yes, possibly yes, cannot say, possibly no and certainly no. It is conceptually easy to understand and is flexible and is tolerant of imprecise data.

© Springer Nature Switzerland AG 2019
C. Benavente-Peces et al. (Eds.): SEAHF 2019, SIST 150, pp. 150–164, 2019.
https://doi.org/10.1007/978-3-030-22964-1_16

The non-local means algorithm does not make any assumptions about noisy image as other denoising methods. The assumptions are the noise contained in the image is white noise and the true image (image without the noise) is smooth. It assumes that image contains an intensive amount of self-similarity.

## 2  Literature Review

Magnetic resonance imaging is a powerful technique to create picture of human analogy so as to diagnose the disease. However MRI images are not provided by any hospitals as these are very confidential. Thus MRI database need to be generated. Few authors have done work in this area.

MR simulators are used to create simulated brain database which encompasses various features as in noise, intensity inhomogeneity as well as slice thickness. A set of brain images can be simulated by varying MRI parameters, hence image database is created which can be further used to test pattern recognition and medical image processing. The creation of SBD provide an environment for measuring the effects of MRI parameters on pattern recognition and image processing (Kollokian 1996).

After few years ago, the tissue parameters i.e. T1, T2 and PD is evaluated simultaneously to test the whole potential of magnetic resonance (MR) imaging for quantitative tissue characterization. MR tissue parameters were determined. Pathologic entities are characterized by wide ranges of T1, T2, and proton density values while normal brain tissues showed only small inter-individual variations of tissue parameters. The significant overlapping of tissue parameters which are analyzed in a three-dimensional T1/T2/proton density space specifies that a reliable diagnosis can't be made on the basis of a quantitative assessment of T1, T2, and proton density alone (Just and Thelen 2000).

Uncertainty may arise due to information which is not fully reliable or due to partial information about the problem. Fuzzy set theory is a mathematical tool to tackle the uncertainty arising due to vagueness. Vagueness (inexact) is nothing but fuzziness. Fuzzy logic is able tolerate imprecise as well as noisy data. Some of the authors have done work on fuzzy logic.

Various uncertainty are involved in image processing and relevant fuzzy theory in handling these uncertainty are discussed. Image ambiguity based on fuzzy entropy as well as choosing of different membership functions are mentioned. Image processing operation like edge detection, feature extraction along with its significance and characteristics are being illustrated. Various real life application are explained like image face modelling and image analysis (Pal 2001). Uncertainty is handled by fuzzy logic. Collection of different fuzzy approaches to image processing is called fuzzy image processing. It is the collection of all approaches that understand, represent and process the images, segments and features as fuzzy sets. The representation and processing rely on the selected fuzzy technique and on the problem to be solved. It consists of image fuzzification, membership functions and image defuzzification. The first and the last steps are due to the fact that we do not own fuzzy hardware. Therefore, the coding of image data is nothing but fuzzification and decoding of the results is nothing but the defuzzification are steps that made possible to process images with fuzzy techniques.

The main phase of this is the modification of membership values. After the image data are transmuted from gray level plane to the membership plane that is fuzzification then fuzzy techniques modify the membership values accordingly. The most important advantage of a fuzzy methodology lies is that the fuzzy membership function delivers a natural means to model the uncertainty dominant in an image scene. Consequently, fuzzy logic results can be applied in feature extraction and object recognition phases of image processing and successive computer vision. (Narang) MRI images can be degraded by some noise which hide some information which is very important to diagnose the disease. Noise is a random variation of image intensity. Hence denoising is very essential to obtain precise images to facilitate the accurate observations as well as to retain the quality of image. There are many ways to denoised image. The main feature of a good image denoising method is to remove noise while preserving edges. Conventionally linear models have been used like mean filter, Gaussian filter and many more. The big advantage of linear models is the speed but at the same time one drawback of the linear models is that they are not able to preserve edges. On the other hand non-linear models can handle edges in a better way than linear models can like non local means filter.

The denoising is very essential but at the same time it is very challenging task. Two noise models are rician noise which is present in MRI images and speckle noise which is present in Ultrasound images. Two ways of denoising are linear (mean filter) and non-linear filter (median filter). Linear filter is not able to preserve edges but at the same time the big advantage of this filter is speed (Nobi and Yousuf 2011). The median filter is able to preserve the edges. Hence the combination of median and mean filter are used to determine more accurate value of pixels of noisy image. The quality of output images is measured by the statistical quantity measures like peak signal to noise ratio (PSNR), signal to noise ratio (SNR) and root mean square error (RSME). They experimental results show that combination of both these filters perform much better than filtering method. The another way to remove noise in the Magnetic Resonance images is Order statics filter which is based on ranking the pixels contained in an image area encompassed by filter (Sarode and Deshmukh 2011). This method demonstrations an optimal assessment result which is more accurate in improving the true signal from Rician noise. This method proposed specially for Rician noise reduction, but as Rician noise can be approached to Gaussian noise when SNR is high, consequently, expected that proposed algorithm also has advantage in denoising of complex MR images. Before processing images for further analysis, attaining an effective method of removing noise from the images, is a great task. Noise can degrade the image at the time of capturing or transmission of the image. Plenty algorithms are available, but they have their own conventions, advantages and disadvantages. The comparison of various noise removal techniques are done (Verma and Ali 2013). The results of applying different noise types to an image model are presented and explored the results of applying various noise reduction techniques. The various image filtering algorithms and techniques used for image filtering/smoothing. Image smoothing is one of the most widely used operation in image processing. Various algorithms and techniques and which algorithm is best, especially concentrated on non-linear filtering algorithms i.e. median filtering is very important in edge preserving. Image may be corrupted because of poor contrast, variations in intensity (Chandel and Gupta 2013).

MRI images suffer from bias, signal dependent which reduce contrast as introduced by Rician distributed noise. Acquisition technique is used to diminish the acquisition time which give rise to correlated noise assumed to be white. The author proposed a two-step denoising procedure where bias is eliminated from the squared magnitude image and denoising itself is then completed on the square root of image in the wavelet domain. When distinguishing significant wavelet coefficients from insignificant ones, then this denoising step is taken into account. The expected statistics of these two classes of wavelet coefficients are used within a bayesian estimator. The experimental results demonstrate that the proposed technique is more effected at removing correlated noise rather than existing MRI denoising techniques. The bias removal technique improve contrast as well as provide a large increase in PSNR (Aelterman et al. 2008). The non-parametric IIH-correction strategy by fusing multiple Gaussian surfaces in MRI images. This non parametric method works directly on spatial domain using local image gradients rather than requiring prior knowledge of intensity probability distribution. Firstly, image histogram is taken into account while extracting different tissue regions. Secondly, a bias field is estimated by fitting Gaussian surface on the gradient map of each of the homogeneous tissue regions. Thirdly, by fusing these bias field, the intensity non-uniformity field of the entire image is obtained. Lastly, this field is removed from the image iteratively so as to obtain the corrected image. The experimental results of the proposed method is efficient in removing bias (Adhikari et al. 2015).

## 3   Summary of Literature Review

There are broadly two filters to denoised image. The main feature of a good image denoising method is to remove noise while preserving edges. Conventionally linear models have been used like mean filter, Gaussian filter and many more. The big advantage of linear models is the speed but at the same time one drawback of the linear models is that they are not able to preserve edges. On the other hand non-linear models can handle edges in a better way than linear models can like non local means filter. The non-local means based filters use all the non-local neighbors for the estimation of noise-free value without region analysis. It is highly feasible that near the image edge information, non-local neighbors belonging to different regions can disturb the accurate estimation of noise free value than the central noisy pixel. Hence by using fuzzy based similarity mechanism non local regions are being analyzed for the estimation noise free pixels. As rician noise introduces bias hence bias correction method must be employed.

## 4   Existing Technique

In the Existing Technique, fuzzy similarity based non local mean FSNLM method is used to denoise MRI Data. The motive of the proposed approach is to remove the low and high categories of Rician noise while preserving the detail information. For the estimation of a noise free pixel's value, the required components are prepared. The fuzzy based similarity mechanism is used to find out the similar and non-similar non-local windows of the central pixel. And finally the restoration of noisy pixel is performed. The existing technique is divided in following steps. The block diagram is shown in Fig. 1 (Figs. 2, 3, 4).

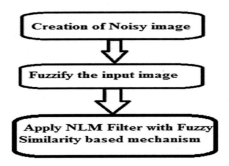

**Fig. 1.** Block diagram of existing technique

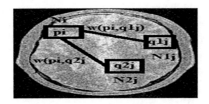

**Fig. 2.** Creation of SBD

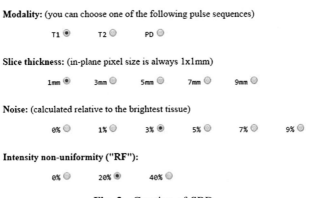

**Fig. 3.** Determining weights for similar windows

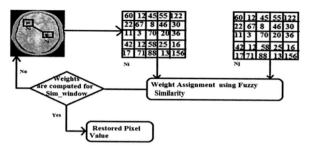

**Fig. 4.** Flow chart of FSNLM

## 4.1    Research Methodology

1. Database collection

MR simulators are used to create simulated brain database which encompasses various features as in modality, noise, and intensity inhomogeneity as well as slice thickness. A set of brain images can be simulated by varying MRI parameters, hence image database is created which can be further used to test medical image processing.

### 4.1.1    Creation of the Database

It is necessary to establish the conditions of MRI parameters prior to construct the database.

- Estimating pulse sequence
- Estimating noise levels
- Estimating RF inhomogeneity
- Estimate different levels of slice thickness.

There are three different modalities T1 weighted, T2 weighted and proton density weighted which reflect amount of nuclei present in body. The size of pixel determines resolution. Smaller is the pixel, higher will be the resolutions. Increasing the slice thickness will effectively increase the pixel size and hence decreases the resolution. The noise percent is given by standard deviation of the Gaussian to be added to real and imaginary channels. Intensity non-uniformity states INU levels.

2. Image fuzzification

The process of making a crisp quantity fuzzy is fuzzification. By simply recognizing that many of the quantities that require to be considered to be crisp and deterministic are actually not deterministic at all. They have some kind of uncertainty. The variable is possibly fuzzy which can be represented by a membership function if the uncertainty happens to arise because of ambiguity, imprecision, or vagueness. In this step, fuzzified values are obtained by applying non-parametric membership function to get the images in fuzzy domain (pixels having values in the range (0, 1)). This process enables us to deal effectively with uncertainties and inexactness present in the images.

## 4.2    Using Parametric Membership Function

Fuzzified values are obtained by applying trapezoidal membership function. Firstly, image is read and then applying the formula of trapezoidal membership function over it. The trapezoidal curve is a function of a vector x that depends on four parameters a, b, c, and d as given by

$$f(x; a, b, c, d) = \begin{cases} 0, & x \leq a \\ \frac{x-a}{b-a}, & a \leq x \leq b \\ 1, & b \leq x \leq c \\ \frac{d-x}{d-c}, & c \leq x \leq d \\ 0, & d \leq x \end{cases}$$

3. Apply Non Local Means Filter with Fuzzy Similarity mechanism

The Basic idea of the NLM filter is considering the data redundancy among the windows of a noisy image and replace the noise free pixel using weighted average of non-local pixels. The weight estimates the similarity between Ni, N1j and N2j neighborhoods which is centered on pixels pi, q1j and q2j called patches or similarity window. The non-local means filter reflects the pixel intensities of the whole image in the weighted average for theoretical reasons while for practical reasons the pixel intensity is restricted to a neighborhood called search window (Binaee and Hasanzadeh 2011).

The non-local means algorithm has two parameters. The first parameter, Rsim, is the radius of the neighborhoods used to find the similarity between two pixels. No similar neighborhoods will be found if Rsim is too large but too many similar neighborhoods will be found if it is too small.

The second parameter, Research, is the radius of a search window. Due to the inadequacy of calculating the weighted average of every pixel for every pixel, it will be diminished to a weighted average of all pixels in a window.

Fuzzy Similarity Criteria: After getting fuzzified values, next step is to identify the similar windows. Similar windows/patches/regions are identified by using these fuzzified values. Window is a process of selecting a particular part of an image. Windowing is needed as processing the whole image is computationally very expensive. Hence smaller size of area is being used. A better noise free image could be estimated by finding the similar pixels in non-local neighborhood. Obtaining the similar window is a very-very challenging task due to uncertainty present in MRI images. And that's why fuzzified values are being used to find similar window. Search area is basically used to obtain pixel j similar to pixel i. Out of the whole image, 5 * 5 and 21 * 21 windows are taken to obtain similar regions. Consider a region of a gray scale image to explain how similar neighborhoods are evaluated in detail as shown in above figure. Considering pi as the central pixel. The central pixel q1ji and q2j of window N1j and N2j respectively are similar to central pixel pi of window Ni.

The different similarity windows are going to be evaluated by the fuzzy similarity criteria. Search area is used to find a pixel j similar to the pixel i. Rsim represents the radius of local and nonlocal regions or windows Wi and Wj centered at pixel i and pixel j (Weken et al. 2005). The region based comparison is performed to find the similarity of pixel i with all non-local neighboring pixels pixel j separately. The area of radius Research (21 * 21), is divided into overlapping windows of radius Rsim (5 * 5) In order to compute the similarity, if the difference between the fuzzified values of these two windows is less than $sim_t$, similarity threshold then windows are similar otherwise non-similar. Smaller the differences, greater will be similarity.

$$W_{sim} = F_{ws} - F_{wnl}$$

where

| | |
|---|---|
| $W_{sim}$ | = Similar windows |
| $F_{ws}$ | = Fuzzified value of local window |
| $F_{wnl}$ | = Fuzzified values of non-local window |

Assigning weights to the pixel and restore the noisy pixel.

The basic principle of the NLM filter is to replace the noisy gray value I (i) of pixel i with a weighted average of the gray-values of all the pixels on the image. The pixel to be denoised is denoted by i and likewise the pixels in the neighborhood of I by j and use them to denoise i. The NLM algorithm computes a weighted average of non-local pixels and restores the pixels which are in a similar environment. The weight is computed based on the similarity mechanism between the neighborhood windows of the pixels of interest and contributing pixels. Buades et al. introduced the NLM filter that averages pixel intensities weighted by the similarity of image neighborhoods which are usually defined as $5 \times 5$, $7 \times 7$ or $9 \times 9$ square windows or patches of pixels which has 25, 49 or 81 elements respectively. After computing similar locations firstly calculate the mean of each neighborhood of entire image. The weight is calculated by using the fuzzified values and mean of those similar windows:

$$weight = \left\{ (1 - (F_{ws} - F_{wnl})) * \left( \frac{Ms}{Mnl} \right) \right\}$$

where $Ms$ denotes mean of local window, $Mnl$ denotes mean of non-local window.

After assigning the weights, calculate the weighted average. The central pixel value is to be replaced by average weighted.

$$Average\_weight = \frac{\sum_{i}^{n} weight * Mnl}{n}$$

where n = no. of counts weight is assigned.

**Drawbacks of the Existing Method**

In existing method, trapezoidal membership function is used for fuzzification which is based on four parameters. Hence specify all parameters makes it complex. Moreover, Rician noise introduce bias which need to be removed as it affects contrast and decrease PSNR.

## 5  Proposed Work

In proposed method to overcome the drawbacks of previous work following modifications have been done.

- Non-parametric membership function is used instead of trapezoidal membership function for fuzzification.
- Bias correction method is used to remove bias which improve contrast and PSNR.

The whole process can be visualized in the flow diagram of the method shown next. As shown in Fig. 5, one more steps has been added to the previous method and one previous step has been modified (Fig 6).

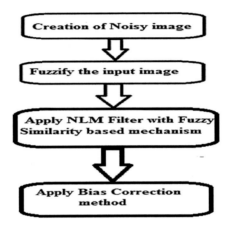

**Fig. 5.** Block diagram of proposed method

| 1 | 2 | 3 | 4 | 5 | 6 | 7 | 8 | 9 |
|---|---|---|---|---|---|---|---|---|
| 0 | 0 | 0 | 0 | 0 | 0 | 0 | 0 | 0 |
| 0 | 0 | 0 | 0 | 0 | 0 | 0 | 0 | 0 |
| 0 | 0 | 0.6119 | 0.5522 | 0.3433 | 0.2985 | 0.2923 | 0.9474 | 0.7544 |
| 0 | 0 | 0.2388 | 0 | 1 | 0.1642 | 0.0179 | 0.7308 | 0.7037 |
| 0 | 0 | 0.1642 | 0.1343 | 0.6269 | 0.1940 | 0.2857 | 0.3878 | 0.2353 |
| 0 | 0 | 0.2537 | 0.5075 | 0.4925 | 0.5970 | 0.5556 | 0.1750 | 0.6739 |
| 0 | 0 | 0.3704 | 0.3148 | 0.0732 | 0.5556 | 0.3333 | 0.4545 | 0.3261 |
| 0 | 0 | 0 | 0.3455 | 0.4714 | 0.1194 | 0.1493 | 0.2344 | 0.3030 |
| 0 | 0 | 0.2778 | 0.4727 | 0.4857 | 0.1940 | 0.1343 | 0.6563 | 0.3750 |
| 0 | 0 | 0.5714 | 0.8036 | 0.2000 | 0.7656 | 1 | 0.2813 | 0.3788 |
| 0 | 0 | 0.8113 | 0.1698 | 0.4853 | 0.3088 | 0.1912 | 0.1176 | 0.2794 |
| 0 | 0 | 0.5303 | 0.0455 | 0.7941 | 0.3529 | 0.4118 | 0.3088 | 0.4265 |
| 0 | 0 | 0.5152 | 0.3182 | 0.0299 | 0.3148 | 0 | 0.6744 | 0.2791 |
| 0 | 0 | 1 | 0.3182 | 0.0149 | 0.5893 | 0.1250 | 0.1250 | 0.4419 |
| 0 | 0 | 0.3333 | 0.4394 | 0.1493 | 0.5714 | 0.7143 | 0.5357 | 0.8500 |
| 0 | 0 | 0.3380 | 0.6479 | 0.6029 | 1 | 0.1167 | 0.2105 | 0.1500 |
| 0 | 0 | 0.3662 | 0.1333 | 0.7368 | 0.2982 | 0.3333 | 0 | 0.1500 |

**Fig. 6.** Fuzzified value using non parametric membership function

## 5.1   Image Fuzzification

The process of making a crisp quantity fuzzy is fuzzification. By simply recognizing that many of the quantities that require to be considered to be crisp and deterministic are actually not deterministic at all. They have some kind of uncertainty. The variable is possibly fuzzy which can be represented by a membership function if the uncertainty happens to arise because of ambiguity, imprecision, or vagueness. In this step, fuzzified values are obtained by applying non-parametric membership function to get the images in fuzzy domain (pixels having values in the range (0, 1)). This process enables us to deal effectively with uncertainties and inexactness present in the images.

## 5.2   Using Non-parametric Equations

Fuzzified values are obtained by applying non-parametric membership functions as shown in figure. Firstly, image is read and then get the minimum and maximum pixel values. And then divide the whole image with maximum value.

## 5.3   Apply Bias Correction Method

After restoring the pixels values, the image we get will be in the de-fuzzy domain. Then apply rician bias estimator to get final denoised image. MRI noise introduces a bias. Bias is a low frequency, smooth and undesirable signal which corrupts MRI images due to the presence of inhomogeneity, change in intensity values of image pixels, in MRI machine. Bias reduction method, a two-step approach is used for post processing of the denoised image obtained by applying Fuzzy similarity based non local means algorithm to improve the denoised image. By such two-step approach edges are preserved in the first step and bias reduction is accomplished in second step. Bias correction is based on variance and variance is calculated by using Bessel correction. Bias is removed by square root of the difference of variance from the result from FSNLM of the image. The bias correction formula:

$$m_1(i,j) = \sqrt{m^2(i,j) - \sigma^2}, \text{ CITATION Muk13\l 1033 (Mukherjee and Qiu 2013)}$$

where $m_1(i,j)$ denoted the improved denoised image. Basically MRI images suffer from bias, a signal dependent which reduce contrast.

# 6   Results and Discussion

## 6.1   Materials and Quantitative Metrics

The experiments and comparative analysis are performed on simulated MRI datasets. The simulated MR data is obtained from Brain-Web. These simulated datasets have been used in the comparative analysis.

## 6.2  Simulated MR Data

In experiments, three types of modalities T1-weighted, T2-weighted, and PD-weighted of simulated MRI volumes for normal brain are analyzed. The size of each simulated MRI image is $181 \times 181$. But for experimental purpose each simulated MRI images are resized to $67 \times 67$. The slice thickness of these datasets is kept 1 mm$^3$.

## 6.3  Quantitative Metrics

In order to measure the performance, most widely used qualitative measure is peak signal to noise ratio (PSNR).

PSNR: This term is the ratio between the power i.e. maximum possible value of a signal and the power of corrupting noise which affects the quality. It is expressed in terms of decibel. It is calculated as:

$$PSNR = 10 * log10 \left( \frac{MAX * MAX}{MSE} \right)$$

where MAX = maximum possible value of the image and MSE is Mean Square Error which is the average of square of errors.

$$MSE = 1/mn \sum_0^{m-1} \sum_0^{n-1} ||f(i,j) - g(i,j)||^2$$

where $f(i,j)$ is the original image, $g(i,j)$ is the degraded image, m represents no of rows and i is index of that row and n represents no of column and j is the index of that column.

MAE: This term is the average difference between reference image or original image and modified image. It is calculated as:

$$MAE = 1/MN \sum_{i=1}^{M} \sum_{j=1}^{N} |x(i,j) - y(i,j)|$$

where $x(i,j)$ is referenced image and $y(i,j)$ is distorted (modified) image.

IEF: This term is Image Enhancement factor. This factor is an important factor in any subjective evaluation of image quality. It is calculated as

$$IEF = \sum_{i,j} (n(i,j) - Y(i,j))^2 / \sum_{i,j} (Y1(i,j) - Y(i,j))^2$$

where $n(i,j)$ is noisy image, $Y(i,j)$ is reference image and $Y1(i,j)$ is modified image.

SSIM: This term is Structural Similarity Index. SSIM measures image quality. It is used to compare the visual quality of image obtained from proposed image and original image. The measure between two windows x and y of common size M * M is:

$$SSIM = \left(2\mu_x\mu_y + c_1\right)\left(2\sigma_{xy} + c_2\right)/\left(\mu_x^2 + \mu_y^2 + c_1\right)\left(\sigma_x^2 + \sigma_y^2 + c_2\right)$$

where $\mu_x$, $\mu_y$ are the average of x and y respectively, $\sigma_x^2$, $\sigma_y^2$ are the variance of x and y respectively. $\sigma_{xy}$ is covariance of x and y respectively. $c_1$ and $c_2$ are two variables to stabilize the division with weak denominator.

To measure the quality of the proposed scheme the detailed experimentation is given on simulated and real MR images. All the experiments were performed using Matlab R2015b.

Table 1 shows the quality measure results for the output obtained from the proposed method of PD-weighted data sets. The PSNR, MAE, IEF and SSIM values of the proposed algorithm are comparing with other existing algorithms by varying noise percent from 3 to 18%. The PSNR value of proposed method is not that up to the mark but it is high as compared with existing method. The MAE value is low as well which shows difference between reference image and test image is low. The IEF value is high which shows image is enhanced. The SSIM value is high which shows both reference image and test image are quite similar.

**Table 1.** Quality metric results comparison of PD-weighted image

| PD-weighted data sets | PSNR | MAE | IEF | SSIM |
|---|---|---|---|---|
| Sample image 1 | 32.15 dB | 3.99 | 1.42 | 0.77 |
| Sample image 2 | 30.97 dB | 6.06 | 1.11 | 0.50 |
| Sample image 3 | 31.27 dB | 4.24 | 1.48 | 0.73 |
| Sample image 4 | 29.23 dB | 4.32 | 1.49 | 0.64 |
| Sample image 5 | 30.92 dB | 6.71 | 1.43 | 0.60 |

Table 2 shows the quality measure results for the output obtained from the proposed method of T1-weighted data sets. The PSNR, MAE, IEF and SSIM values of the proposed algorithm are comparing with other existing algorithms by varying noise percent from 3 to 18%. The PSNR value of proposed method is not that up to the mark but it is high as compared with existing method. The MAE value is low as well which shows difference between reference image and test image is low. The IEF value is high which shows image is enhanced. The SSIM value is high which shows both reference image and test image are quite similar.

**Table 2.** Quality metric results comparison of T1-weighted image

| T1-weighted data sets | PSNR | MAE | IEF | SSIM |
|---|---|---|---|---|
| Sample image 1 | 30.42 dB | 5.88 | 1.08 | 0.53 |
| Sample image 2 | 30.12 dB | 8.81 | 1.33 | 0.58 |
| Sample image 3 | 29.42 dB | 7.02 | 1.08 | 0.32 |
| Sample image 4 | 29.67 dB | 4.45 | 0.97 | 0.31 |
| Sample image 5 | 30.11 dB | 3.45 | 1.04 | 0.44 |

# 7    Conclusions

Fuzzy similarity based non-local means filter has been presented which has mainly three components. Firstly Fuzzification has been done. After that fuzzy similarity mechanism is used to identify similar windows or regions for the estimation of noise free pixel. FSNLM calculates the weighted average of the surrounding similar pixels so as to restore a noisy pixel in the image which gives better result. But somehow, rician noise introduce bias which reduces contrast and hence decreases PSNR value as image quality decreases. So the proposed method has been introduced so as to increase the contrast and PSNR value. Experiments results shows that the proposed method is good at increasing contrast and PSNR values.

After analyzing the statistical and visual results, it is to be concluded that, the output obtained from proposed method has better visual quality and contrast as compared to the image obtained from existing method.

## References

Agrawal, N., Sinha, D.: A survey on fuzzy based image denoising methods. Int. J. Eng. Res. Technol. (IJERT) **4**(5), 528–531 (2015)

Adhikari, S.K., Sing, J.K., Bbasu, D.K., Nasipuri, M., Saha, P.K.: A nonparametric method for intensity inhomogeneity correction in MRI brain images by fusion of Gaussian surfaces. SIViP **9**, 1945–1954 (2015)

Aelterman, J., Goossens, B., Pizurica, A., Philips, W.: Removal of correlated rician noise in magnetic resonance imaging. In: 16th European Signal Processing Conference (EUSIPCO), pp. 1–5 (2008)

Amza, C.G., Cicic, D.T.: Industrial image processing using fuzzy-logic. In: 25th DAAAM International Symposium on Intelligent Manufacturing and Automation, DAAAM, pp. 492–498 (2015)

Binaee, K., Hasanzadeh, R.P.: A non local means method using fuzzy similarity criteria for restoration of ultrasound images. In: IEEE Machine Vision and Image Processing (MVIP), pp. 1–5 (2011)

Borkar, A.D., Atulkar, M.: Fuzzy inference system for image processing. Int. J. Adv. Res. Comput. Eng. Technol. (IJARCET) **2**(3), 1007–1010 (2013)

Brar, A.K., Wasson, V.: Image denoising using improved neuro-fuzzy based algorithm: a review. Int. J. Adv. Res. Comput. Sci. Softw. Eng. **4**(4), 1072–1075 (2014)

Brinkmann, B., Manduca, A., Robb, R.: Optimized homomorphic unsharp masking for MR grayscale inhomogeneity correction. IEEE Trans. Med. Imaging **17**(2), 161–171 (1998)

Buades, A., Coll, B., Morel, J.M.: A non-local algorithm for image denoising. IEEE Comput. Soc. Conf. **2**, 60–65 (2005)

Buades, A., Coll, B., Morel, J.M.: Nonlocal image and movie denoising. Int. J. Comput. Vis. **76**, 123–139 (2008)

Chandel, R., Gupta, G.: Image filtering algorithms and techniques: a review. Int. J. Adv. Res. Comput. Sci. Softw. Eng. **3**(10), 198–202 (2013)

Cocosco, C.A., Kollokian, V., Kwan, R.K., Pike, G.B., Evans, A.: Brainweb: online interface to a 3D MRI simulated brain database. In: NeuroImage, pp. 1–1 (1997)

D, S. R., M, S., & M.H.M, K. P.: Quality Assessment parameters for iterative image fusion using fuzzy and neuro-fuzzy logic and applications. In: 8th International Conference inter Disciplinarity in Engineering, pp. 889–895 (2015)

D, S. R., M, S., & Prasad, K.: Comparison of fuzzy and neuro fuzzy image fusion techniques and its applications. Int. J. Comput. Appl., 31–37 (2012)

Juntu, J., Sijbers, J., Dyck, D.V., Gielen, J.: Bias field correction for MRI images. In: IEEE International Conference on Computer Graphics, Vision and Information Security (CGVIS), pp. 1–8 (2015)

Just, M, Thelen, M.: Tissue characterization with T1, T2, and proton density values: results in 160 patients with brain tumors. Radiology, 779–785 (1988)

K, M. P., Rai, D.: Applications of fuzzy logic in image processing—a brief study. Int. J. Adv. Comput. Technol. **4**(3), 1555–1559 (2015)

K, M. P., Rai, D S.: Fuzzy logic—a comprehensive study. Int. J. Adv. Found. Res. Comput. (IJAFRC), **1**(10), 1–6 (2014)

Kaur, A., Kaur, A.: Comparison of Mamdani-type and Sugeno-type fuzzy inference systems for air conditioning system. Int. J. Soft Comput. Eng. (IJSCE) **02**(02), 323–325 (2012)

Kaur, J., Sethi, P.: Evaluation of Fuzzy inference system in image processing. Int. J. Comput. Appl. **68**, 1–4 (2013)

Kollokian, V.: Performance analysis of automatic techniques for tissue classification in magnetic resonance images of the human brain (1996)

Kozlowska, E.: Basic principles of fuzzy logic (2012, 08 01). http://access.fel.cvut.cz/rservice.php?akce=tisk&cisloclanku=2012080002

Kumar, B.K.: Image denoising based on non local-means filter and its method noise thresholding. Signal Image Video Process., 1–12 (2013)

Metin Ertas, A.A.: Image denoising by using non-local means and total variation. Signal Process. Commun. Appl. Conf. (SIU), 2122–2125 (2014)

Narang, S.: Applying Fuzzy Logic to Image Processing Applications: A Review (n.d.)

Nobi, M., Yousuf, M.: A new method to remove noise in magnetic resonance and ultrasound images. J. Sci. Res., 81–89 (2011)

Pal, S.K.: Fuzzy image processing and recognition: uncertainty handling and applications. Int. J. Image Graph. **01**, 169–195 (2001)

Pathak, M., Sinha, D.: A survey of fuzzy based image denoising techniques. J. Electron. Commun. Eng. (IOSR-JECE), 27–36 (2014)

Pereza, M.G., Concib, A., Morenoc, A.B., Andaluz, V.H., Hernández, J.A.: Estimating the Rician Noise Level in Brain MR Image. IEEE, pp. 1–6 (2014)

Quality Assessment parameter for iterative image fusion using fuzzy and neuro fuzzy logic and applications. In: 8th International Conference Inter Disciplinarity in Engineering, pp. 888–894 (2015)

Sarode, M.V., Deshmukh, D.R.: Performance evaluation of noise reduction algorithm in magnetic resonance images. Int. J. Comput. Sci. **8**(3), 198–201 (2011)

Sharif, M., Hussain, A., Jaffar, M.A., Choi, T.S.: Fuzzy similarity based non local means filter for Rician noise removal. Multimed. Tools Appl., 5533–5556 (2015)

Sijbers, J., Dekker, A.D., Scheunders, P., Dyck, D.V.: Maximum-likelihood estimation of rician distribution parameters. IEEE Trans. Med. Imaging **17**(3), 357–361 (1998)

Sijbers, J., Dekker, A.D., Audekerke, J.V., Verhoye, M., Dyck, D.V.: Estimation of the noise in magnitude MR images. IEEE, pp. 87–90 (2014)

Styner, M., Brechbühler, C., Szekely, G., Gerig, G.: Parametric estimate of intensity inhomogeneities applied to MRI. IEEE Trans. Med. Imaging **19**(3), 153–165 (2000)

Tasdizen, T.: Principal neighborhood dictionaries for non-local means image denoising. IEEE Trans. Image Process., 1–12 (2009)

Vaidya, S.D., Hanchate, V.: Implementation of NLM for denoising of MRI images by using FPGA mechanism. Int. J. Adv. Res. Electr. Electron. Instrum. Eng. 5343 –5352 (2016)

Verma, R., Ali, D.: A comparative study of various types of image noise and efficient noise removal techniques. Int. J. Adv. Res. Comput. Sci. Softw. Eng. 3(10), 617–622 (2013)

Wang, X., Wang, H., Yang, J., Zhang, Y.: A new method for nonlocal means image denoising using multiple images. PloS One, 1–9 (2016)

Weken, D.D., Nachtegael, M., Witte, V., Schulte, S., Kerre, E.: A survey on the use and the construction of fuzzy similarity measures in image processing. In: IEEE International Conference on Computational Intelligence for Measurement Systems and Applications, pp. 187–192 (2005)

# Augmented Reality in Children's Education in the Republic of Macedonia

Marija Minevska[⊠] and Smilka Janeska-Sarkanjac

Faculty of Computer Science and Engineering, University 'Ss. Cyril and Methodious', 1000 Skopje, Republic of Macedonia
marijaminevska@gmail.com, smilka.janeska.
sarkanjac@finki.ukim.mk

**Abstract.** The aim of the research presented in this paper is to support the hypothesis that augmented reality is a good educational tool compared to the traditional ways of learning. In order to understand the level of knowledge, attitude and preferences towards the use of augmented reality (AR) in education in the Republic of Macedonia, we have conducted a survey where we designed several questions and examples that helped us to collect initial insights on the topic. The second part of the study was a practical experiment showing us that regardless of the environment, children prefer this visual way of studying compared to the traditional one. The AR applications used in this experiment were Skyview, Augment and Anatomy 4D, introduced to two classes of 6–8 and 10–12 years old. Additionally, AR applications were used by two children with dyslexia, 9 and 12 years old.

**Keywords:** Augmented reality · Education · Republic of Macedonia

## 1 Introduction

Augmented reality (AR) is a technology that allows virtual, visual and descriptive computer-generated information to be overlaid onto a live direct or indirect real-world environment in real time [6]. With AR the information about the surrounding real world of the user becomes interactive and digitally manipulatable. AR is variation of a Virtual Reality (VR). VR technologies completely immerse a user inside a synthetic environment and while immersed, the user cannot see the real world around him. In contrast, AR takes digital computer generated information, whether they be images, audio, video, and touch or haptic sensations and overlaying them in a real-time environment [4]. Research indicates that student participation in an AR-integrated learning environment could lead to enhancing students' spatial abilities and motivational factors [2, 7]. Augmented reality has pedagogical potential to improve critical thinking, creativity and critical analysis [1, 5]. AR contents developed for teaching science were evaluated by teachers and students as effective one year ago [3]. There are many AR based applications that can be used for education, starting with simple games through which children could learn, and later they can be used in primary school and high school education in subjects such as Mathematics, Geography, Biology, Chemistry, as well as in higher education such as the Medical Faculty. The era of making hybrid

© Springer Nature Switzerland AG 2019
C. Benavente-Peces et al. (Eds.): SEAHF 2019, SIST 150, pp. 165–168, 2019.
https://doi.org/10.1007/978-3-030-22964-1_17

mobile AR application for education is emerging, and it is expected that in a short period of time the number of AR applications will grow rapidly, covering different areas.

## 2    Research

### 2.1    Pre-research Study

In order to understand the level of knowledge, attitude and preferences towards the use of AR in education in Macedonia, we conducted a survey where we designed several questions and examples that helped us to collect initial insights on topic. The survey was conducted on 229 respondents and 95% of them think that use of AR in education will have motivating effect on children and will develop attraction for science. The respondents' explanations were that the child's attention is better retained in comparison with traditional education method, it is fun to study the material and the creativity grows. Parents would encourage schools that implement augmented reality lectures (partially or completely). Visual learning has been effective for generations. The most common explanation given by the interviewees about visualization and 3D projection is that it is more realistic, children do not study by heart and students acquire quality knowledge, it is faster and better to remember and easier to understand the material. Since visual learning is effective, the use of AR would be a great help for children. The number of positive opinion is greater than negative, so a conclusion is made that the interest for augmented reality is rapidly increasing and that it can transfer knowledge in a way that is more closely and immediately related to the world around us. Negative comments are also taken in consideration. Children nowadays are using tablets and phones constantly for playing, because for them that is more interesting than reading a book. But, if we combine the technologies with education, knowledge will be fun and challenging.

### 2.2    Research

The study was conducted in three different cities in the Republic of Macedonia: Skopje, Kavadarci and Strumica, to see that no matter what kind of childhood or environment kids might have, they all prefer this visual way of studying compared to the traditional one. The primary schools "Dimo Hadzi Dimov", "Petar Pop Arsov" and "Aleksandar Makedonski" located in Skopje were part of this study with kids 6–8 years old, attending first and second grade. Firstly, the children were taught in the traditional way, with explanation on a board. Afterwards, when they were asked about the topic, there was little response in answering and remembering. Then, the lesson was taught with help of AR application, such as Skyview, Augment and Anatomy 4D, with help from an author of this paper with phones and tablets. 30 kids from each school participated in our project. Skyview was the first application, showing the planets and some stars. Children interacted with the application by clicking on stars or planets they liked and information about them was popping out, helping them to learn. Following that Augment was presented. Children were able to see how Earth looks like, animals, human's skeleton or some geometric figures like a sphere. Anatomy 4D makes the children understand in an interactive way the human body and heart through AR, providing fun facts to keep their attention. After two days, children were used to the applications and they were able to use them on their own. In the end we were asking them for things taught the previous days to check what they had remembered. We were also giving them instructions to them to show us a particular thing, to rotate it, to zoom it, etc. In Kavadarci, a volunteer teacher assisted in conducting this study in the school "Dimkata Angelov Gaberot", in two classes with children 10–12 years old, who worked on the applications Augment and Skyview. They were observing galaxy and learning about planets with Skyview. After that they were supposed to tell some information from the application that they understood, for example, what they learned about the planet Pluto, while the teacher was assisting them. The rest of the days they were using Augment. During the experiment, kids reacted with curiosity and interest using the AR applications. Afterwards they could repeat all the information they received by observing. The attention was constantly retained and it was fun to study the material. It saved time of the lecture because there was no need to explain abstract terms from the textbook. Retained knowledge was good, as the child observed in detail and visually remembered each part. In the end, the pupils from both cities answered affirmatively to the question if they liked this way of learning. When children were asked which is their favorite part, most of them answered they liked Augment 4D the most thanks to a heart beating representation. From here we conclude that children like the new technology and they want to use it, especially for natural sciences. Teaching them in this way saves time and helps them to remember better, instead of learning by heart. Children learning with AR easily master the material and acquire knowledge that will be used in forming them as individuals and professionals. They were amazed when they saw how the heart pumps, excited when they were able to see galaxy around them and nearly all of them stayed breathless when they were observing human's body only with muscles and turning it.

## 2.3  Using Augmented Reality for Education of Children with Dyslexia

In the city of Strumica, all applications were used by a specialist working with two children 9 and 12 years old who suffer from dyslexia in the school "Vidoe Podgorec". The aim of this study is to prove that it helps those children to learn without problems, without reading and to start using it for that purpose in Strumica. The specialist and the children were using Augment 4D, she was translating the explanation and they were writing down the things they have learnt. Afterwards, they used Skyview for learning the planets, same as the children from Skopje and Kavadarci. Learning with AR helps a lot because the children see what is supposed to learn, self-confidence increases and they are able to attend normal classes with children their age. This kind of learning will help children with dyslexia to have the same knowledge and everyday life as everyone else.

## 3 Conclusion

The aim of this paper is to support the hypothesis that augmented reality is a good educational tool, compared to the traditional way of learning. As an introduction to the study, descriptive research was done with a survey and according to the responses, a conclusion was made that visual learning is more effective, easier to remember, inspiring innovation, fun and interaction. Moreover, 94% of the interviewees supported using AR application for education. According to the research, by involving children in the augmented reality learning process, they master the material more easily and acquire quality knowledge that will be used in forming them as individuals and professionals. The lessons are modern, interesting and attract children's attention, thus saving teachers' time. The visualization gives a better idea of what is being taught and increases children's imagination. Also, AR helps children with dyslexia so they can fit in normal classes, progress intellectually like everyone else and be accepted by the society. Benjamin Franklin said: "Tell me something and I will forget. Teach me something and I will remember. Involve me and I will learn." Augmented reality is this future that Benjamin Franklin described, changing the way we learn by involving us.

## References

1. Bower, M., Howe, C., McCredie, N., Robinson, A., Grover, D.: Augmented reality in education—cases, places and potentials. Educ. Media Int. **51**(1), 1–15 (2014)
2. Han, J., Jo, M., Hyun, E., So, H.J.: Examining young children's perception toward augmented reality-infused dramatic play. Educ. Tech. Res. Dev **63**(3), 455–474 (2015)
3. Karagozlu, D., Ozdamli, F.: Student opinions on mobile augmented reality application and developed content in science class. Tem J.-Technol. Educ. Manag. Inform. **6**(4), 660–670 (2017)
4. Kipper, G., Rampolla, J.: Augmented Reality: An Emerging Technologies Guide to AR. Elsevier (2012)
5. Koutromanos, G., Sofos, A., Avraamidou, L.: The use of augmented reality games in education: a review of the literature. Educ. Media Int. **52**(4), 253–271 (2015)
6. Lee, K.: Augmented reality in education and training. TechTrends **56**(2), 13–21 (2012)
7. Zhou, F., Duh, H.B.L., Billinghurst, M.: Trends in augmented reality tracking, interaction and display: a review of ten years of ISMAR. In: Proceedings of the 7th IEEE/ACM International Symposium on Mixed and Augmented Reality, pp. 193–202. IEEE Computer Society (2008)

# Impact of Quality Management on Green Innovation: A Case of Pakistani Manufacturing Companies

Tahir Iqbal[✉]

Imam Abdulrahman Bin Faisal University, Dammam, Saudi Arabia
timuniruddin@iau.edu.sa

**Abstract.** Quality management is regarded as a one of the significant practices that are incorporated by the management of the organization in order to improve quality, enhance performance and productivity, and reducing the cost. Furthermore, in order for the effective implementation of quality management, Green innovation is one of the most potent tool to inculcate long term sustainability. The aim of the research is to analyze the role of quality management practices on green innovation. This research paper has contributed to the existing body of literature by providing focus on green innovation. The researcher has used a secondary quantitative research where the data has been collected with the help of secondary sources. 10 manufacturing companies from Pakistan are selected from different sectors for the time period ranging from 2010 to 2015. The data for the selected companies has been obtained from annual reports, sustainability reports, press releases, and other online sources. The results of correlation analysis show that green management innovation has a relationship with quality management while the green technology innovation does not have an impact on quality management. The results of the research have shown that green management innovation and green technology innovation is influenced by quality management practices that are implemented by the manufacturing companies of Pakistan.

**Keywords:** Quality management · Green management innovation · Green technology innovation

## 1 Introduction

The welfare of human being is greatly dependent on the environmental quality. People became more aware of the quality due to Japanese products that were of extremely high quality during the time of late 20th century. However, it should be noted that extreme environmental deterioration also played a major role in regard to raising the awareness among the consumers and businesses regarding the need of environmental quality. Pakistan being one of the rapidly developing nations in the world is experiencing environmental pollution, energy shortages, and other environmental problems which have been affecting the quality of lives of people in an adverse manner. There are a number of challenges which are faced by Pakistan and its people in terms of environmental corrosion (Huma 2018). Pakistani organizations have been facing pressure

© Springer Nature Switzerland AG 2019
C. Benavente-Peces et al. (Eds.): SEAHF 2019, SIST 150, pp. 169–179, 2019.
https://doi.org/10.1007/978-3-030-22964-1_18

from different stakeholders such as regulatory bodies, consumers, media, and non-governmental organizations to use strategies for managing the environment and different quality management practices for their operational activities.

From the analysis of past literature, it has been found that total quality management is one of the tools of quality management that helps the company to achieve sustainable development (Jackson et al. 2016). Irrespective of the significant amount of literature found in this regard there are conflicting conclusions. Certain studies have found that quality improvement practices foster green innovation however Prajogo and Sohal (2001) have concluded through their research that such practices tend to obstruct the green innovation. Presently, the literature has focused on exploring the impact of quality improvement practices on general innovation which also shows mixed results. There are hardly any studies which have focused on finding the impact of quality management practices on green innovation. Thus, the aim of this research paper is to analyze the role of quality management practices on green innovation. This research paper has contributed to the existing body of literature by providing focus on green innovation.

## 2  Literature Review

### 2.1  Quality Management and Green Innovation

Quality management refers to different practices that are incorporated by the management of the organization in order to improve quality, enhance performance and productivity, and reducing the cost (Kaynak 2003). There is a significant amount of literature which is available in regard to the importance of quality management. The synthesis of literature shows that there are certain important elements which contribute towards quality management such as: planning, leadership, stakeholder management, and process management (Garvare and Johansson 2010). Some of the aforementioned elements foster environmental management and green innovation (Costantini and Mazzanti 2012). According to the definition of sustainable development provided by World Commission on Environment and Development (Brundtland 1987), economic development deals with strategies and measures for meeting the needs of present generation in such a manner that the ability of future generations to meet their needs is not compromised. On the other hand, green innovation is defined as the creation of new products and services, process, procedures, or organizational structure that contributed towards environmental betterment (Chen et al. 2006).

From the critical analysis of the definitions of green innovation & sustainable development it has been found that broadly sustainable development deals with designing strategies and practices which foster economic development however, in order to achieve sustainable development green innovation is used as a tool. According to Bon and Mustafa (2013), environment deterioration and corrosion green innovation and total quality management have emerged as significant tools in aiding the organizations to attain sustainable development. Green innovation is distinctive from the general innovation because it deals with ensuring sustainability by incorporating such strategies that contributes towards the environmental well-being.

Generally, the firms are reluctant to invest in green innovation alone and hence they seek investments from government and other regulatory bodies in order to ensure sustainable development in their operational procedures. An important point that is highlighted by the study conducted by Teece (2000) has shown that there are two aspects which should be considered while achieving sustainable advantage by an organization: administrative innovation and the technological innovation. Hence green innovation is divided into two components as well: green technology innovation & green management innovation (Qi et al. 2010). The green technology is concerned with the use of advanced technology in order to protect the environment. It is related with the usage of scientific technology to save resources and raw materials to make the operational process more efficient (Siva et al. 2016).

On the other hand, green management innovation is defined as the adoption of improved and innovation organizational structure and process which intends to improve the management processes. Such management processes tend to contribute towards reduction in environmental deterioration. Energy conservation and use of comprehensive environmental management systems are considered to be the best examples for green innovation (Manders et al. 2016). In an environment constrained society, quality management and green innovation are considered as tools for attaining competitive advantage however from the review of past researches no consensus has been reached in terms of the relationship among the two variables (Prajogo and Sohal 2001). There are certain studies that have concluded that there is a positive relationship between quality management and promotion of sustainable development. The study carried out by Zeng et al. (2017) shows that green innovation is an important quality management tool which tends to positively impact the sustainable development.

There are also certain researches which have particularly followed a pessimistic approach where it has concluded that QM methodologies such as ISO9001 are extremely formal and systematic rules which tends to hinder the green innovation because the bureaucracy is increased (Castillo-Rojas et al. 2012). From the review of literature available regarding quality management and green innovation it can be argued that quality management is implemented as a cost-effective strategy rather than differentiation strategy which tends to hinder the green innovation. In this regard it has been pointed out quality management can hinder the creativity of the organizations because of implementation of heavy formal policies such as ISO 9001. On contrary to this, it has been pointed out by the research carried out by Steiber and Alänge (2013) that radical green innovation can be hindered by continuous improvement of quality management practices. Following hypotheses are developed based on the literature:

H1a    Quality management have a negative impact on corporate green technology innovation in manufacturing companies of Pakistan

H1b    Quality management have a negative impact on corporate green management innovation in manufacturing companies of Pakistan.

## 2.2 Environmental Regulation

Interdependence of quality management (QM) and green innovation (GI) can be best understood with the help of a moderating variable which is environmental regulations. It has been found by the research carried out by Prajogo and Sohal (2001) that there are certain contingent factors, which can weaken the pessimistic impact of QM on GI. In this research paper, the researcher has analyzed the role of government organizations in regard to environmental regulations. In this regard, it is important to use the institutional theory which states the way in which social influence towards conformity impacts the actions undertaken by an organization (Berrone et al. 2013).

## 3   Research Design

The researcher has used a secondary quantitative research where the data has been collected with the help of secondary sources. Ten manufacturing companies from Pakistan are selected from different sectors for the time period ranging from 2010 to 2015. The data for the selected companies has been obtained from a4nnual reports, sustainability reports, press releases, and other online sources. Figure 1 shows the model which is used by the researcher to assess the aim and test the hypothesis of the study.

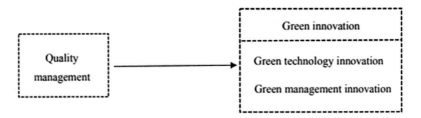

**Fig. 1.** Research model used for the study

The research method shows that quality management is the independent variable and the dependent variable of the research is green innovation. In accordance to the analysis of literature review green innovation (GI) is divided into: green technology innovation (GTI) & green management innovation (GMI). The independent variable of the research has been assessed by the firms that have passed ISO 9001. On the other hand, the dependent variable of the research is bifurcated into following sub-variables: green technology innovation (GTI) & green management innovation (GMI). The former is measured with the help of presence of green patents in the selected organizations and the latter is assessed with the help of firms that have passed ISO 14001.

The researcher has used e-views for the purpose analyzing the data in order to fulfil the aim and objectives of the research. In order to analyze the statistical properties of the data the researcher has used descriptive analysis. Moreover, relationship between predicting variable and dependent variable is judged with the help of correlation analysis. In order to find the impact of quality management on green innovation the researcher has used Ordinary Least Square (OLS) method. Following is the regression equation:

$$GTI = \alpha + \beta1QM + \beta2Size + \beta3Own + \beta4Ind + \beta5FP + \beta6Lev + \varepsilon \qquad (1)$$

$$GMI = \alpha + \beta1QM + \beta2Size + \beta3Own + \beta4Ind + \beta5FP + \beta6Lev + \varepsilon \qquad (2)$$

where,

GTI    Green Technology Innovation
Size    Firm Size (Total Assets)
GMI    Green Management Innovation
Own    Ownership type (State-owned enterprises and Private enterprises)
Ind    Industry type (environmentally sensitive industries and environmentally non-sensitive industries)
QM    Quality Management
FP    Financial Performance (ROA)
Lev    Leverage.

## 4    Empirical Analysis

### 4.1    Descriptive Analysis

Descriptive analysis has been used to highlight the key features of the data. Table 1 provides the summary about the data collected for the completion of the study. Important values are mean and standard deviation. The value of mean shows the average of each variable that has been selected for the research. The mean value of firm size shows that on an average the total assets of the manufacturing firms selected are 30502969 units and based on the value of standard deviation it can be said that for the sample size selected on an average firm size is deviated by 56007979. The mean value of leverage shows that on an average the leverage of the manufacturing firms selected are 0.5983 units and based on the value of standard deviation it can be said that for the sample size selected on an average leverage is deviated from 0.916 units. The mean value of financial performance shows that on an average the ROA of the manufacturing firms selected is 15.16 and based on the value of standard deviation it can be said that for the sample selected on an average ROA is deviated from 11.07 units.

### 4.2    Correlation Analysis

Correlation analysis focuses on highlighting the existing relationship between the variables, while elaborating on the strength of association and the nature of relationship. This was quite appropriate in this study to explore the possible relationship between quality management, and green innovation.

**Table 1.** Descriptive statistics

| | FIRMSIZE | GREEN_M. | GREEN_TE | INDUSTRY | LEVERAGE | OWNERSI- | QUALITY J | ROA |
|---|---|---|---|---|---|---|---|---|
| Mean | 30502969 | 0.S33333 | 0.5 | 0.7 | 0.593357 | 0.4 | 0.8 | 15.16424 |
| Median | 14328697 | 1 | 0.5 | 1 | 0 | 0 | 1 | 12.94415 |
| Maximum | 2.52E + 08 | 1 | 1 | 1 | 3 | 1 | 1 | 43.84254 |
| Minimum | 0 | 0 | 0 | 0 | 0 | 0 | 0 | 0 |
| Std. Dev. | 56007979 | 0.375823 | 0.504219 | 0.462125 | 0.916904 | 0.494032 | 0.403376 | 11.07214 |
| Skewness | 2.933486 | −1.78885 | 0 | −0.87287 | 1.128192 | 0.408248 | −1.5 | 0.866418 |
| Kurtosis | 10.58199 | 4.2 | 1 | 1.761905 | 2.81987 | 1.166667 | 3.25 | 2.8842 |
| Jarque–Bera | 233.5859 | 35.6 | 10 | 11.45125 | 12.80929 | 10.06944 | 22.65625 | 7.540334 |
| Probability | 0 | 0 | 0.006738 | 0.003261 | 0.001654 | 0.006508 | 0.000012 | 0.023048 |
| Sum | 1.83 E + 09 | 50 | 30 | 42 | 35.90141 | 24 | 48 | 909.8543 |
| Sum Sq. Dev. | 1.85E + 17 | 8.333333 | 15 | 12.6 | 49.60203 | 14.4 | 9.6 | 723 2.9 38 |
| Observations | 60 | 60 | 60 | 60 | 60 | 60 | 60 | 60 |

Table 2 is extracted from e-views where the correlation analysis has been carried out in order to determine the relationship among quality management (QM) and green technology innovation (GTI) and green management innovation (GMI). Correlation matrix also shows the interdependency of other control variables with the dependent variables of the research. Based on the data that has been collected for the research it is evident that there is no correlation between quality management and green technology innovation as the value of coefficient is 0. Another dependent variable of the research is green management innovation. The correlation coefficient of this variable against quality management is calculated to be 0.4472. This value can be interpreted as moderate correlation among both the variables.

**Table 2.** Correlation analysis

| | FIRM_SIZE | GREEN_M. | GREEN_TE | INDUSTRY | LEVERAGE | OWNERSH | QUALITY_ | ROA |
|---|---|---|---|---|---|---|---|---|
| FIRM_SIZE | 1 | 0.142465 | 0.341776 | 0.26625 | −0.0296 | 0.264025 | 0.110373 | 0.392109 |
| GREEN_MANAGEMENT_INNOVAT | 0.142465 | 1 | 0.447214 | 0.68313 | 0.087603 | 2.03E-17 | 0.447214 | 0.11377 |
| GREEN_TECHNOLOGY_INNOVAT | 0.341776 | 0.447214 | 1 | 0.654654 | −0.15588 | −0.40825 | 0 | 0.56286 |
| INDUSTRY_TYPE | 0.26625 | 0.68313 | 0.654654 | 1 | −0.13729 | −0.35635 | 0.218218 | 0.237903 |
| LEVERAGE | −0.029604 | 0.087603 | −0.15588 | −0.13729 | 1 | 0.061343 | −0.00826 | −0.26651 |
| OWNERSHIP | 0.264025 | 2.03E-17 | −0.40825 | −0.35635 | 0.061343 | 1 | 0.408248 | 0.026099 |
| QUALITY_MANAGEMENT | 0.110373 | 0.447214 | 0 | 0.218218 | −0.00826 | 0.408248 | 1 | 0.222304 |
| ROA | 0.392109 | 0.11377 | 0.56286 | 0.237903 | −0.26651 | 0.026099 | 0.222304 | 1 |

## 4.3   Regression Analysis

The purpose of regression analysis was to highlight the impact of quality management on green innovation, especially in reference to the manufacturing companies operating in Pakistan.

The first regression analysis was carried out in order to assess the influence of quality management on green management innovation (see Table 3). In order to make the model more statistically efficient the researcher has made use of certain the control variables viz. ROA, ownership, leverage, industry type, and firm size. From the individual p-value of the control variables, it has been found that only industry type tends to impact the green management innovation. This means that environmental sensitive industry tends to explain the green management innovation in the manufacturing companies of Pakistan. The remaining control variables do not have an impact on green management innovation on an individual basis. In order to understand the model in an inclusive manner the values of R-square and adjusted R-square are interpreted. The former value shows that the independent variables and control variables incorporated in the model shows that quality management has 61.96% ability to explain the variations that occur in green management innovation. After eliminating the impact of insignificant factors in the model, quality management can explain 57.66% of the changes that occur in green management innovation. To conclude the model the p-value of F-statistic can be interpreted which is 0.0000 hence the null hypothesis is rejected. The statement of alternate hypothesis is accepted stating that Quality management have a negative impact on corporate green management innovation in manufacturing companies of Pakistan.

The second regression analysis was carried out in order to assess the influence of quality management on green technology innovation (see Table 4). In order to make the model more statistically efficient the researcher has made use of certain the control variables. The control variables of the research are: ROA, ownership, leverage, industry type, and firm size. From the individual p-value of the control variables it has been found that only industry type and financial performance tends to impact the green management innovation. This means that environmental sensitive industry and financial performance of companies tends to explain the green management innovation in the manufacturing companies of Pakistan. The remaining control variables do not have an impact on green management innovation on an individual basis. In order to understand the model in an inclusive manner the values of R-square and adjusted R-square are interpreted. The former value shows that the independent variables and control variables incorporated in the model shows that quality management has 68.8% ability to explain the variations that occur in green technology innovation. After eliminating the impact of insignificant factors in the model, quality management can explain 65.27% of the changes that occur in green technology innovation. To conclude the model the p-value of F-statistic can be interpreted which is 0.0000 hence the null hypothesis is rejected. The statement of alternate hypothesis is accepted stating that Quality management have a negative impact on corporate green technology innovation in manufacturing companies of Pakistan.

**Table 3.** First regression analysis

Dependent Variable: GREEN_MANAGEMENT_INNOVAT
Method: Least Squares
Date: 09/04/18   Time: 19:36
Sample: 1 60
Included observations: 60

| Variable | Coefficient | Std. Error | t-Statistic | Prob. |
|---|---|---|---|---|
| QUALITY_MANAGEMEN T | 0.196840 | 0.100394 | 1.960673 | 0.0552 |
| ROA | -0.000937 | 0.003370 | -0.277872 | 0.7822 |
| OWNERSHIP | 0.163801 | 0.091618 | 1.787878 | 0.0795 |
| LEVERAGE | 0.070422 | 0.036449 | 1.932070 | 0.0587 |
| INDUSTRY_TYPE | 0.632917 | 0.091383 | 6.926010 | 0.0000 |
| FIRM_SIZE | -8.66E-10 | 7.11E-10 | -1.217045 | 0.2290 |
| C | 0.165769 | 0.093596 | 1.771108 | 0.0823 |

| | | | | |
|---|---|---|---|---|
| R-squared | 0.619666 | Mean dependent var | | 0.833333 |
| Adjusted R-squared | 0.576609 | S.D. dependent var | | 0.375823 |
| S.E. of regression | 0.244542 | Akaike info criterion | | 0.130424 |
| Sum squared resid | 3.169449 | Schwarz criterion | | 0.374764 |
| Log likelihood | 3.087294 | Hannan-Quinn criter. | | 0.225998 |
| F-statistic | 14.39188 | Durbin-Watson stat | | 0.606474 |
| Prob(F-statistic) | 0.000000 | | | |

**Table 4.**  Second regression analysis

Dependent Variable: GREEN_TECHNOLOGY_INNOVAT
Method: Least Squares
Date: 09/04/18  Time: 19:37
Sample: 1 60
Included observations: 60

| Variable | Coefficient | Std. Error | t-Statistic | Prob. |
|---|---|---|---|---|
| QUALITY_MANAGEMEN T | -0.141003 | 0.121984 | -1.155922 | 0.2529 |
| OWNERSHIP | -0.250580 | 0.111320 | -2.250984 | 0.0286 |
| LEVERAGE | 0.024493 | 0.044287 | 0.553060 | 0.5825 |
| INDUSTRY_TYPE | 0.500262 | 0.111034 | 4.505471 | 0.0000 |
| ROA | 0.020458 | 0.004095 | 4.995812 | 0.0000 |
| FIRM_SIZE | 1.10E-09 | 8.64E-10 | 1.272280 | 0.2088 |
| C | 0.004419 | 0.113724 | 0.038854 | 0.9692 |

| | | | | |
|---|---|---|---|---|
| R-squared | 0.688053 | Mean dependent var | | 0.500000 |
| Adjusted R-squared | 0.652739 | S.D. dependent var | | 0.504219 |
| S.E. of regression | 0.297131 | Akaike info criterion | | 0.519993 |
| Sum squared resid | 4.679200 | Schwarz criterion | | 0.764333 |
| Log likelihood | -8.599790 | Hannan-Quinn criter. | | 0.615568 |
| F-statistic | 19.48347 | Durbin-Watson stat | | 0.456881 |
| Prob(F-statistic) | 0.000000 | | | |

# 5  Discussion

In the light of the existing studies that are conducted, it was determined that quality management does tend to influence Green Innovation (GI) for the manufacturing companies in Pakistan however in a negative manner as revealed by the findings. However, some of the studies had also revealed that quality management standards of ISO 9001 are helpful in controlling the accidents. Quality management (QM) & Green innovation (GI) are some related concepts that show that QM significantly has the ability to explain the changes in GI. In manufacturing companies of Pakistan, green innovation and green management are taken into consideration because green management innovation (GMI) and green technology innovation (GTI) are used as a tool to measure sustainability (Prajogo and Sohal 2001). Some of the studies that are conducted in this domain have revealed that environmental regulations reduce the impact of negative quality management on green innovation which further indicates that Pakistan's government is playing a vital role to curb environmental issues (Berrone et al. 2013; Kaynak 2003; Prajogo and Sohal 2001). The environmental regulations that are placed investigate the uncertainty of the issues caused in the environment. Quality management refers to the practices that are incorporated that are incorporated by the management of the organization in order to improve quality, enhance performance and productivity, and reducing the cost (Kaynak 2003). In addition to this it was also found that important elements which contribute towards quality management such as planning, leadership, stakeholder management, and process management play a major role in ensuring highest quality of products. Therefore, it has been concluded from the findings of the study that the researcher was successful in achieving the aim of the study along with answering the research questions and determining the role that is played by QM in terms of affecting GI in terms of both management and technology.

## 6 Conclusion

Overall, from the detailed analysis of findings obtained from this research as well as the secondary findings it has been found that quality management significantly impacts the usage of GI. The quantitative results of the study show that green management innovation (GMI) and green technology innovation (GTI) is influenced by quality management (QM) practices that are carried out by the manufacturing companies of Pakistan. There are certain practical implications of this research the first one is that the quality management practices that are implemented by the manufacturing companies of Pakistan tends to restrict their focus on development of existing management processes instead of green innovation. The limitation of the research is that the sample size selected for conducting the research is very small hence the methodological issue of generalizability of the data is present.

## References

Berrone, P., Fosfuri, A., Gelabert, L., Gomez-Mejia, L.R.: Necessity as the mother of 'green'inventions: Institutional pressures and environmental innovations. Strateg. Manag. J. **34**(8), 891–909 (2013)

Bon, A.T., Mustafa, E.M.: Impact of total quality management on innovation in service organizations: Literature review and new conceptual framework. Procedia Eng. **53**, 516–529 (2013)

Brundtland, G.: Our common future: Report of the 1987 World Commission on Environment and Development. United Nations, Oslo **1**, 59 (1987)

Castillo-Rojas, S.M., Casadesús, M., Karapetrovic, S., Coromina, L., Heras, I., Martín, I.: Is implementing multiple management system standards a hindrance to innovation? Total Quality Management & Business Excellence **23**(9–10), 1075–1088 (2012)

Chen, Y.-S., Lai, S.-B., Wen, C.-T.: The influence of green innovation performance on corporate advantage in Taiwan. J. Bus. Ethics **67**(4), 331–339 (2006)

Costantini, V., Mazzanti, M.: On the green and innovative side of trade competitiveness? The impact of environmental policies and innovation on EU exports. Res. Policy **41**(1), 132–153 (2012)

Garvare, R., Johansson, P.: Management for sustainability–a stakeholder theory. Total Qual. Manag. **21**(7), 737–744 (2010)

Huma, Z.: Pakistan's environmental challenges, https://dailytimes.com.pk/247550/pakistans-environmental-challenges/. Last accessed 09 Apr 2018

Jackson, S.A., Gopalakrishna-Remani, V., Mishra, R., Napier, R.: Examining the impact of design for environment and the mediating effect of quality management innovation on firm performance. Int. J. Prod. Econ. **173**, 142–152 (2016)

Kaynak, H.: The relationship between total quality management practices and their effects on firm performance. J. Operat. Manag. **21**(4), 405–435 (2003)

Manders, B., de Vries, H.J., Blind, K.: ISO 9001 and product innovation: a literature review and research framework. Technovation **48**, 41–55 (2016)

Prajogo, D.I., Sohal, A.S.: TQM and innovation: a literature review and research framework. Technovation **21**(9), 539–558 (2001)

Qi, G., Shen, L.Y., Zeng, S., Jorge, O.J.: The drivers for contractors' green innovation: an industry perspective. J. Clean. Prod. **18**(14), 1358–1365 (2010)

Siva, V. Gremyr, I., Bergquist, B., Garvare, R., Zobel, T., Isaksson, R.: The support of quality management to sustainable development: a literature review. J. Clean. Product. 1–10 (2016)

Steiber, A., Alänge, S.: Do TQM principles need to change? Learning from a comparison to Google Inc. Total Quality Management & Business Excellence **24**(1–2), 48–61 (2013)

Teece, D.J.: Strategies for managing knowledge assets: the role of firm structure and industrial context. Long Range Plan. **33**(1), 35–54 (2000)

Zeng, J., Zhang, W., Matsui, Y., Zhao, X.: The impact of organizational context on hard and soft quality management and innovation performance. Int. J. Prod. Econ. **185**, 240–251 (2017)

# Hamming Distance and K-mer Features for Classification of Pre-cursor microRNAs from Different Species

Malik Yousef[(⊠)] [ⓘ]

Community Information Systems, Zefat Academic College, 13206 Safed, Israel
malik.yousef@gmail.com

**Abstract.** MicroRNAs (miRNAs), are short RNA sequences involved in targeting post transcriptional gene regulation. These mature miRNAs are derived from longer sequence precursors (pre-miRNAs) (70nt-100nt in mammalian) and have been shown to integrate multiple genes into biologically networks. Previously, we have shown that pre-miRNAs can be categorized into their species of origin using sequence-based features (such as frequency of k-mer) and machine learning.

In this study, we introduce a new set of features which are extracted from the precursor sequence that based Hamming distance between k-mer and pre-miRNAs sequence. These new set of features reveal an interesting result where in some cases it outperforms the k-mer frequency.

In the Hamming distance, we consider k-mers words with k = 4 and k = 5 while in k-mer frequency we consider k = 1, 2, 3. Hamming distance allows mismatches (flexible match) while k-mer frequency require the appearance of the whole word with length k. The Hamming flexibility allows getting more accurate representation to some clades and results in improving the performance.

This study suggests that there is no one universal feature set that applicable to all microRNA clades, so one needs to examine a different set of features and apply a function that associates the best set of feature to each clade.

**Keywords:** microRNA · Features · Hamming distance · Classification

## 1 Background

MicroRNAs have been described for a variety of species ranging from viruses [1] to plants [2–4]. MicroRNAs by themselves are not functional but when co-expressed with their targets [5] are involved in modulating the protein abundance of their targets. Since this, for many cases, only occurs in response to internal or external stresses it may not be possible to experimentally determine all miRNAs, their targets, and their interactions. Therefore, there is reliance on computational approaches to detect miRNAs and their target genes, and many approaches have been developed [6–9]. Many such approaches are based in machine learning and these, with few exceptions [10–12], perform two class classification.

Saçar Demiric et al. [13] and miRNAfe [14] implemented almost all of the published features categorized into sequence, structural, thermodynamic, probabilistic

© Springer Nature Switzerland AG 2019
C. Benavente-Peces et al. (Eds.): SEAHF 2019, SIST 150, pp. 180–189, 2019.
https://doi.org/10.1007/978-3-030-22964-1_19

based ones or a mixture of these types which can further be normalized by other features like stem length, number of stems, or similar. The tool, izMiR, also evaluated the previously published approaches in terms of their selected feature sets [13].

Similarly, in the microRNA target aspect designation of the feature is a very important step for using computation approaches. Peterson et al. has [15] reviewed different computational tools for microRNA target predictions and in comparison across all miRNA target prediction tools four main aspects of the miRNA:mRNA target interaction emerge as common features on which most target prediction is based: seed match, conservation, free energy, and site accessibility.

Short nucleotide sequences (*k*-mers) have been used early on for the machine learning-based ab initio detection of pre-miRNAs [16]. Additionally, we have recently conducted different studies to answer the question whether the pre-miRNA sequence (ignoring the secondary structure) can be differentiated among species and may, therefore, contain a hidden message that could influence recognition via the protein machinery of the miRNA pathway. We further investigated whether there is a consistent difference among species taking into account their evolutionary relationship. Similarly, we have conducted sa imilar study that consider the 3'UTR target sites [17].

In order to answer these questions, we established random forest machine learning models using two class classification with the positive class being pre-miRNAs from one species/clade and the negative pre-miRNAs from a different species/clade [18] and found that distantly related species can be distinguished on this basis. In another recent study [19], we corroborated on this approach and introduced information-theoretic features but found that k-mers were sufficient for this type of analysis. Here, we have established novel features based on Hamming k-mers distance (k = 4, 5) and compare the performance results with k-mer frequency (k = 1, 2, 3).

This is the first time that we notice that there is the influence of k-mer word length (k greater than 3). In previous studies the k was chosen to be 1, 2, 3 to form 84 features, also we have tested k-mer with k more than 3 to yield no clear improvement in the results.

Interestingly, in this study we see that using Hamming distance and k greater than 3, the results improved for some of the clades. Hamming distance allows flexibility in the appearance of the k-mer in the pre-microRNA sequences. With Hamming distance, we don't require that the whole word k-mer appears instead we allow some mismatches.

## 2    Methods

### 2.1    Parameterization of Pre-miRNAs

The first step in applying machine learning to the current data is to transform the data into vector space, where each component v relates to a specific feature (k-mer frequency or k-mer Hamming Distance) and where n is the number of features. One simple way of representing sequences that consist of 4 nucleotide letters is by employing k-mers. We have couple of studies that have shown that k-mers are sufficient to categorize pre-miRNAs into species [3, 19–21].

## 2.2    *K*-mer Features

Many studies performing pre-cursor miRNA analysis used sequence-based features. Sequence-based features can be words or short sequence of nucleotides over the alphabet {A, U, C, G,} with the length k (so-called k-mers or n-grams). For example, 1-mers are the 'words' A, U, C, and G; 2-mers are the words AA, AC, ..., UU, and 3-mers lead to 64 (43) short nucleotide sequences ranging from AAA to UUU. In general, the number of all k-mers up to and including length k is $\sum_1^k 4^i$.

Higher *k* have also been used [22], but here we chose 1-, 2-, and 3-mers as features. The *k*-mer counts in a given sequence were normalized by the length of the sequence (i.e., len(sequence) - k + 1). Hence, for *k*-mers with $k = \{1, 2, 3\}$, 84 features were calculated per example. The *k*-mer frequency ranges between 0 (if the k-mer is not present in the sequence) and 1 (if the sequence is a repeat of a mononucleotide which is not observed since such a sequence does not fold into secondary structures).

## 2.3    K-mer Hamming Distance Features

The traditional Hamming distance between two sequences is the number of mismatch between them. In this study, we have defined other metric that based on the merit of Hamming distance. We define the distance between k-mer and pre-microRNA sequences in a specific position to be the number of matches in the current window. The length of the windows is the length of the k-mer. The following equation is considered to compute the distance between a k-mer and sequence:

d(k-mer, pre-miRNA seq) = maximum distance between k-mer and all k-mers in sequences.

In order to compute the d(k-mer, pre-miRNA seq) value we use a sliding window of length k-mer. The window starts from the first position of the sequences and computes the number of matches between this window and the k-mer. The window is moving one nucleotide where we compute the number of matches in the new window. In the end, we consider the maximum value as the distance between the k-mer and the pre-microRNA sequence.

For example, d(GATTCTCA, GCTACTCCGCTGACCAA) is 5. We perform sliding windows of the k-mer GATTCTCA over the sequence by moving one nucleotide each time. For each sliding widows we compute the number of matches and at the end, we consider the maximum value (Fig. 1).

In the current study, we have considered all the k-mer with length 4 and 5 to form a feature vector of length 1280. The following are examples of k-mer of length 4 and 5.

**Fig. 1.** Example of matches between the k-mer and the sequence. The number of matches for the first windows is 5.

4-mer: ACAG, GACA, GCGT, ATTC, TGGA, GGAG, TCTT, TGAT, GTTT, CCAA, ATCG, ATAC, GCTG, GTTA,.........
5-mer: TGTTC, GCCAT, ACTCC, CGCGG, AAGGT, TAGAA, GTAAT, TGTCG, TGAAC, GGAGG,.....

# 3    Datasets and Methods

## 3.1    Pre-processing the Data

The data consists of information from 15 clades. The sequences of Homo sapiens were taken out of the data of its clade Hominidae. The process of removing homology sequences (keeping just one representative) consisted of combining all clades and Homo sapiens sequences into one dataset and then applying the USEARCH [23] to clean the data by removing similar sequences. The USEARCH tool clustered the sequences by similarity. From each cluster, one representative was chosen to form a new dataset with non-homologous sequences. The new dataset was then broken into clades without similar sequences between each pair of clades. Cleaning the data ensured that the results were accurate. The list of clades and associated number of precursors and unique pre-cursor are listed in Table 1.

**Table 1.** The list of clades used in the study, the first column presents the name of the clade, the second column the number of pre-cursors available on miRBase, and the third column is a number of precursors after preprocessing the data. The organism Homo sapiens is taken out from its clade Hominidae. Clades in parentheses refer to the name available in the NCBI taxonomy while the other name option is the provided by miRBase.

| Clade/Species name | Number of precursors | Number of unique precursors |
|---|---|---|
| Hominidae | 3629 | 1326 |
| Brassicaceae | 726 | 535 |
| Hexapoda | 3119 | 2050 |
| Monocotyledons (Liliopsida) | 1598 | 1402 |
| Nematoda | 1789 | 1632 |
| Fabaceae | 1313 | 1011 |
| Pisces (Chondricthyes) | 1530 | 682 |
| Virus | 306 | 295 |
| Aves | 948 | 790 |
| Laurasiatheria | 1205 | 675 |
| Rodentia | 1778 | 993 |
| *Homo sapiens* | 1828 | 1223 |
| Cercopithecidae | 631 | 503 |
| Embryophyta | 287 | 278 |
| Malvaceae | 458 | 419 |
| Platyhelminthes | 424 | 381 |

### 3.2    Feature Vector and Feature Selection

In this study, we considered two kinds of features, k-mer frequency-based features, and the Hamming distance features. We also used information gain measurement [24] as implemented in KNIME (version 3.1.2) [25] for feature selection when we combined a different kind of features.

### 3.3    Classification Approach

Following the study of [18], we used the random forest (RF) classifier implemented by the platform KNIME [25]. The classifier was trained and tested with a split into 80% training and 20% testing data. Negative and positive examples were forced to equal amounts while performing a 100-fold Monte Carlo cross-validation (MCCV) [26] for the model establishment.

## 4    Model Performance Evaluation

For each established model, we calculated a number of statistical measures like the Matthews's correlation coefficient (MCC) [27], sensitivity, specificity, and accuracy for evaluation of model performance. The following formulations were used to calculate the statistics (with TP: true positive, FP: false positive, TN: true negative, and FN referring to false negative classifications):

1. Sensitivity (SE, Recall) = TP/(TP + FN)
2. Specificity (SP) = TN/(TN + FP)
3. Accuracy (ACC) = (TP + TN)/(TP + TN + FP + FN)

All reported performance measures refer to the average of 100-fold MCCVs.

## 5    Results and Discussion

We have conducted different studies shown that k-mers may be sufficient to allow the categorization of miRNAs into species [17, 19, 20]. Additionally, we have shown that the usage of motif sequences features able to reach sufficient performance [3, 28]. For this study, we selected pre-miRNAs of a number of species and/or clades (Table 1) to analyze the ability of a new set of features to aid the categorization of pre-miRNAs into their species/clades. The selected data come from a range of clades at various evolutionary distances and their placement in the tree of life is shown in Fig. 2.

For each pair of species/clades, we trained a classifier using 100-fold MCCV. First, we evaluated the known k-mer features and how accurately they can categorize the given data into the tested species and clades (Table 2). The average performance of 100-fold MCCV was recorded for all pairs of clades, with one of the clades used as the positive class and the other as the negative one. For example, Aves vs. Hexapoda leads to an average accuracy of 0.87 using 100-fold MCCV which is a comparably good

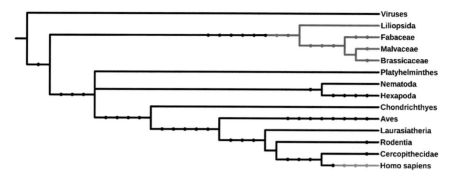

**Fig. 2.** Phylogenetic relationships among organisms and clades used in this study, (Chondrichthyes replaces Pisces which is deprecated but used by miRBase). This subset of the tree of life was generated using the phylot website (http://phylot.biobyte.de/)

categorization and about 5% better than the general average for all categorization attempts.

The results presented in Table 2 for k-mer features confirm our previous observation [18, 19] that with increasing evolutionary distance the average model accuracy also increases.

**Table 2.** Average accuracy for 100-fold MCCV using a random forest classifier and a split of 80% training and 20% testing employing only k-mer features. Yellow shades indicate lower accuracy while red shades show higher accuracy.

| | Viruses | Monocotyledons | Fabaceae | Embryophyta | Brassicaceae | Malvaceae | Platyhelminthes | Nematoda | Hexapoda | Pisces | Aves | Laurasiatheria | Rodentia | Hominidae | Homo sapiens | Cercopithecidae | Average |
|---|---|---|---|---|---|---|---|---|---|---|---|---|---|---|---|---|---|
| Viruses | 0.10 | 0.85 | 0.90 | 0.90 | 0.92 | 0.90 | 0.88 | 0.82 | 0.79 | 0.81 | 0.84 | 0.86 | 0.82 | 0.83 | 0.83 | 0.81 | 0.85 |
| Monocotyledons | 0.84 | 0.10 | 0.74 | 0.69 | 0.77 | 0.74 | 0.80 | 0.85 | 0.81 | 0.84 | 0.85 | 0.84 | 0.86 | 0.88 | 0.88 | 0.83 | 0.81 |
| Fabaceae | 0.89 | 0.74 | 0.10 | 0.81 | 0.69 | 0.69 | 0.75 | 0.83 | 0.80 | 0.85 | 0.86 | 0.89 | 0.87 | 0.87 | 0.87 | 0.82 | 0.82 |
| Embryophyta | 0.90 | 0.70 | 0.80 | 0.10 | 0.83 | 0.82 | 0.86 | 0.86 | 0.84 | 0.88 | 0.88 | 0.85 | 0.88 | 0.90 | 0.91 | 0.87 | 0.85 |
| Brassicaceae | 0.93 | 0.78 | 0.69 | 0.84 | 0.10 | 0.74 | 0.80 | 0.86 | 0.83 | 0.89 | 0.91 | 0.92 | 0.91 | 0.91 | 0.92 | 0.89 | 0.85 |
| Malvaceae | 0.89 | 0.74 | 0.68 | 0.81 | 0.74 | 0.10 | 0.85 | 0.87 | 0.83 | 0.87 | 0.88 | 0.87 | 0.87 | 0.88 | 0.89 | 0.87 | 0.84 |
| Platyhelminthes | 0.88 | 0.80 | 0.75 | 0.87 | 0.80 | 0.84 | 0.11 | 0.74 | 0.69 | 0.81 | 0.87 | 0.90 | 0.88 | 0.86 | 0.86 | 0.80 | 0.82 |
| Nematoda | 0.82 | 0.85 | 0.83 | 0.86 | 0.86 | 0.87 | 0.74 | 0.10 | 0.72 | 0.79 | 0.88 | 0.90 | 0.87 | 0.89 | 0.89 | 0.83 | 0.84 |
| Hexapoda | 0.79 | 0.82 | 0.81 | 0.84 | 0.83 | 0.83 | 0.69 | 0.72 | 0.10 | 0.79 | 0.87 | 0.88 | 0.87 | 0.88 | 0.88 | 0.81 | 0.82 |
| Pisces | 0.82 | 0.85 | 0.84 | 0.88 | 0.89 | 0.86 | 0.81 | 0.79 | 0.80 | 0.10 | 0.72 | 0.83 | 0.76 | 0.81 | 0.81 | 0.70 | 0.81 |
| Aves | 0.83 | 0.85 | 0.87 | 0.88 | 0.91 | 0.87 | 0.87 | 0.88 | 0.87 | 0.71 | 0.10 | 0.80 | 0.69 | 0.71 | 0.72 | 0.66 | 0.81 |
| Laurasiatheria | 0.86 | 0.84 | 0.88 | 0.84 | 0.92 | 0.88 | 0.90 | 0.90 | 0.88 | 0.83 | 0.80 | 0.10 | 0.77 | 0.81 | 0.83 | 0.76 | 0.85 |
| Rodentia | 0.83 | 0.86 | 0.87 | 0.88 | 0.91 | 0.88 | 0.87 | 0.87 | 0.88 | 0.76 | 0.69 | 0.77 | 0.10 | 0.63 | 0.64 | 0.63 | 0.80 |
| Hominidae | 0.82 | 0.87 | 0.87 | 0.90 | 0.91 | 0.88 | 0.86 | 0.89 | 0.88 | 0.81 | 0.71 | 0.81 | 0.64 | 0.10 | 0.14 | 0.63 | 0.77 |
| Homo sapiens | 0.83 | 0.88 | 0.88 | 0.91 | 0.92 | 0.88 | 0.86 | 0.89 | 0.89 | 0.81 | 0.72 | 0.82 | 0.64 | 0.14 | 0.10 | 0.64 | 0.78 |
| Cercopithecidae | 0.81 | 0.82 | 0.82 | 0.86 | 0.88 | 0.86 | 0.80 | 0.83 | 0.81 | 0.70 | 0.67 | 0.77 | 0.63 | 0.63 | 0.65 | 0.10 | 0.77 |

The Hamming distance k-mer features were tested and the results presented in (Table 3). Overall the results are comparable to the k-mer results. Interestingly, is that if we examine the difference for each cell of the two tables, Table 2 and Table 3 (see Table 4 = Table 3 − Table 2). For simplicity, positive values in Table 4 indicates that hamming features performance approach is greater than k-mer approach. One can observe that in some cases the hamming features improving the performance more than 3% (33 times), For example, we reach an improvement of 11% with the pair Cerco-pithecidae- Laurasiatheria, and 12% for the pair Laurasiatheria- Cercopithecidae. Those results are very encouraging to examine in more details the influence of each feature set on different clades. It seems that the hamming distance where is applied to k-mer greater than the regular k-mer is providing more information about some of the clades. It can be explained that the k-mer with length 1–3 is for some clades is not sufficient to carry all the required information for classification. However, with k (k-mer) bigger than 3 we are making the model more complex as the number of features is increasing to 1280 with k = 4 and 5 while we have 84 features with k = 1, 2 and 3.

**Table 3.** Average accuracy for 100-fold MCCV using a random forest classifier and a split of 80% training and 20% testing employing hamming features only. The results are for the top 100 selected features from the total of 1280 features. Yellow shades indicate lower accuracy while red shades show higher accuracy.

| | Viruses | Monocotyledons | Fabaceae | Embryophyta | Brassicaceae | Malvaceae | Platyhelminthes | Nematoda | Hexapoda | Pisces | Aves | Laurasiatheria | Rodentia | Hominidae | Homo sapiens | Cercopithecidae | Average |
|---|---|---|---|---|---|---|---|---|---|---|---|---|---|---|---|---|---|
| Viruses | 0.10 | 0.89 | 0.89 | 0.94 | 0.95 | 0.94 | 0.82 | 0.80 | 0.80 | 0.76 | 0.78 | 0.91 | 0.76 | 0.74 | 0.72 | 0.74 | 0.83 |
| Monocotyledons | 0.89 | 0.10 | 0.64 | 0.63 | 0.67 | 0.64 | 0.81 | 0.84 | 0.78 | 0.85 | 0.84 | 0.82 | 0.81 | 0.85 | 0.86 | 0.85 | 0.79 |
| Fabaceae | 0.90 | 0.64 | 0.10 | 0.71 | 0.62 | 0.62 | 0.74 | 0.85 | 0.77 | 0.84 | 0.86 | 0.87 | 0.83 | 0.85 | 0.86 | 0.84 | 0.79 |
| Embryophyta | 0.93 | 0.63 | 0.72 | 0.10 | 0.74 | 0.69 | 0.87 | 0.89 | 0.84 | 0.90 | 0.90 | 0.82 | 0.86 | 0.91 | 0.91 | 0.91 | 0.83 |
| Brassicaceae | 0.95 | 0.66 | 0.62 | 0.73 | 0.10 | 0.62 | 0.83 | 0.90 | 0.83 | 0.91 | 0.91 | 0.88 | 0.89 | 0.91 | 0.92 | 0.91 | 0.83 |
| Malvaceae | 0.94 | 0.64 | 0.62 | 0.69 | 0.62 | 0.10 | 0.85 | 0.89 | 0.83 | 0.91 | 0.91 | 0.85 | 0.87 | 0.91 | 0.91 | 0.91 | 0.82 |
| Platyhelminthes | 0.83 | 0.80 | 0.74 | 0.87 | 0.83 | 0.84 | 0.10 | 0.75 | 0.66 | 0.73 | 0.80 | 0.88 | 0.74 | 0.75 | 0.76 | 0.74 | 0.78 |
| Nematoda | 0.79 | 0.84 | 0.85 | 0.90 | 0.90 | 0.89 | 0.74 | 0.10 | 0.72 | 0.73 | 0.81 | 0.91 | 0.81 | 0.85 | 0.85 | 0.77 | 0.82 |
| Hexapoda | 0.80 | 0.78 | 0.77 | 0.84 | 0.84 | 0.83 | 0.67 | 0.72 | 0.10 | 0.70 | 0.76 | 0.85 | 0.72 | 0.74 | 0.75 | 0.72 | 0.77 |
| Pisces | 0.77 | 0.85 | 0.84 | 0.91 | 0.91 | 0.91 | 0.73 | 0.73 | 0.70 | 0.10 | 0.66 | 0.89 | 0.71 | 0.72 | 0.71 | 0.65 | 0.78 |
| Aves | 0.79 | 0.84 | 0.86 | 0.90 | 0.92 | 0.91 | 0.80 | 0.81 | 0.75 | 0.66 | 0.10 | 0.87 | 0.68 | 0.68 | 0.69 | 0.63 | 0.79 |
| Laurasiatheria | 0.91 | 0.82 | 0.87 | 0.82 | 0.88 | 0.86 | 0.87 | 0.91 | 0.85 | 0.89 | 0.87 | 0.10 | 0.83 | 0.87 | 0.89 | 0.88 | 0.87 |
| Rodentia | 0.76 | 0.82 | 0.84 | 0.86 | 0.89 | 0.88 | 0.74 | 0.81 | 0.72 | 0.71 | 0.68 | 0.82 | 0.10 | 0.59 | 0.61 | 0.64 | 0.76 |
| Hominidae | 0.73 | 0.85 | 0.86 | 0.91 | 0.91 | 0.91 | 0.75 | 0.85 | 0.75 | 0.72 | 0.68 | 0.87 | 0.58 | 0.10 | 0.15 | 0.61 | 0.74 |
| Homo sapiens | 0.72 | 0.86 | 0.87 | 0.91 | 0.92 | 0.91 | 0.75 | 0.85 | 0.75 | 0.71 | 0.68 | 0.89 | 0.61 | 0.15 | 0.10 | 0.63 | 0.75 |
| Cercopithecidae | 0.74 | 0.85 | 0.84 | 0.91 | 0.91 | 0.91 | 0.74 | 0.77 | 0.72 | 0.65 | 0.63 | 0.87 | 0.63 | 0.61 | 0.63 | 0.10 | 0.76 |

Interestingly some clades are getting improved results (see row results of clades), for example, the clades Laurasiatheria has improved its results dramatically in most cases, see the red shaded on the row of Laurasiatheria clades.

**Table 4.** The difference between a pair of results from Table 3 and Table 2. The values are Table 3-Table 2 corroding cells. Red shades (positive values) shows the higher accuracy of the hamming features than the k-mer features. Yellow shades (negative values) shows the lower accuracy of hamming features than k-mer features.

| | Viruses | Monocotyledons | Fabaceae | Embryophyta | Brassicaceae | Malvaceae | Platyhelminthes | Nematoda | Hexapoda | Pisces | Aves | Laurasiatheria | Rodentia | Hominidae | Homo sapiens | Cercopithecidae |
|---|---|---|---|---|---|---|---|---|---|---|---|---|---|---|---|---|
| Viruses | 0.00 | 0.04 | 0.00 | 0.03 | 0.03 | 0.04 | -0.01 | 0.01 | 0.02 | -0.01 | -0.04 | 0.05 | -0.01 | -0.04 | -0.03 | -0.04 |
| Monocotyledons | 0.05 | 0.00 | -0.01 | -0.02 | 0.00 | -0.01 | 0.01 | -0.01 | -0.03 | 0.01 | -0.01 | -0.02 | -0.04 | -0.03 | -0.02 | 0.02 |
| Fabaceae | 0.00 | 0.00 | 0.00 | 0.01 | -0.08 | -0.06 | 0.00 | 0.02 | -0.03 | -0.01 | 0.00 | 0.02 | -0.04 | -0.01 | -0.01 | 0.02 |
| Embryophyta | 0.03 | -0.02 | 0.02 | 0.00 | 0.01 | -0.02 | 0.01 | 0.04 | -0.01 | 0.02 | 0.02 | -0.02 | -0.02 | 0.00 | 0.01 | 0.05 |
| Brassicaceae | 0.02 | -0.01 | -0.07 | 0.00 | 0.00 | -0.03 | 0.02 | 0.04 | 0.01 | 0.02 | 0.01 | 0.00 | -0.03 | 0.00 | 0.00 | 0.02 |
| Malvaceae | 0.04 | -0.01 | -0.05 | -0.01 | -0.03 | 0.00 | 0.00 | 0.03 | 0.00 | 0.04 | 0.04 | 0.02 | 0.00 | 0.02 | 0.02 | 0.04 |
| Platyhelminthes | -0.01 | 0.00 | 0.00 | 0.00 | 0.02 | 0.00 | -0.01 | 0.01 | 0.01 | -0.01 | -0.01 | 0.01 | -0.03 | -0.04 | -0.04 | -0.02 |
| Nematoda | 0.01 | -0.01 | 0.02 | 0.04 | 0.04 | 0.03 | 0.02 | 0.00 | 0.00 | 0.00 | -0.02 | 0.01 | 0.00 | 0.00 | 0.00 | -0.01 |
| Hexapoda | 0.01 | -0.04 | -0.04 | -0.01 | 0.01 | 0.00 | 0.00 | 0.00 | 0.00 | -0.02 | -0.01 | -0.03 | -0.01 | -0.02 | -0.02 | -0.02 |
| Pisces | -0.02 | 0.00 | 0.01 | 0.02 | 0.02 | 0.05 | -0.01 | -0.01 | -0.03 | 0.00 | -0.05 | 0.06 | -0.02 | -0.02 | -0.02 | -0.02 |
| Aves | -0.03 | -0.01 | -0.01 | 0.02 | 0.01 | 0.04 | -0.02 | -0.02 | -0.01 | -0.04 | 0.00 | 0.07 | -0.01 | -0.04 | -0.04 | 0.00 |
| Laurasiatheria | 0.05 | -0.02 | 0.03 | -0.02 | 0.00 | 0.01 | 0.01 | 0.01 | -0.03 | 0.06 | 0.07 | 0.00 | 0.05 | 0.06 | 0.06 | 0.12 |
| Rodentia | -0.02 | -0.04 | -0.04 | -0.02 | -0.03 | 0.00 | -0.02 | 0.00 | -0.02 | -0.01 | -0.01 | 0.06 | 0.00 | -0.02 | -0.02 | 0.01 |
| Hominidae | -0.03 | -0.02 | -0.02 | 0.01 | 0.00 | 0.03 | -0.04 | 0.00 | -0.02 | -0.01 | -0.03 | 0.06 | -0.02 | 0.00 | 0.01 | -0.02 |
| Homo sapiens | -0.03 | -0.02 | -0.02 | 0.00 | 0.00 | 0.03 | -0.05 | 0.00 | -0.03 | -0.02 | -0.03 | 0.07 | -0.01 | 0.01 | 0.00 | 0.00 |
| Cercopithecidae | -0.04 | 0.03 | 0.02 | 0.06 | 0.03 | 0.05 | -0.01 | -0.01 | -0.01 | -0.01 | -0.01 | 0.11 | 0.00 | -0.01 | -0.02 | 0.00 |

We have combined the set of features, k-mer frequency, and the Hamming distance features and apply the information, gain to sleet the top 100 and top 500 features. Unfortunately, the results on average are not improved.

## 6   Conclusions

MicroRNAs are involved in post-transcriptional gene regulation and have been described for many species covering most of the tree of life. MicroRNA evolution is subject to investigations [29]. Previously, we have shown that microRNAs can be categorized into their species of origin [18, 30].

Table 4 shows the difference performance between the two methods indicates that we need to explore in more depth the influence of the features set on each pair of clades rather than considering just one set of features. It is the first time that we see that there is the influence of the length of the k-mer word (k greater than 3). In previous studies, the k was chosen to be 1, 2, 3 to form 84 features and also we have tested k-mer with k more than 3 and we don't see any improvement in the results. Moreover, in this study, we see that with k greater than 3 we can reach improved results for some of the clades. We explain those improvements due to the hamming distance that considers distance and allows some mismatches, while k-mer frequency requires the appearance of the whole work with length k.

This study suggests that one need to examine a different set of features and apply a function that associates the best set of feature to each clade. Additionally, we might need to develop a model that able to combine a different set of features and provide some weights that associated with the clades. We have tried to pull together all the k-mer and the hamming features and perform feature selection, let say the top 100 top features. The results still not improved. That's is, one need to develop a more sophisticated approach for combing those set of features as it seems that there is no one universal feature set that applicable to all microRNA clades.

**Acknowledgments.** The work was supported by Zefat Academic College to MY.

# References

1. Grey, F.: Role of microRNAs in herpesvirus latency and persistence. J. Gen. Virol. **96**, 739–751 (2015)
2. Zhang, B., Pan, X., Cobb, G.P., Anderson, T.A.: Plant microRNA: a small regulatory molecule with big impact. Dev. Biol. [Internet] **289**, 3–16 (2006). http://www.sciencedirect.com/science/article/pii/S0012160605007645
3. Yousef, M., Allmer, J., Khalifa, W.: Sequence motif-based one-class classifiers can achieve comparable accuracy to two-class learners for plant microRNA detection. J. Biomed. Sci. Eng. [Internet] **08**, 684–94 (2015). http://www.scirp.org/journal/PaperDownload.aspx?DOI= 10.4236/jbise.2015.810065
4. Yousef, M., Saçar Demirci, M.D., Khalifa, W., Allmer, J.: Feature selection has a large impact on one-class classification accuracy for MicroRNAs in plants. Adv. Bioinform. [Internet] **2016**, 1–6 (2016). https://www.researchgate.net/publication/301244460_Feature_ Selection_Has_a_Large_Impact_on_One-Class_Classification_Accuracy_for_MicroRNAs_ in_Plants
5. Saçar, M.D., Allmer, J.: Current limitations for computational analysis of miRNAs in cancer. Pak. J. Clin. Biomed. Res. **1**, 3–5 (2013)
6. Yousef, M., Jung, S., Kossenkov, A.V., Showe, L.C., Showe, M.K.: Naive Bayes for microRNA target predictions machine learning for microRNA targets [Internet], pp. 2987–2992 (2007). http://bioinformatics.oxfordjournals.org/cgi/content/abstract/23/22/2987
7. Yousef, M., Nebozhyn, M., Shatkay, H., Kanterakis, S., Showe, L.C., Showe, M.K.: Combining multi-species genomic data for microRNA identification using a Naive Bayes classifier. Bioinformatics [Internet] **22**, 1325–1334 (2006). http://bioinformatics. oxfordjournals.org/cgi/content/abstract/22/11/1325
8. Krek, A., Grün, D., Poy, M.N., Wolf, R., Rosenberg, L., Epstein, E.J., et al.: Combinatorial microRNA target predictions. Nat. Genet. **37**, 495–500 (2005)
9. Lim, L.P., Lau, N.C., Weinstein, E.G., Abdelhakim, A., Yekta, S., Rhoades, M.W., et al.: The microRNAs of Caenorhabditis elegans. Genes Dev. **17**, 991–1008 (2003)
10. Dang, H.T., Tho, H.P., Satou, K., Tu, B.H.: Prediction of microRNA hairpins using one-class support vector machines. In: 2nd International Conference on Bioinformatics and Biomedical Engineering, iCBBE 2008, pp. 33–36 (2008)
11. Khalifa, W., Yousef, M., Sacar Demirci, M.D., Allmer, J.: The impact of feature selection on one and two-class classification performance for plant microRNAs. PeerJ **4**, e2135 (2016) (United States)

12. Yousef, M., Jung, S., Showe, L.C., Showe, M.K.: Learning from positive examples when the negative class is undetermined–microRNA gene identification. Algorithms Mol. Biol. **3**, 2 (2008)
13. Saçar Demirci, M.D., Baumbach, J., Allmer, J.: On the performance of pre-microRNA detection algorithms. Nat. Commun. **8**, 330 (2017)
14. Yones, C.A., Stegmayer, G., Kamenetzky, L., Milone, D.H.: miRNAfe: a comprehensive tool for feature extraction in microRNA prediction. Biosystems **138**, 1–5 (2015) (Elsevier Ireland Ltd.)
15. Peterson, S.M., Thompson, J.A., Ufkin, M.L., Sathyanarayana, P., Liaw, L., Congdon, C.B.: Common features of microRNA target prediction tools. Front. Genet. (2014)
16. Lai, E.C., Tomancak, P., Williams, R.W., Rubin, G.M.: Computational identification of Drosophila microRNA genes. Genome Biol. **4**, R42 (2003)
17. Yousef, M., Levy, D., Allmer, J.: Species categorization via MicroRNAs—based on 3'UTR target sites using sequence features. In: Proceedings of the 11th International Joint Conference on Biomedical Engineering Systems and Technology, Bioinformatics, vol. 4, pp. 112–118. SciTePress (2018)
18. Yousef, M., Khalifa, W., İlhan Erkin, A., Allmer J.: MicroRNA categorization using sequence motifs and k-mers. BMC Bioinform. [Internet] **18**, 170 (2017). http://dx.doi.org/10.1186/s12859-017-1584-1
19. Yousef, M., Nigatu, D., Levy, D., Allmer, J., Henkel, W.: Categorization of species based on their MicroRNAs employing sequence motifs, information-theoretic sequence feature extraction, and k-mers. EURASIP J. Adv. Signal Process (2017)
20. Nigatu, D., Sobetzko, P., Yousef, M., Henkel, W.: Sequence-based information-theoretic features for gene essentiality prediction. BMC Bioinform. [Internet] **18**, 473 (2017). https://doi.org/10.1186/s12859-017-1884-5
21. Yousef, M., Khalifa, W., Acar, E., Allmer, J.: MicroRNA categorization using sequence motifs and k-mers. BMC Bioinform. **18** (2017)
22. Cakir, M.V., Allmer, J.: Systematic computational analysis of potential RNAi regulation in Toxoplasma gondii. In: 2010 5th International Symposium on Health Informatics and Bioinformatics (HIBIT), pp. 31–38. IEEE, Ankara, Turkey (2010)
23. Edgar, R.C.: Search and clustering orders of magnitude faster than BLAST. Bioinformatics **26**, 2460–2461 (2010)
24. Shaltout, N.A.N., El-Hefnawi, M., Rafea, A., Moustafa, A.: Information gain as a feature selection method for the efficient classification of Influenza-A based on viral hosts. In: Proceedings of the World Congress on Engineering, pp. 625–631. Newswood Limited (2014)
25. Berthold, M.R., Cebron, N., Dill, F., Gabriel, T.R., Kötter, T., Meinl, T., et al.: KNIME: The Konstanz Information Miner. SIGKDD Explor. 319–326 (2008)
26. Xu, Q.-S., Liang, Y.-Z.: Monte Carlo cross validation. Chemom. Intell. Lab. Syst. **56**, 1–11 (2001)
27. Matthews, B.W.: Comparison of the predicted and observed secondary structure of T4 phage lysozyme. BBA—Protein Struct. **405**, 442–451 (1975)
28. Yousef, M., Allmer, J., Khalifaa, W.: Plant MicroRNA Prediction employing Sequence Motifs Achieves High Accuracy (2015)
29. Tanzer, A., Stadler, P.F.: Evolution of microRNAs. Methods Mol. Biol. **342**, 335–350 (2006)
30. Yousef, M., Nigatu, D., Levy, D., Allmer, J., Henkel, W.: Categorization of species based on their microRNAs employing sequence motifs, information-theoretic sequence feature extraction, and k-mers. EURASIP J. Adv. Signal Process. **2017** (2017)

# Hybrid Feature Extraction Techniques Using TEO-PWP for Enhancement of Automatic Speech Recognition in Real Noisy Environment

Wafa Helali[✉], Zied Hajaiej, and Adnen Cherif

Research Unite of Processing and Analysis of Electrical and Energetic Systems, Faculty of Sciences of Tunis, University Tunis El-Manar, 2092 Tunis, Tunisia
wafa.helali@yahoo.com

**Abstract.** In recent years, most research areas have focused their attention on the exactitude of Speech Recognition (SR). Despite being reasonably performant in quiet conditions, these systems are indeed far too ineffective in distorted conditions or malformed channels. Given these observations, finding functional feature extraction methods capable of improving the capacities of those systems in non-optimal conditions is more than an indispensable requirement. The present paper presents an investigation that was carried out on those Speech Recognition systems in noisy conditions, with many combinations of new three hybrid feature extraction algorithms such as Teager-Energy Operator-Perceptual Wavelet Packet (TEO-PWP), Mel Cepstrum Coefficient (MFCC) and Perceptual Linear Production (PLP). A (HMM) was also used to classify the extracted features. Our model was tested on TIMIT database that contains both clean and noisy speech files recorded at different level of Speech-to-Noise Ratio (SNR). The analytic bases for speech processing and classification procedures were exhibited and the recognition results were given depending on speech recognition rates.

**Keywords:** Teager-Energy Operator · TEO-PWP · Enhancement Speech · MFCC · PLP · RASTA-PLP · HMM

## 1 Introduction

Speech enhancement becomes a crucial stage due to the increased development in many signal processing applications and recent technological devices. In general, there are many improvements of speech algorithms that are relied on linear processing techniques such as spectral subtraction, Wiener filter, linear prediction, and signal subspace approach [1]. Very promising results have been achieved in reduction of noise while preserving speech quality when compressive sensing based methods has been tested in frequency domain, as recent methods in [2–4]. In speech recognition systems, MFCC [5] and PLP [6] are mostly the used commonly acoustic features. In matched environments, both MFCC and PLP front-ends are so promised to be used as speech data since they are originated from reasonably clean environments. So, the objective of this study is to test the effectiveness of the proposed model in real noises

© Springer Nature Switzerland AG 2019
C. Benavente-Peces et al. (Eds.): SEAHF 2019, SIST 150, pp. 190–195, 2019.
https://doi.org/10.1007/978-3-030-22964-1_20

by using different combination of feature extraction algorithms such as PLP, RASTA-PLP, GF, MFCC, AMS and TEO-PWP through using HMM as a classifier.

This paper is organized as follows: In Sect. 2, the speech pre-processing steps are defined. A description of different feature extraction techniques is given in Sect. 3. The proposed model and the statistical modeling classifier are presented in Sect. 4. Analyzes and discussions of different obtained results are exhibited in Sect. 5. In Sect. 6, conclusion and perspectives are included.

## 2  Speech Pre-processing

To make the speech data that we are going to set and then take the different acoustic characteristics, the following common steps are firstly performed: sampling, pre-emphasis, frame blocking and windowing [7].

### 2.1  Pre-emphasis

Many values of the realistic disturbing sounds at different SNRs values between −15 and 15 dB are added in order to disturb and corrupt digitally the input speech signal.

### 2.2  The Expected Rating of Signal-to-Noise

To make the speech spectrum straight and replace the unwanted elevated frequency section of speech signal, first order High-pass Filter (FIR) was applied [8]. Indeed, the transfer function of this filter is expressed as follows:

$$y[n] = x[n] - Ax[n-1] \qquad (1)$$

where $x[n]$ and $x[n-1]$ are the input and previous speech signal, respectively, and $A$ is the pre-emphasized factor which is set to 0.98.

### 2.3  Frame Blocking and Windowing

The window size and window shifting size are set respectively to 25 ms and 10 ms in order to block the pre-emphasized signal into samples and ensure a good and easy conversion of expected specifications from frame to frame.

## 3  Feature Extraction Techniques

### 3.1  Mel Frequency Cepstral Coefficients (MFCC)

To extract spectral features, we have used MFCC as it is the most dominant method for this task. The inverse Fourier transform of logarithm of that spectrum is computed [9].

## 3.2   Perceptual Linear Prediction (PLP)

The steps of the extraction of PLP feature are as follows: A spectral analysis is carried out as a first step and then a frequency band analysis is done in a second step. In third, fourth, fifth steps, an equal-loudness pre-emphasis, intensity-loudness power law, and autoregressive modeling are successively applied [10].

## 3.3   RASTA-PLP

The application of a band pass filter to filter the time path of each one of the converted spectral motif using Eq. (2), the simulation of the power law of listening, and at the end the computation of an all-pole representation of the spectrum, like in PLP process [11].

$$H(z) = 0.1 \times \frac{2 + z^{-1} - z^{-3} - 2z^{-4}}{z^{-4}(1 - 0.98z^{-1})} \tag{2}$$

## 3.4   Gammatone Filters (GFs)

The form of impulsive response of Gammatone filter in time domain is formally expressed as follows:

$$g(t) = at^{n-1}e^{-2\pi bt}\cos(2\pi f_c t + \phi) \tag{3}$$

where $f_c$ represents the central frequency of the filter and $\phi$ is the phase which is usually set to be 0. For $a$, it is a constant which used to control the gain. Concerning $n$, it represents the order of filter which is usually set to be equal or less than 4 [12]. For $b$, it is a decay factor which is related to $f_c$ and it is given by the following expression:

$$b = 1.019 * 24.7 * \frac{(4.37 * f_c)}{(1000 + 1)} \tag{4}$$

## 3.5   Amplitude Modulation Spectrogram (AMS)

The extraction of AMS features is then performed using 30-ms frames by applying Hamming window and 256-point FFT, as in [13].

# 4   The Proposed Method

## 4.1   PWP Transform and TE Operation

To make difference between speech and noise, PWP transform does not offer enough frequency resolution [14]. Let's consider $W_{k,m}$ the $m^{th}$ PWP coefficient in the $k^{th}$ subband, the expression of TE operated coefficient $t_{k,m}$ corresponding to $W_{k,m}$ can be given as follows:

$$t_{k,m} = T\left(W_{k,m}\right) \tag{5}$$

where the definition of discrete TE operator $T\left(W_{k,m}\right)$ is given as [15]:

$$T\left(W_{k,m}\right) = W_{k,m}^2 - W_{k,m+1}W_{k,m-1} \tag{6}$$

### 4.2 Erlang-2 PDF for Modeling of TE Operated PWP Coefficients

Thresholding in wavelet, wavelet packet, or PWP domain [14–16], it is not reasonable to consider a unique threshold for all sub-bands in [17]. In this paper, a common PDF was used to model the TE operated PWP coefficients of noisy speech and noise. Indeed, the symmetric K-L divergence is expressed as follows:

$$SKL(p,q) = \frac{KL(p,q) + KL(q,p)}{2} \tag{7}$$

where p and q represent 2-PDFs.

### 4.3 Algorithm of Our Proposed Method

The algorithm of our proposed method is performed by following steps (Fig. 1).

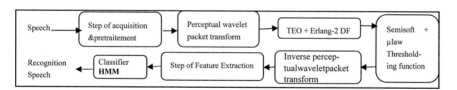

**Fig. 1.** Bloc diagram of our proposed method

## 5 Results and Analysis

### 5.1 Results for Speech Signals

In general, we can say that the mixture of PWP, MFCC, and RASTA-PLP has shown a good capacity in achieve good recognition rates in clean speech condition (100%) and noisy speech condition (98.8 and 92.8 at −15 and 15 dB of SNR, respectively) (Figs. 2 and 3).

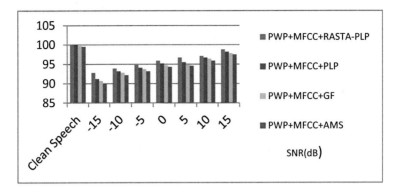

**Fig. 2.** Recognition rate in clean and street noisy speech conditions using different combinations of feature extraction techniques.

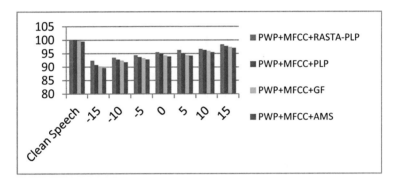

**Fig. 3.** Recognition rate in clean and car noisy speech conditions using different combinations of feature extraction techniques.

## 6   Conclusion and Perspectives

In this research, we have tested the performance of HMM-based speech recognition system using four combinations of feature extraction techniques such as, PWP, MFCC and RASTA-PLP; PWP, MFCC, and PLP; PWP, MFCC and GF; and PWP, MFCC and AMS. To be sure about the performance of the proposed method, we suggest testing it with other combinations of feature extraction techniques (combinations of more than three feature extraction techniques) other than PWP, MFCC, and RASTA-PLP, and to use neural networks as classifiers such as, Conventional Neural Network (CNN) and Recurrent Neural Network (RNN).

# References

1. Lei, S.F., Tung, Y.K.: Speech enhancement for nonstationary noises by wavelet packet transform and adaptive noise estimation. In: Intelligent Signal Processing and Communication Systems (ISPACS 2005), Proceedings of 2005 International Symposium on IEEE, pp. 41–44 (2005)
2. Firouzeh, F.F., Ghorshi, S., Salsabili, S.: Compressed sensing based speech enhancement. In: 2014 8th International Conference on Signal Processing and Communication Systems (ICSPCS). IEEE, pp. 1–6 (2014)
3. Donoho, D.: De-noising by soft-thresholding. IEEE Trans. Inf. Theory **41**, 613–627 (1995)
4. Wu, D., Zhu, W.-P., Swamy, M.: The theory of compressive sensing matching pursuit considering time-domain noise with application to speech enhancement. IEEE ACM Trans. Audio Speech Lang. Process. **22**(3), 682–696 (2014)
5. Davis, S., Mermelstein, P.: Comparison of parametric representations for monosyllabic word recognition in continuously spoken sentences. IEEE Trans Acoust. Speech Signal Process. **28**(4), 357–366 (1980)
6. Hermansky, H.: Perceptual linear prediction analysis of speech. J. Acoust. Soc. Am. **87**(4), 1738–1752 (1990)
7. Zhu, Q., Alwan, A.: On the Use of variable frame rate analysis in speech recognition. In: 2000 IEEE International Conference on Acoustics, Speech, and Signal Processing, vol. 3, pp. 1783–1786 (2000)
8. Rabiner, L.R., Juang, B.-H.: Fundamentals of Speech Recognition, vol. 14. PTR Prentice Hall, EnglewoodCliffs (1993)
9. Chetouani, M., Gas, B., Zarader, J.: Discriminative training for neural predictive coding applied to speech features extraction. In: Proceedings of the 2002 International Joint Conference on Neural Networks, vol. 1, pp. 852–857 (2002)
10. Hermansky, H.: Perceptual Linear Predictive (PLP) analysis of speech. J. Acoust. Soc. Am. **87**, 1738–1752 (1990)
11. Hermansky, H., Morgan, N., Bayya, A., Kohn, P.: The Challenge of Inverse-E: The RASTA-PLP Method. In: 1991 Conference Record of the 25th Asilomar Conference on Signals, Systems and Computers, Pacific Grove, 4–6 Nov 1991, pp. 800–804 (1991)
12. Patterson, R.D., Moore, B.C.J.: Frequency Selective in Hearing, Chapter Auditory Filters and Excitation Patterns as Representations of Frequency Resolution, pp. 123–177. Academic Press Ltd., London (1986)
13. Kim, G., Lu, Y., Hu, Y., Loizou, P.C.: An algorithm that improves speech intelligibility in noise for normal-hearing listeners. J. Acoust. Soc. Am. **126**, 1486–1494 (2009)
14. Islam, M.T., Shahnaz, C., Zhu, W.-P., Ahmad, M.O.: Speech enhancement based on student modeling of teager energy operated perceptual wavelet packet coefficients and a custom thresholding function. IEEE ACM Trans. Audio Speech Lang. Process. **23**(11), 1800–1811 (2015)
15. Sanam, T., Shahnaz, C.: Enhancement of noisy speech based on a custom thresholding function with a statistically determined threshold. Int. J. Speech Technol. **15**, 463–475 (2012)
16. Sanam, T.F., Shahnaz, C.: A combination of semisoft and -law thresholding functions for enhancing noisy speech in wavelet packet domain. In: 2012 7th International Conference on Electrical & Computer Engineering (ICECE), pp. 884–887. IEEE (2012)
17. Islam, M.T., Shahnaz, C., Zhu, W.-P., Ahmad, M.O.: Modeling of teager energy operated perceptual wavelet packet coefficients with an Erlang-2 PDF for real time enhancement of noisy speech. Prepr. Submitt. J. LATEX Templates (2018)

# Investigation on the Behavior of Detonation Gun Sprayed Stellite-6 Coating on T-22 Boiler Steel in Actual Boiler Environment

Yogesh Kumar Sharma[1] , Anuranjan Sharda[2(✉)] ,
Rahul Joshi[2] , and Ketan Kakkar[1]

[1] GNA University, Phagwara, Punjab, India
[2] Faculty of Engineering Design and Automation, GNA University, Phagwara,
Punjab, India
anuranjansharda@gmail.com

**Abstract.** Detonation gun thermal spraying widely is used as erosion, wear and corrosion protective coatings to the various industrial applications. In present study, ASTM SA213 T-22 boiler tube steel sample was coated with Stellite-6 using Detonation Gun (D-Gun) thermal spraying method. Coated and uncoated samples were kept in the super-heater zone of a coal fired boiler at 900 °C up-to 1000 h (Hrs.) for 10 cycles each of 100 h. in boiler followed by 1 h. cooling at ambient temperature. An approach was made to improve the properties of the substrate material to withstand high temperature corrosion and erosion. The thermogravimetric technique was used to establish kinetics of corrosion. X-ray diffraction (XRD), scanning electron microscopy/energy-dispersive analysis (SEM/EDAX) and X-Ray Mapping techniques were used to analyze the corrosion products. The results reveal that coating on the boiler steel provided the substrate with necessary protection against erosion -corrosion at high temperature as compared to the uncoated one.

**Keywords:** Erosion-Corrosion · T-22 Boiler Steel · Stellite-6 Coating · Detonation gun

## 1 Introduction

Material failure due to degradation at high temperature erosion and corrosion is serious problem of components in the hot sections of gas turbines, boilers etc. [1]. Hot corrosion and erosion are recognized as serious problems in coal-based power generation plants in India. The coal used in Indian power stations has large amounts of ash (about 50%) which contain abrasive mineral species such as hard quartz (up to 15%) which increase the erosion propensity of coal [2]. The atmosphere in power plant boilers has a sufficient free oxygen content to account for a combined erosion–corrosion process, consisting of an oxidizing gas at elevated temperature carrying erosive fly ashes which impact against metallic surfaces. These erodent particles can deposit or embed

© Springer Nature Switzerland AG 2019
C. Benavente-Peces et al. (Eds.): SEAHF 2019, SIST 150, pp. 196–210, 2019.
https://doi.org/10.1007/978-3-030-22964-1_21

themselves on the test surface in quite significant quantities, and mixtures of erodent particles and oxide scales can form on the eroding surface, so that quantification of the erosion rate is a difficult task, and the oxidation kinetics and erodent particle embedment have to be taken into account [3]. The annual direct cost of metallic corrosion in the U.S. is $276 billion-which is 3.1% of the Nation's Gross Domestic Product (GDP) [4]. Thermal spraying is an effective and low-cost method to apply thick coatings to change surface properties of the component. Coatings are used in a wide range of applications including automotive systems, boiler components, and power generation equipment, chemical process equipment, aircraft engines, pulp and paper processing equipment, bridges, rollers and concrete reinforcements, orthopedics and dental, land-based and marine turbines, ships [5]. Coating is deposition of layer of material in organized way with aim to obtain required anti-corrosive properties [6]. The composition and structure of the coating are decided by role that they have to play in various application [7]. Although the material withstands high temperature without coating, the coating enhances the lifetime of the material [8]. One among the various coating techniques is detonation Gun (D-Gun) spray technique. The essence of D-Gun spraying is to use the energy of the gas mixture detonation to heat and relay high kinetic energy of the particles of powdered coating materials [9]. The D-Gun deposited coatings are formed by collision with the substrate material and deformation of grains of the powder which together with detonation products, form metallic spray [10]. Detonation gun (D-gun) spray coating process is a thermal spray process, which gives an extremely good adhesive strength, low porosity and coating surfaces with compressive residual stresses. The porosity values of D-gun sprayed coatings are very much lower (<0.8%) than that of HVOF sprayed $Cr_3C_2$–NiCr coating as reported in the literature [11]. Due to presence of elements like Cobalt and Chromium in Stellite-6 offer excellent resistance towards abrasion, corrosion, galling and cavitation without altering the base metal properties. [12, 13] In the present study, a boiler steel ASTM SA213 T-22 has been investigated for high temperature erosion-corrosion behavior in the uncoated as well as coated condition. Detonation Gun thermal spraying process was used for coating of Stellite-6 on boiler steel. The coated as well as uncoated T-22 specimens were exposed to super-heater zone of coal fired boiler environment at 900 ° C for 10 cycles. Each of 100 h heating followed by 1-h cooling at ambient conditions. SEM/EDAX, XRD and X-ray mapping analysis done to reveal the microstructure, compositional features of boiler steel and the corrosion products. Thermogravimetric techniques are used to investigate the erosion-corrosion behavior at 900 °C for 1000 h.

## 2  Experimental Procedure

### 2.1  Substrate Material

2.25Cr-1Mo steel ASTM-SA213 T-22 boiler steel with nominal chemical composition 0.15C- 0.3 to 0.6 Mn- 0.5Si- 0.03max S- 0.03 max P- 1.9 to 2.6 Cr- 0.87 to 1.13 Mo- balance Fe. has been used as a substrate material in the experiment. The material for the study was made available from Guru Nanak Dev Thermal Plant, Bathinda (Punjab), India. The Specimens (20 mm × 15 mm × 5 mm) were cut from the boiler tubes.

They were polished by using 180, 220, 400, 600, and 1000 grade SiC emery papers and grit blast with alumina powder ($Al_2O_3$) Grit 30 before spraying with coating by Detonation Gun (D-Gun) Process. After that samples were washed with acetone.

## 2.2    Development of Coating

Commercially available Stellite-6 (ST-6) in powder form, made by H.C. Starck of Germany was deposited on all six surfaces (sides) of boiler steel samples by the Detonation Gun (D-Gun) spraying apparatus, AWAAZ spray Detonation system (ARCI Developed) at M/S SVX POWDER M SURFACE ENGINEERING (PVT.) LIMITED, Greater Noida, (U.P.) India. Oxy-acetylene gas was used as fuel gas. Nitrogen gas was used as carrier gas. The coating powder characteristics and standard spray parameters used for deposition of coating as supplied by manufacturer are given in the Tables 1 and 2 respectively.

**Table 1.**  Chemical composition and average particle size of coating powder

| Powder name | Chemical composition (wt.%age) | Particle size |
|---|---|---|
| Stellite-6 | Co Base, Cr 27–32%, W 4–6%, C 0.9–1.4% balance Ni, Fe, Si, Mn, Mo | 53/15 μm |

**Table 2.**  Detonation Gun spray parameters for the coating

| Variant | $O_2$ flow rate | Fuel ($C_2H_2$) flow rate | Carrier gas ($N_2$) flow rate | Spray distance | Flame temp. | Detonation frequency |
|---|---|---|---|---|---|---|
| ST-6 coated | 3120 SLPH | 2400 SLPH | 1040 SLPH | 165 mm | 3900 °C | 3 Shots/s |

The SEM/EDAX and XRD analysis of the coating powder was conducted to confirm the chemical composition and average particle size. It is observed from the Fig. 1 that the powder particles have round in shape. Chromium (Cr) and Cobalt (Co) are the main elements presents in the Stellite-6. EDAX as well as XRD analysis confirm the present of the above elements. The highest peak corresponds to the Cr and Co. Along with that SEM/EDAX and XRD of the fly ash was also conducted to find out various elements presents in the ash. It is observed from the Fig. 2 that Silicon (Si) and Aluminum (Al) are main elements presents in the ash. The XRD analysis confirms presence of $SiO_2$, $Al_2O_3$, $Fe_2O_3$, CaO and MgO.

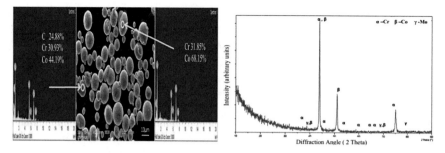

**Fig. 1.** SEM/ EDAX analysis along with X-ray Diffraction pattern of "Stellite-6" powder at different points, (1200X).

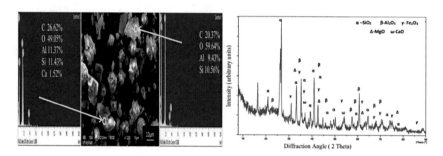

**Fig. 2.** SEM/EDAX analysis along with X-ray Diffraction pattern for "Fly-ash" at different points, (1200X).

## 2.3   Characterization of As-Sprayed Coating

Field-emission-scanning electron microscopy/energy dispersive spectroscopy (FE-SEM/EDS) analysis of the deposited coating was studied using Scanning Electron Microscope (JEOL JSM-6610 LV) with EDAX Genesis software attachment at IIT Ropar (Punjab), India. SEM indicates the surface morphology and EDAX genesis software indicates the elemental composition (wt.%) present at point/area of interest. To identify the various phases formed on the surface, the XRD analysis of the coated specimens were carried out with X-Ray Diffractometer (X-Pert PRO, PANalytical PW 3050/60) at Punjab Agricultural University, Ludhiana (Punjab), India. The specimens were scanned with a scanning speed of 2°/min in 2θ range of 20°–90° and the intensities were recorded. The diffractometer interface with X-Pert PRO, PANalytical X-Ray diffraction software provides the "d" values directly on the diffraction pattern. After the surface characterization, the cross-sectional analyses of the samples were carried out. EDAX analysis was carried out at different points of interest along the cross-section of D-Gun sprayed coated samples. X-ray mapping of different elements presents across the coated samples was revealed using SEM. Firstly specimens were cut with the help of Diamond Cutter (Model: Scientific Precimet-1 Low Speed Precision Saw) across its cross-section at IIT Ropar (Punjab), India and subsequently mounted in epoxy resin. The mounted specimens were polished manually using SiC emery papers of 400, 600, 800, 1000, 1500 grit sizes and subsequently on 1/0, 2/0, 3/0 and 4/0 grade. Finally, specimens were mirror polished on a cloth polishing wheel machine with 0.5 μm alumina powder. The specimens were washed thoroughly with flowing water, and dried in air to remove any moisture. After that the specimens were washed with the help of Acetone solution in order to remove any impurity. The analysis was carried out by Scanning Electron Microscope (JEOL JSM-6610 LV) with EDAX Genesis software attachment at IIT Ropar (Punjab), India.

## 2.4    E-C Studies in Actual Boiler Environment

The coated as well as uncoated specimens were exposed to the platen super-heater zone of the coal fired boiler unit-2 (height 27 m) at Guru Nanak Dev Thermal Plant, Bathinda, (Punjab), India. A hole of approximate 1.5 mm diameter was drilled in all the specimens to hang them in the boiler for experimentation. The average temperature at that point was about 900° C with variations of ±10° C. The specimens were exposed to the combustion environment for 10 cycles. Each cycle consisted of 100 h heating followed by 1-h cooling at ambient conditions. After the end of each cycle, the specimens were clean properly with a gentle brush to remove any ash or loose contaminants, etc. They have been washed by acetone, then visually observed for any change in the surface texture and weight of the specimens were measured subsequently using an electronic balance (Sartorius BP-210S accuracy 0.0001 g) at Guru Nanak Dev Thermal Power Plant, Bathinda (Punjab), India. In case of the specimens exposed to the industrial environment, the actual working conditions of the coal fired boiler in a thermal power plant, the spalled scale could not be collected and incorporated in the mass change. In this case, the weight change consists of a weight gain owing to the formation of the oxide scales and a weight loss due to the suspected spalling and fluxing of the oxide scales. Therefore, the net weight change in the industrial environment represents the combined effects of these two processes. The mass change data along with physical observation to approximate the kinetics of E-C conditions. XRD, FE-SEM/EDS and X-ray mapping techniques were used to characterize the oxide scales formed on the samples after exposure to the above said environments. In case of actual boiler environment, the weight change data alone could not be of much use for predicting E-C behavior because of suspected spalling and ash deposition on the samples. Although the specimens were washed with acetone after every cycle before weight measurement in order to remove ash deposited yet it was difficult to remove the ash completely. Therefore, the extent of E-C has also been evaluated by measuring the thickness of the scale lost due to erosion, spalling or evaporation by finding the difference in thickness of the specimens before and after 1000 h exposure to the actual environment.

# 3  Experimental Results

## 3.1  SEM/EDS and XRD Analysis of As-Sprayed Coating:

The Surface morphology and EDAX analysis of Stellite-6 coated "ASTM SA-213 T-22" is as shown in Fig. 3. The microstructure indicates uniform and dense coating. The coating contains some un-melted and semi melted particles. Chromium (Cr), Cobalt (Co) and Molybdenum (Mo) are the main elements presents in the Stellite-6 coating. Oxygen might have penetrated into the coating during coating process. The XRD analysis also confirms the presence of above elements.

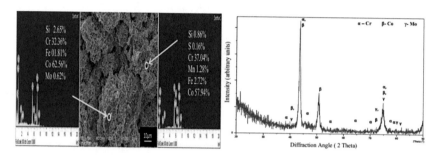

**Fig. 3.** SEM/ EDAX analysis along with X-ray Diffraction pattern for Stellite-6 coated "ASTM SA-213 T-22" boiler steel at different points, (1200X).

## 3.2  Thickness of Coating

The thickness of the coating has been measured from the Back Scattered Electron Images (BSEI) Fig. 4 and its average values were found to be 166.8 μm for the coating. It is observed that the coating has formed a continuous and homogeneous bond with the substrate.

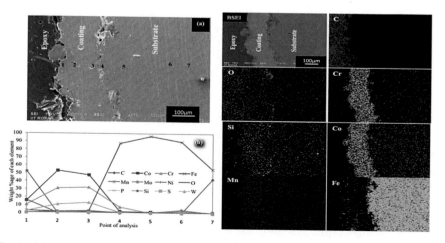

**Fig. 4.** (a) Cross-Sectional Morphology, (b) Variation of elemental composition and (c) X-ray mapping of the cross-section of Stellite-6 coated "ASTM-SA213 T-22" boiler steel

### 3.3    Cross-Sectional Analysis of As-Sprayed Coating

Back Scattered Electron Image (BSEI) micrograph and X-ray mapping of the different elements in the Stellite-6 coated "ASTM-SA213 T-22" boiler steel is as shown in the Fig. 4. The figure illustrates that Chromium (Cr) and Cobalt (Co) distributed uniformly throughout the coating. A small amount of oxygen might have penetrated into the coating during spraying process.

### 3.4    Visual Examination

The macrographs of Stellite-6 coated "ASTM-SA213 T-22" boiler steel exposed to super-heater zone of the coal fired boiler environment at 900 °C for 1000 h are shown in Fig. 5. In this case no more damage to the coating occurs up to the $7^{th}$ cycle. At the end of the $1^{st}$ cycle welded pool is developed near the hole with the deposition of the small amount of the ash. After the end of $3^{rd}$ cycle, the specimen start going to be blackish and a small welded pool of carbon particles is developed on the side of the specimen. The coating protects the specimen up to the end of the cycle. On the end of $7^{th}$ cycle the welded pool of carbon particles is developed on the two sides of the specimen and the specimen remain no longer rectangular. After each cycle the welded pool near the hole goes on developed up to $10^{th}$ cycle. After end of $9^{th}$ cycle the color of the specimen changes with the bottom of the specimen turn more blackish with more deposition of the ash on the specimen.

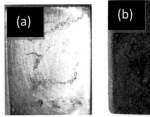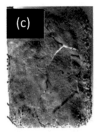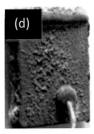

**Fig. 5.** Macrographs of the uncoated and coated "ASTM-SA213 T-22" boiler steel before exposure: (a) Uncoated (b) Stellite-6 coated and after 1000 h exposure in the super-heater zone of the coal fired boiler environment at 900 °C: (c) Uncoated (d) Stellite-6 coated.

### 3.5    Erosion-Corrosion Rates

Figure 6 shows weight change in mg/cm$^2$ as a function of times expressed in hours for Uncoated, and Stellite-6 coated "ASTM-SA213 T-22" boiler steel for high temperature erosion-corrosion at 900 °C for 1000 h in coal fired boiler environment. The weight change data is used to established kinetics of oxidation process. Uncoated boiler steel show higher rate of oxidation as compare to coated one. The cumulative weight gain after the completion of the 10 cycles of erosion-corrosion for the Uncoated, and Stellite-6 coated "ASTM-SA213 T-22" boiler steel is 332.2179 mg/cm$^2$, and 122.3947 mg/cm$^2$ respectively. Figure 7 shows the gain in overall thickness for

Uncoated, and Stellite-6 coated "ASTM-SA213 T-22" boiler steel after the completion of the 10 cycles are 5.696 mm, and 0.848 mm respectively. Also Fig. 8 shows erosion-corrosion rate in terms of mils per year after the completion of the 10 cycles for the Uncoated and Stellite-6 coated "ASTM-SA213 T-22" boiler steel is 1965.79mpy and 292.66mpy respectively. Average scale thickness after the completion of 10 cycles for the Uncoated and Stellite-6 coated "ASTM-SA213 T-22" boiler steel is 912.0 µm and 419.2 µm respectively.

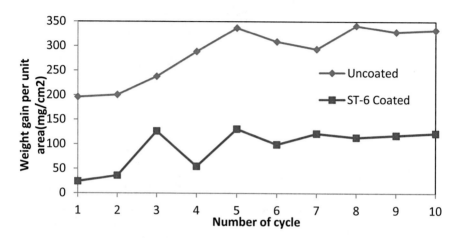

**Fig. 6.** Weight gain per unit area (mg/cm$^2$) Vs Number of cycles for Uncoated, and Stellite-6 coated "ASTM-SA213 T-22" boiler steel for high temperature erosion-corrosion at 900° C for 1000 h in coal fired boiler environment

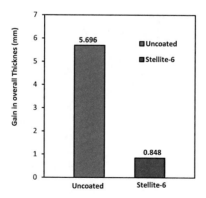

**Fig. 7.** Gain in overall thickness (mm) for Uncoated, and Stellite-6 coated "ASTM-SA213 T-22" boiler steel for high temperature erosion-corrosion at 900° C for 1000 h in coal fired boiler environment

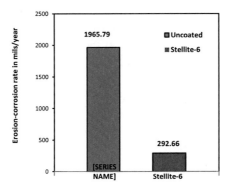

**Fig. 8.** Erosion-corrosion rate in mils/year (mpy) for Uncoated, and Stellite-6 coated "ASTM-SA213 T-22" boiler steel for high temperature erosion-corrosion at 900° C for 1000 h in coal fired boiler environment

### 3.6    SEM/EDAX and XRD Analysis

SEM micrograph, EDAX analysis along with the XRD diffractograms for the uncoated, and Stellite-6 coated "ASTM-SA213 T-22" boiler steel for high temperature erosion-corrosion at 900 °C for the 1000 h are shown in the Figs. 9 and 10 respectively. SEM micrograph mainly indicates grey phases in Fig. 9 for Uncoated T-22. EDAX analysis indicates the presence of Al, O, Si and Fe. The presence of significant amount of oxygen and Iron, Aluminum and Silicon point out the possibility of $Fe_2O_3$, $Al_2O_3$ and $SiO_2$ respectively. The XRD analysis in Fig. 9 also confirms the presence of above elements.

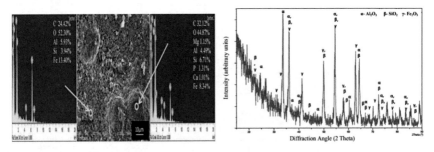

**Fig. 9.** SEM/ EDAX analysis along with X-Ray Diffraction Pattern for Uncoated "ASTM SA-213 T-22" boiler steel after high temperature erosion corrosion at 900° C for 1000 h in coal fired boiler environment, (1200X).

In the case of Stellite-6 coated "ASTM-SA213 T-22" boiler steel for high temperature erosion-corrosion at 900 °C for the 1000 h, SEM micrograph mainly indicates grey with dark phases in Fig. 10. EDAX analysis indicates the presence of Al, O and Si and Fe. Along with the above elements some amount of carbon is also present which may be from the ash contents which stick with the specimen. The presence of significant amount of oxygen, Iron, Aluminum and Silicon point out the possibility of $Fe_2O_3$, $Al_2O_3$ and $SiO_2$ respectively. The XRD analysis in Fig. 10 also confirms the presence of above elements.

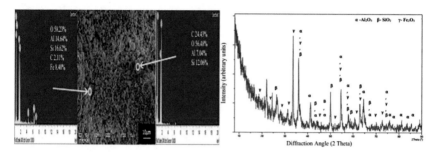

**Fig. 10.** SEM/ EDAX analysis along with X-Ray Diffraction Pattern for Stellite-6 Coated "ASTM SA-213 T-22" boiler steel after high temperature erosion corrosion at 900° C for 1000 h in coal fired boiler environment, (1200X).

### 3.7    Scale Thickness

The average oxide scale thickness of uncoated, and Stellite-6 coated "ASTM-SA213 T-22" boiler steel for high temperature erosion-corrosion at 900 °C for the 1000 h is 912 μm and 419.2 μm respectively.

### 3.8    Cross-Sectional Morphology and X-Ray Mapping

Back Scattered Electron Image (BSEI) micrograph and elemental variation across the cross-section of uncoated, Stellite-6 coated, "ASTM-SA213 T-22" boiler steel after high temperature erosion-corrosion at 900 °C for 1000 h in the coal fired boiler environment are as shown in the Fig. 11, Fig. 12, respectively. In the case of uncoated "ASTM-SA213 T-22" boiler steel Fig. 11 micrograph shows that there is formation of thick scale. The scale is fragile and indicating cracking. EDAX analysis reveals the presence of Fe, Mo and O throughout the scale along with some amount of Aluminum (Al). The existence of significant amount of oxygen points out the possibility $Fe_2O_3$ in the oxide scale. At point 3, Scale is rich in Aluminum (Al) and Silicon (Si) along with significant amount of oxygen which point out the possibility of formation of $Al_2O_3$ and $SiO_2$. BSEI and X-ray mapping for a part of oxide scale for uncoated "ASTM-SA213 T-22" boiler steel after high temperature erosion corrosion at 900 °C for 1000 h in the coal fired boiler environment is shown in Fig. 11. It indicates the significant amount of oxygen (O) and Iron (Fe) in the scale. This mean the oxygen penetrated the specimen. At the point.3 dense concentration of the Aluminum (Al) is observed. Silicon (Si) is also present throughout the specimen. BSEI and X-ray mapping for Stellite-6 coated "ASTM-SA-213 T-22" boiler steel after high temperature erosion-corrosion, which shows the thick bands of oxygen (O) and iron (Fe) in the scale. Also, thick bands of Chromium (Cr) and Cobalt (Co) are observed.

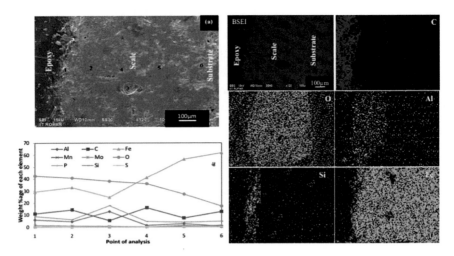

**Fig. 11.** (a) Oxide scale Morphology (b) Variation of elemental composition across the cross section and (c) BSEI and X-ray mapping of the cross-section of Uncoated "ASTM-SA213 T-22" boiler steel after high temperature erosion-corrosion at 900° C for 1000 h in the coal fired boiler environment.

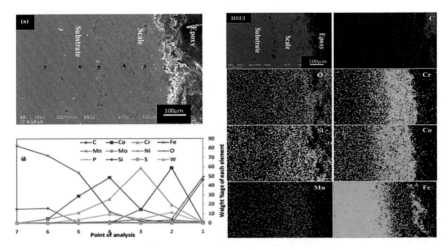

**Fig. 12.** (a) Oxide scale Morphology (b) Variation of elemental composition across the cross section and (c) BSEI and X-ray mapping of the cross-section of Stellite-6 Coated "ASTM-SA213 T-22" boiler steel after high temperature erosion-corrosion at 900° C for 1000 h in the coal fired boiler environment

**Discussion**: Thermal spraying techniques have been gaining attention as they present a cost-effective way to combat degradation without sacrificing properties of the material. Thermal spraying is one of the most versatile coating techniques and find wide ranging applications in numerous industry sectors. The coating of Stellite-6 has been formulated successfully by D-Gun Thermal spraying technique on "ASTM- SA213 T-22" boiler tube steel. Chromium (Cr), Cobalt (Co) and Molybdenum (Mo) are the main elements presents in the Stellite-6 coating. The SEM/EDAX and XRD analysis of powder and coating surface confirms the presence of above elements. Nearby spherical particles are observed during SEM of Stellite-6 and may be un-melted particles as suggested by [14]. The coating has ordered and granular structure. The X-ray mapping of the coating shows (Fig. 4) that Chromium (Cr) and Cobalt (Co) distributed uniformly throughout the coating and a small amount of oxygen have penetrated into the coating during spraying process. Coating-substrate interface shows no gaps or cracks feature good adhesion between them. X-Ray mapping also confirms the clearly visible and thick bands of protective elements uniformly distributed throughout the coating. Thick bands of CO, Cr and Mn in stellite-6 coating provide protection to substrate. Grain boundary can be clearly seen in the coating. All the elements are uniformly distributed throughout the coating supported by cross-sectional morphology as shown in Fig. 4. Samples of Fly ash shows the presence of Al, Si, Ca, Mg, C and O from coal fired boiler as indicated by SEM/EDAX analysis in Fig. 2, and XRD analysis Figure confirms the availability of $Al_2O_3$, $Fe_2O_3$ and $SiO_2$. There is deposition of ash (Si, Al, Ca etc) on the surface of uncoated and Stellite-6 coated "ASTM-SA213 T-22" boiler steel which is supported by SEM/EDAX analysis. Uncoated and coated boiler steel shows the formation of $Al_2O_3$, $Fe_2O_3$ and $SiO_2$ at the surface of specimens exposed to actual environment of coal fired boiler as indicated by SEM/EDAX analysis and equally supported by XRD analysis as shown in Figs. 9 and 10. Most metals are unstable and react spontaneously with oxygen at high temperature, to form the thermodynamically favoured metal oxide. However, once a layer of oxide has developed on the metal surface, it acts as a barrier between the metal and the environment; further reaction takes place only if one or both of the reactants are able to penetrate the layer, leading to formation of new oxide at either the scale-gas or scale-metal interface, [15], whereas in the case of uncoated boiler steel, there is no thermal barrier coating available. The weight change data for the D-gun spray Stellite-6 coated and uncoated boiler steels in actual boiler environment are shown in Figs. 6 and 8. Uncoated boiler steel gain much higher weight as compared to the coated one when exposed to high temperature erosion corrosion boiler environmental. Which may be due to rapid formation of oxides at splat boundaries and within the open pores due to the penetration of the oxidizing species. There is continuous formation of thin oxide scale with subsequent depletion by spallation and erosion under cyclic test conditions. The top surface may contain inclusions, which leads to vertical cracks through which the corrosive species might have penetrated along the crack and between the metallic layers as reported by [16, 17]. Formation of $Fe_2O_3$ as shown by XRD might be due to the reaction of iron with oxygen since iron is the main constituent of boiler steel. Formation of such type of oxides has also been analyzed by [18, 19] during the failure analysis of superheater tubes caused by fireside corrosion. The formation of $Al_2O_3$ and $SiO_2$ might be due to the deposition of ash on the eroded-corroded tubes. The presence of such phases in slag has also been reported by john in his study on slag, gas and deposit thermo-chemistry in a coal gasifier [20]. Thermogravimetric analysis shows high rate of erosion corrosion rate in terms of mils per year is achieved for the uncoated boiler steel as compare with the coated one. Many alloys based on iron, nickel or cobalt, because these metals have relatively high melting points and can be fabricated into any shape without much difficulty; however, the metal oxides are not very protective at high temperatures. Hence, addition of sufficient amounts of other elements to alloys may ensure establishment of a more resistant external oxide scale, usually containing a protective and healing layer of $Cr_2O_3$, $Al_2O_3$ and occasionally $SiO_2$ at the scale/alloy interface [21]. It is commonly reported that as a result of the oxidation process under isothermal conditions a protective Cr-containing oxide and Fe-containing oxide is developed on the surface of the steel causing a decrease of the oxidation rate with time. Oxide scale is constituted by a layered structure with compositional and microstructural variations from the substrate to the outer interface [22]. That is why presence of hard phases of Cr, Co, Ni and Mn might become barrier to the further oxidation of coated specimens and simultaneously provide the protection to the substrate. Also X-Ray mapping (Fig. 12) confirms thick bands of Co and Cr in stellite-6 coating, which provide high protection to substrate as amount of oxygen and carbon is very small after exposure to industrial boiler. Clear grain boundaries of all elements in stellite-6 coating are shown. Thermogravimetric analysis shows the gain in overall thickness and erosion-corrosion rate in terms of mills per year (mpy) for the uncoated and Stellite-6 coated "ASTM-SA213 T-22" boiler steel tube follows the sequence:

Uncoated > Stellite-6 coated

# 4  Conclusion

The particles present in the Stellite-6 powder have rounded in shape having approximate size ±53/15 μm. The main elements are Chromium (Cr) and Cobalt (Co). Whereas Silicon (Si) and Aluminium (Al) are the main elements present in the ash. Stellite-6 coating have been successfully deposited with the Detonation Gun (D-Gun) thermal spraying process. The present range of coating thickness is between 150 μm to 200 μm and during the coating process oxygen is diffused into the coating. Thick bands of the Chromium (Cr), Cobalt (Co) and Manganese (Mn) in the Stellite-6 coated ASTM SA-213 T-22" boiler steel is visible in the X-ray mapping. Simultaneously thick band of Iron (Fe), the substrate, is also shown. Thus, the coating protects the substrate as the thick bands of the elements prevent the oxygen to pass through and thus resist oxidation after the completion of the 10 cycles of hot corrosion. Uncoated "ASTM SA-213 T-22" boiler steel specimen undergoes more weight gain, with severe spalling with visible cracks, as compare to the coated "ASTM SA-213 T-22" boiler steel specimens. Uncoated as well as coated "ASTM SA-213 T-22" boiler steel has shown ash deposit on the surface. Thus, final thickness is contributed by scale formation, erosion and ash deposition. In addition to Detonation Gun thermal spraying coating process, Stellite-6 coating may be deposited by other thermal spraying methods like High velocity-oxy fuel (HVOF), High velocity-air fuel (HVAF), Plasma Thermal spraying or Cold Spraying. These comparative studies may be considered as future scope of this paper.

# References

1. Singh, G., Goyal, K., Bhatia, R.: Hot corrosion studies of plasma-sprayed chromium oxide coating on boiler tube steel at 850 °C in simulated boiler environment. Iran. J. Sci. Technol. Trans. Mech. Eng. 149–159 (2017)
2. Chawla, V., Chawla, A., Prakash, S., Puri, D., Gurbuxani, P.G., Sidhu, B.S.: Hot corrosion & erosion problems in coal based power plants in india and possible solutions—a review. J. Miner. Mater. Charact. Eng. 10(4), 367–385 (2011)
3. Hidalgo, V.H., Varela, J.B., Menéndez, A.C., Martınez, S.P.: High temperature erosion wear of flame and plasma-sprayed nickel–chromium coatings under simulated coal-fired boiler atmospheres. Wear 247, 214–222 (2001)
4. Kumar, M., Singh, H., Singh, N.: Study of Ni–20Cr coatings for high temperature applications–a review. Arch. Metall. Mater. 58(2), 523–528 (2013). https://doi.org/10.2478/amm-2013-0030
5. Kumar, N., Prashar, G., Dhawan, R.K.: To check the feasibility of Cr2O3 coating on boiler steel tubes simulated coal fired boiler conditions to prevent the erosion. Int. J. Emerg. Technol. 3(1), 126–129 (2012)
6. Kumar, A., Srivastava, V., Mishra, N.K.: Oxidation resistance of uncoated & detonation gun sprayed WC-12Co and Ni-20Cr coating on T-22 boiler steel at 900 °C. In: International Conference on Mechanical, Materials and Renewable Energy (2018). https://doi.org/10.1088/1757-899x/377/1/012076
7. Goyal, K., Sidhu, V.P.S., Goyal, R.: Comparative study of corrosion behaviour of HVOF-coated boiler steel in actual boiler environment of a thermal power plant. J. Aust. Ceram. Soc. 925–932 (2017)
8. Singh, R., Goyal, A., Singh, G.: Study of high-temperature corrosion behaviour of D-Gun spray coatings on ASTM-SA213, T-11 steel in molten salt environment. Mater. Today: Proc. 4, 142–151 (2017)
9. Kaushal, G., Kaur, N., Singh, H., Prakash, S.: Oxidation behaviour of D-Gun spray Ni-20Cr coated ASTM A213 347H steel at 900 °C. Int. J. Surf. Eng. Mater. Technol. 2(1), 33–38 (2012). ISSN: 2249-7250
10. Murthy, J.K.N., Ventakaraman, B.: Surf. Coat. Technol. 200, 2642 (2006)
11. Jain, A., Sharma, Y.K., Singh, B., Chawla, V.: Combating erosion-corrosion behavior of detonation gun sprayed satellite-6 coating on grade A-1 boiler steel in a coal fired boiler. CORSYM-2014 (20–21 Feb 2014), pp. 186–200. IIT, Bombay (2014)
12. Singh, G., Kaushal, G., Krishan, B.: Characterization of as-sprayed stellite-6 coating deposited by D-gun spray process for perspective wear application. Int. J. Latest Trends Eng. Technol. 131–136 (2017)
13. Sidhu, B.S., Puri, D., Prakash, S.: Mechanical and metallurgical properties of plasma sprayed and laser remelted Ni–20Cr and Stellite-6 coatings. J. Mater. Process. Technol. 159, 347–355 (2005)
14. Ak, N.F., Tekmen, C., Ozdemir, I., Soykan, H.S., Celik, E.: Ni-Cr coatings on stainless steel by HVOF technique. Surf. Coat. Technol. 1070–1073 (2003)
15. Lawless, K.: The oxidation of metals. Rep. Prog. Phys. 37:231, 173–174 (1974)
16. Sidhu, B.S., Prakash, S.: Evaluation of the corrosion behaviour of plasma sprayed Ni3Al coatings on steel in oxidation and molten salt environments at 900 °C. Surf. Coat. Technol. 166, 89–100 (2003)
17. Niranatlumpong, P., Ponton, C.B., Evans, H.E.: The failure of protective oxides on plasma-sprayed NiCrAlY overlay coatings. Oxid. Met. 53(3–4), 241–256 (2000)

18. Prakash, S., Singh, S., Sidhu, B.S., Madeshia, A.: Tube failures in coal fired boilers. In: Proceeding of the National Seminar on Advances in Material and Processing, Nov 9–10, pp. 245–253. IITR, Roorkee, India (2001)
19. Srikanth, S., Ravikumar, B., Das, S. K., Gopala Krishna, K., Nandakumar, K., Vijayan, P.: Analysis of Failures in Boiler Tubes due to Fireside Corrosion in a Waste Heat Recovery Boiler, Engineering Failure Analysis, vol. 10, pp. 59–66 (2003)
20. John, R.C. et al.: Mater. High Temp. 11(1–4), 124 (1993)
21. Wood, G.C., Whittle, D.P.: Mechanisms of oxidation of Fe -16.4% Cr at high temperature. Corros. Sci. 4(1–4), 263–292 (1964)
22. Chawla, V., Gond, D., Puri, D., Prakash, S.: Oxidation studies of T-91 and T-22 boiler steels in air at 900 °C. J. Miner. Mater. Charact. Eng. 9(8), 749–761 (2010)

# ICI Reduction Using Enhanced Data Conversion Technique with 1 × 2 and 1 × 4 Receive Diversity for OFDM Systems

Vaishali Bahl$^{(\boxtimes)}$, Vikrant Sharma, Gurleen Kaur,
and Ravinder Kumar

GNA University, Phagwara, India
er.vaishalibahl@gmail.com, vikrant.sharma@yahoo.com

**Abstract.** The intensifying demand for tremendously high rate data transmission over wireless mediums needs ingenious usage of electromagnetic resources considering restrictions like power incorporation, spectrum proficiency, robustness in disparity to multipath propagation and implementation complication. Orthogonal frequency division multiplexing (OFDM) is a favorable approach for upcoming generation wireless communication systems. However its susceptibility to the frequency offset triggered by frequency difference between local oscillator of transmitter and receiver or due to Doppler shift results to Inter Carrier Interference. This delinquent of ICI results in worsening performance of the wireless systems as bit error rate increases with increase in value of frequency offset. In this paper simulation results are demonstrated for analyzing the effect of varying frequency offset factor on system's error rate performance. Also the paper proposes the reduction in Inter Carrier Interference using the enhanced Data Conversion Technique with receive diversity for OFDM based wireless systems.

**Keywords:** Bit Error Ratio (BER) · Inter-Carrier Interference (ICI) · Additive White Gaussian noise (AWGN) · Carrier Frequency Offset (CFO) · Maximal Ratio Combining (MRC)

## 1 Introduction

Orthogonal frequency division multiplexing (OFDM) is a multicarrier multiplexing technique, in which data is transmitted over several parallel frequency sub channels at a lower rate. It has been standardized in several wireless applications such as Digital Video Broadcasting (DVB), HIPERLAN, IEEE 802.11 (Wi-Fi), and IEEE 802.16 (WiMAX) and is used for wired applications as in the Asynchronous Digital Subscriber Line (ADSL), Digital Audio Broadcasting (DAB) and power-line communications [1, 2]. One of the foremost reasons to use OFDM is to step-up the robustness against narrowband interference or frequency selective fading. As every technique has its inadequacies, this technique also has problem of being sensitive towards frequency mismatch. This mismatch in frequency can either arise because of variation in local oscillator frequencies of transceivers or due to Doppler shift initiating carrier frequency

© Springer Nature Switzerland AG 2019
C. Benavente-Peces et al. (Eds.): SEAHF 2019, SIST 150, pp. 211–220, 2019.
https://doi.org/10.1007/978-3-030-22964-1_22

offset. The CFO upshots in loss of orthogonality of the subcarriers which causes ICI. This paper is outlined in way that Sect. 2 expresses the basic description and issues of OFDM system succeeded by the system portrayal and interference scrutiny and Mathematical description of ICI is given in Sect. 3 succeeded by simulation results in Sect. 4. The conclusion of paper is given in Sect. 5.

## 1.1 Orthogonal Frequency Division Multiplexing

OFDM is a distinct case of multi-carrier modulation. The dictum of OFDM is to split a single high-data rate stream into a number of lesser rate streams that are transferred simultaneously over some narrower sub channels which are orthogonal to each other. Henceforward it is not only a modulation technique nevertheless a multiplexing technique too. The merits of this technique that make it a desired choice over other modulation techniques are its extraordinary spectral efficiency, easier implementation of FFT, lower receiver intricacy, robustness for high-data rate transmission over multipath fading channel, high controllability for link adaptation are few benefits to list. However, it has two elementary impairments: (1) higher peak to average power ratio (PAPR) as paralleled single carrier signal [3]. (2) Susceptibility to phase noise, timing and frequency offsets that acquaint with ICI into the system. The carrier frequency offset is instigated by the disparity of frequencies amongst the oscillators at the transceivers, or from the Doppler spread due to the relative motion between them. The phase noise arises mainly due to imperfections of the LO in the transceiver. The timing offset emerges due to the multipath delay spread and because of it not only inter-symbol interference, but ICI also transpires. However, ICI influenced by phase noise and timing offset can completely be compensated or fixed. But the manifestation of frequency offset due to the Doppler spread or frequency shift resulting in ICI is arbitrary, henceforth only its impact can be lessened. Many diverse ICI mitigation schemes have been extensively reconnoitered to fray the Inter-Carrier Interference in OFDM systems, comprising frequency-domain equalization [4], time-domain windowing [5], and the ICI self-cancellation (SC) schemes [6–12], frequency offset estimation and compensation techniques [13] and so on. Amidst the schemes, the ICI self-cancellation scheme is a modest method for ICI minimization. It is a two-phase approach that uses redundant modulation to overpower ICI with ease for OFDM [18].

## 1.2 System Depiction and ICI Analysis

Figure 1 portraits a distinctive discrete-time base-band equivalent OFDM system model. As presented, a stream of input bit stream is first mapped into symbols using BPSK modulation. The symbols are modulated using IFFT on N-parallel subcarriers succeeding the serial-to-parallel (S/P) conversion. With cyclic prefix (CP) appending, the OFDM symbols are sequential using parallel to serial (P/S) conversion and referred to the channel. At the receiver, the received symbols are recaptured by S/P transformation, CP removal, FFT transformation, P/S conversion and are demapped with equivalent scheme to obtain the anticipated novel bit stream [14, 17, 18].

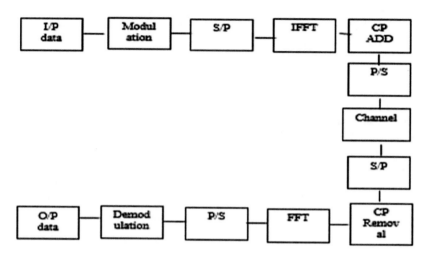

**Fig. 1.** OFDM transceiver

In OFDM systems, the transmitted signal in time domain can be exhibit as:

$$x(n) = \frac{1}{N} \sum_{k=0}^{N-1} X(k) e^{j\frac{2\pi kn}{N}} \tag{1}$$

Where x (n) represents the nth sample of the OFDM transmitted signal, X (k) symbolizes the modulated symbol for the kth subcarrier and k = 0, 1… N − 1, N is the total quantity of
OFDM subcarriers.

The received signal in time domain is specified by:

$$y(n) = x(n) e^{j\frac{2\pi n\in}{N}} + w(n) \tag{2}$$

Where ∈ is the normalized frequency offset known by ∈ = Δf.NTs in which Δf is the frequency difference which is either due to variance in the local oscillator carrier frequencies of transmitter and receiver or due to Doppler shift and Ts is the subcarrier frequency and w (n) is the Additive White Gaussian Noise acquainted in the channel. The outcome of this frequency offset on the received symbol stream can be implicit by considering the received symbol Y (k) on the kth subcarrier. The received signal at subcarrier index k can be stated as

$$Y(k) = X(k)S(0) + \sum_{l=0, l \neq k}^{N-1} X(l)S(l-k) + W(k) \tag{3}$$

where  k = 0, 1… N − 1 and X (k) S (0) is the wanted signal and $\sum_{l=0, l \neq k}^{N-1} X(l)S(l-k)$ is the ICI component of acknowledged OFDM signal.

ICI component S (l-k) can be given as:

$$S(l - k) = \frac{\sin(\pi(l + \epsilon - k))}{N\sin(\pi(l + \epsilon - k)/N)} \exp\left(j\pi\left(1 - \frac{1}{N}\right)(l + \epsilon - k)\right) \tag{4}$$

### 1.3    PROPOSED ICI Self-cancellation Scheme with Receive Diversity

The scheme works in two very simple steps. At the transmitter side, one data symbol is modulated onto a group of adjacent subcarriers with a group of weighting coefficients. The weighting coefficients are designed so that the ICI caused by the channel frequency errors can be minimized. At the receiver side, by linearly combining the received signals on these subcarriers with proposed coefficients, the residual ICI contained in the received signals can then be further reduced.

Various schemes implemented having different weighting coefficients have been described below:-

**Data Conversion Scheme**

In the data-conversion self-cancellation scheme [5] for ICI mitigation, data symbol allocation is as

$$X'(k) = X(k)$$

$$X'(k + 1) = -X(k)$$

Where k = 0, 2, ...., N − 2 in consecutive subcarriers to deal with the ICI. The received signal Y (k) is determined by the difference between the adjacent subcarriers. This self-cancellation scheme relies on the fact that the real and imaginary parts of the ICI coefficients change gradually with respect to the subcarrier index k. Therefore, the difference between consecutive ICI coefficients $S(l - k) - S(l - k + 1)$ is very small. Then the resultant data sequence is used for making symbol decision can be represented

$$Y''(k) = \frac{1}{2}[Y'(k) - Y'(k + 1)] \tag{5}$$

The simulation model of the proposed hybrid scheme for mitigating ICI has been illustrated in the Fig. 2. Now to analyze the performance of system, OFDM signal is generated with the input parameters as per shown in Table 1 which are as per IEEE 802.16n-2013 specifications that are taking 1024000 binary input data bits over 10 MHz bandwidth are reshaped and modulated by BPSK (2-QAM) modulation technique and are mapped as per DC SC mapping scheme on 1024 subcarriers. Then 1024 point IFFT is performed and CP having length of 256 samples is added. This signal is transmitted with transmitting frequency of 2.3 GHz in the environment which is Rayleigh channel. The Normalized Frequency Offset gets added due to Doppler shift caused by moving receiver. The received signal is combined using Maximal Ratio Combining and then fed to Serial to Parallel converter. After CP removal its transformed back into frequency domain by doing fast Fourier Transform on the signal.

After this signal is recovered by doing de-mapping in accordance with linear combination of the data conversion based ICI self-cancellation scheme. The recovered signal is demodulated and output data is received and thereby the performance in terms of Bit Error Ratio is evaluated.

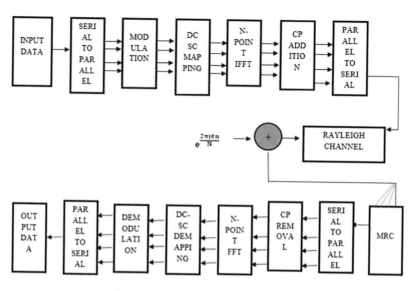

**Fig. 2.** Proposed OFDM transceiver

## 2  Related Work

The performance comparison has been done for the SIMO system using proposed ICI Self-Cancellation Scheme with the OFDM system using Maximal Ratio Combining in terms of Bit Error Ratio Analysis. The input parameters have taken in accordance with those given in Table 1 for the simulations. For all 4 cases the OFDM signal is transmitted at the 2.3 GHz Frequency in bandwidth 10 MHz using 1024 data subcarriers carrying 1024000 input data bits over the AWGN channel. The receiver being in motion results in Doppler shift which causes Normalized frequency offset causing loss of orthogonality. Thereby introducing inter carrier interference. This ICI deteriorates the system performance.

A simulation is directed to evaluate performance of system for input specifications given in Table 2.

The simulation graphs in figures from Figs. 3, 4, 5, 6 and 7 shows Bit Error Ratio Comparison of Standard OFDM without ICI Self-Cancellation Scheme given by curve of black color and OFDM System with Proposed Self-Cancellation based MRC Scheme given by curve of red color over the SNR range for the receiver moving at the speed of 20 kmph corresponding to 0.05 value of Normalized Frequency Offset and 60 kmph corresponding to 0.15 value of Normalized Frequency Offset. The systems are SISO (1 × 1) and SIMO (1 × 2) and (1 × 4). The results shows that for SISO system

**Table 1.** Proposed input parameters for simulations

| Parameters | Values |
|---|---|
| Bandwidth (MHz) | 10 |
| Carrier frequency (GHz) | 2.3 |
| Input data bits | 51200 |
| No. of data subcarriers | 1024 |
| Modulation technique | BPSK |
| FFT size | 1024 |
| No. of symbols | 50 |
| CP length | 256 |
| OFDM symbol length | 1280 |
| Symbol duration (μs) | 91.4 |
| Subcarrier frequency (KHz) | 10.94 |
| Channel | Rayleigh |
| SNR (dB) | 0 : 2 : 20 |
| Normalized frequency offset | 0.05, 0.15 |
| Noise | AWGN |

**Table 2.** Performance comparison

| System | Offset = 0.05 Speed = 20 kmph | | Offset = 0.15 Speed = 60 kmph | |
|---|---|---|---|---|
| | Stand. | Prop. | Stand. | Prop. |
| $1 \times 1$ | $10^{-1}$ | $10^{-2}$ | $10^{-1.4}$ | $10^{-2.8}$ |
| $1 \times 2$ | $10^{-2.5}$ | $10^{-4.4}$ | $10^{-2.3}$ | $10^{-4}$ |
| $1 \times 4$ | $10^{-3.4}$ | $10^{-4.7}$ | $10^{-3}$ | $10^{-4}$ |

the performance is not improvised much but as we increased the antennas upto $1 \times 4$, the BER curve depicts considerable amount of improvement. Therefore the hybrid data conversion self-cancellation based MRC scheme outperforms its counterparts.

Results present demonstrate the performance of proposed hybrid scheme and its contribution in improving the capability of the OFDM receiver even for high speed and hence it improves BER of the OFDM system.

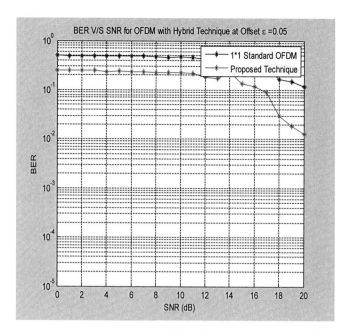

**Fig. 3.** Performance comparison of standard SISO OFDM (1 × 1) without ICI self-cancellation and proposed hybrid scheme of DC self cancellation with MRC for receiver moving at speed of 20 kmph over the SNR range.

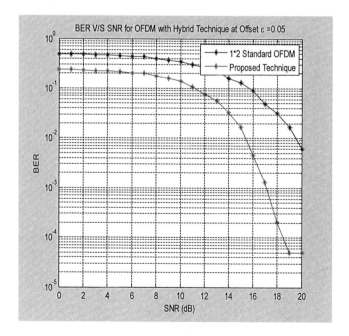

**Fig. 4.** Performance comparison of standard SIMO OFDM (1 × 2) without ICI self-cancellation and proposed hybrid scheme of DC self cancellation with MRC for receiver moving at speed of 20 kmph over the SNR range.

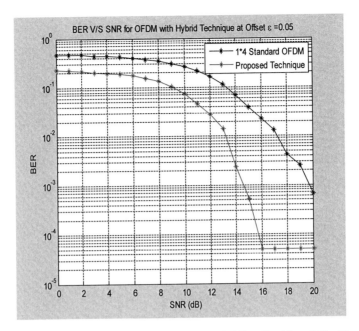

**Fig. 5.** Performance comparison of standard SIMO OFDM ($1 \times 2$) without ICI self-cancellation and proposed hybrid scheme of DC self cancellation with MRC for receiver moving at speed of 20 kmph over the SNR range

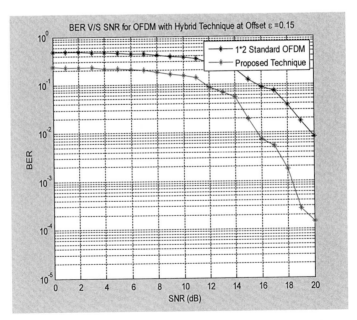

**Fig. 6.** Performance comparison of standard SIMO OFDM ($1 \times 2$) without ICI self-cancellation and proposed hybrid scheme of DC self cancellation with MRC for receiver moving at speed of 60 kmph over the SNR range.

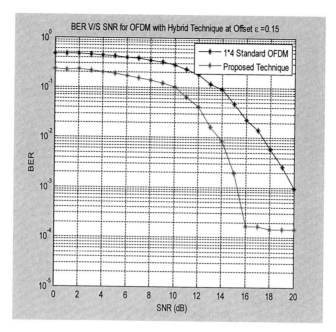

**Fig. 7.** Performance comparison of standard SIMO OFDM (1 × 4) without ICI self-cancellation and proposed hybrid scheme of DC self cancellation with MRC for receiver moving at speed of 60 kmph over the SNR range.

## 3   Conclusion

This paper demonstrates the improvement in the performance of OFDM systems at different speeds of receiver. The offset gets introduced either due to frequency difference or due to Doppler shift system's performance starts worsening in terms of BER of the system. This is because the ICI component gets introduced as soon as offset occurs. The proposed technique of hybridisation of receive diversity and self-cancellation successfully simulated in this paper improvises the performance. It can be further enhanced for higher speed as well.

## References

1. Chang, R.W.: Synthesis of band limited orthogonal signal for multichannel data transmission. Bell Syst. Tech. **45**, 1775–1796 (1996)
2. Salzberg, B.R.: Performance of an efficient parallel data transmission system. IEEE Trans. Commun. **Com-15**, 805–813 (1967)
3. Bahl, V., Dubey, R., Kaur, D.: MIMO-OFDM: foundation for next-generation wireless systems. Int. J. Recent Innov. Trends Comput. Commun. **2**(6), 1692–1695 (2014). ISSN: 2321-8169
4. Jeon, W.G., et al.: An equalization technique for orthogonal frequency division multiplexing systems in time-variant multipath channels. IEEE Trans. Commun. **47**(1), 27–32 (2001)

5. Ryu, H.-G., Li, Y., Park, J.-S.: An improved ICI reduction method in OFDM communication system. IEEE Trans. Broadcast. **51**(3), 395–400 (2005)
6. Armstrong, J.: Analysis of new and existing methods of reducing intercarrier interference due to carrier frequency offset in OFDM. IEEE Trans. Commun. **47**(3), 365–369 (1999)
7. Goyal, R., Dewan, R.: ICI cancellation using self ICI symmetric conjugate symbol repetition for OFDM system. Int. J. Emerg. Trends Signal Process. **1**(6) (2013). ISSN: 2319-9784
8. Fu, Y., Ko, C.C.: A new ICI self-cancellation scheme for OFDM systems based on a generalized signal mapper. In: Proceedings of the 5th Wireless Personal Multimedia Communications, pp. 995–999, Oct 2002
9. Zhao, Y., Häggman, S.: Intercarrier interference self-cancellation scheme for OFDM mobile communication systems. IEEE Trans. Commun. **49**(7), 1185–1191 (2001)
10. Ying-shan, L., et al.: ICI compensation in MISO-OFDM system affected by frequency offset and phase noise. J. Commun. Comput. **5**(12), 32–38 (2008)
11. Peng, Y.-H., et al.: Performance analysis of a new ICI-SC-scheme in OFDM systems. IEEE Trans. Consum. Electron. **53**(4), 1333–1338 (2007)
12. Yeh, H.-G., Chang, Y.-K., Hassibi, B.: A scheme for cancelling intercarrier interference using conjugate transmission in multicarrier communication systems. IEEE Trans. Wirel. Commun. **6**(1) (2007)
13. Shi, Q., Fang, Y., Wang, M.: A novel ICI self-cancellation scheme for OFDM systems. In: IEEE WiCom, pp. 1–4 (2009)
14. Bishnu, A., et al.: A new scheme ICI self-cancellation in OFDM system. In: IEEE ICCSNT (2013)
15. Shim, E.-S., et al.: OFDM carrier frequency offset estimation methods with improved performance. IEEE Trans. Broadcast. **53**(2), 567–573 (2007)
16. van Nee, R., Prasad, R.: OFDM for Wireless Multimedia Communications. Artech House, Inc. (2000)
17. Bahl, V., Kaur, D., Buttar, A.S.: BER analysis of ICI self cancellation schemes for OFDM based wireless systems. Int. J. P2P Trends Technol. **13**(1), 1–5 (2014)
18. Bahl, V., Sehmby, A.S., Singh, N.P.: A perlustration of ICI self cancellation schemes for OFDM systems. J. Netw. Commun. Emerg. Technol. **5**(2), 20–23 (2015)

# Architecture Analysis and Specification of the RacingDrones Platform: Online Video Racing Drones

César Benavente-Peces[✉], David Tena-Ramos, and Ao Hu

Universidad Politécnica de Madrid, ETSIS Telecomunicación, Campus Sur, Calle de Nikola Tesla sn, 28031 Madrid, Spain
cesar.benavente@upm.es
http://www.upm.es

**Abstract.** This research work is aimed at the specification and development of a platform that allows online participation of multiple users and take control of a car (Racing-Drone). The management of the number of participating users will be analyzed according to the quality that the system can offer. Likewise, the different types of connection that can be used to guarantee the highest possible quality of service will be analyzed. This research has as its final objective the specification of the on-line multi-user platform to be developed to support the RacingDrones project. On the one hand, it will be necessary to study the architecture of the system in order to support the platform's engine and user access, taking into account aspects such as the capacity of the system. The software necessary to support the management of the game and the access of the users will also be analyzed. It will be necessary to carry out an analysis of the communications framework that must be developed in order to support multiple on-line players with the appropriate quality, managing the access permissions, the access requests and the maintenance of the connection. The most suitable protocols will be analyzed for the different functionalities of the system. After carrying out the analysis of all the aspects related to the platform (hardware, software and communications), the complete specification of the platform will be made. The devices from which users can access the platform will also be taken into account in the system specification. . . .

**Keywords:** Drones · Video streaming · Video coding · Remote driving · Virtual reality

## 1 Introduction

Videogames, from the beginning, follow the footsteps of the film industry, with scripts increasingly worked, protagonists with their own personality, and a soundtrack performed by a renowned symphony orchestra. But in terms of the visual aspect, despite tending to photo-realism, they have always been somewhat

© Springer Nature Switzerland AG 2019
C. Benavente-Peces et al. (Eds.): SEAHF 2019, SIST 150, pp. 221–230, 2019.
https://doi.org/10.1007/978-3-030-22964-1_23

behind, due to the difficulty in capturing a scene with the same fidelity as a real camera would.

One of the little explored possibilities to provide real sensations to the player is the use of real environments in a videogame, and instead of generating the images by computer, use a direct video recording that generates a competition vehicle. Although the result would be perfect in its photo-realism (it is real), we are faced with the technical problem of video coding in real time without affecting the game-play, which is a challenge and a problem not solved at present. In addition, in the case of mounting a video compressor inside a competition vehicle, the computer would be limited in size and weight and, consequently, in processing capacity, which makes the challenge a difficult problem to solve with the available technologies. The market for this type of technological solution would be very promising and is not limited to video games, as an important market that would have a tractor effect on this technology is the modeling competition market, where realism is not sought that can "simulate" a computer, but the same reality, for which high performance coding technologies are necessary.

Through a high-performance video coding technology solution, modeling competitions (and drones, both terrestrial and aerial) can be taken to the online world and take advantage of the potential of the synergy that both markets offer.

Within the modeling market, there are numerous circuits in cities around the world, and their exploitation is limited to specific meetings where competitions are held. On the contrary, in online games, every day there are ad-hoc competitions of players competing thanks to the online capacity of the games. Uniting both worlds could lead the modeling market much further, but for this a new technology is necessary.

In this context the main goal of this investigation is the analysis and definition of an appropriate architecture to provide real-time gaming facilities with a high QoS and a higher user experience.

The remaining part of this paper is structured as follows: Sect. 2 introduces the constraints and possible solution. The analysis of different solutions is carried out in Sects. 3, 4 and 5. The results obtained along this investigation are discussed in Sect. 6. Finally, some concluding remarks are provided in Sect. 7.

## 2   Analysis of Possible Solutions

In this section, the proposed solutions for the design and implementation of the platform are analyzed, taking into account the different aspects that may affect the quality of the online game service offered. One of the aspects that most impacts on the quality of the video-game service is latency. The aspects of the platform design that most commonly impact, to a greater or lesser extent, the latency of the platform are the following:

- The latency of the camera.
- The latency of the encoder.
- The technology of access to the network and the topology of it.

- The distance between the ends in which the communication is established, as well as the networks or intermediate nodes through which the data packets are routed.
- Computation capacity of the devices that are interconnected to process the data packets and generate the data packets from the video information at the transmitting end and receive the data packets at the receiving end and shape the video.
- The latency of the decoder.
- The latency of the screen.

There are some aspects which can not be influenced by an adequate design of the platform, such as the latency of the camera or the screen, which depend on the kind of device used and the manufacturing technology. Other aspects that influence the global latency will have higher or lower impact depending on how the architecture of the platform will be and what the specifications of the platform will determine to guarantee the best possible quality. There are other aspects such as jitter and lag produced in the transmission that are influenced by the network architecture (communications) that is used.

## 2.1   GAW vs. Other Solutions

GAW vs. other solutions. In order to evaluate GamingAnywhere [1], they have compared the performance of GamingAnywhere with OnLive [2] and Stream-MyGame (SMG) [3]. The experimental configuration consists of a server, a client and a router. The OnLive server resides in the OnLive data centers, while the GamingAnywhere and SMG servers are installed on our own PCs. More specifically, the OnLive client connects to the OnLive server over the Internet, while the GamingAnywhere and SMG clients connect to their servers through a LAN. To evaluate the performance of gaming systems in the cloud under various network conditions, we added a FreeBSD router between the client and the server, and executed dummynet to inject restrictions on delays, packet losses and network bandwidth.

Because the OnLive server is outside our LAN, the quality of the network path between our OnLive client and the server can affect our evaluations. However, according to our observations, the quality of the route was consistently good during all the experiments. The delay of the route network was around 130 ms with few fluctuations. In addition, packet loss rates were measured to less than 10–6 when OnLive transmissions were received at the recommended 5 Mbps. Therefore, the route between the OnLive server and our client can be considered as a communication channel with sufficient bandwidth, zero packet loss rate and a constant latency of 130 ms.

Since the performance of the game systems in the cloud can depend on the game, we consider games of three popular categories: action adventure, first person shooter and real time strategy. A representative game of each category is selected:

- LEGO Batman: The Videogame (Batman) [4] is an action adventure game created by Traveler's Tales in 2008. All the interactive objects in this game are made of Lego bricks. In this game, players control the characters to fight against enemies and solve riddles from a third person perspective.
- F.E.A.R. 2: Project Origin (FEAR) [5] is a first-person shooter game, developed by Monolith Productions in 2009. The combat scenes are designed to be as close as possible to those in real life. In this game, players have a great freedom to interact with environments, for example, they can flip over a desk to take cover.
- Warhammer 40,000: Dawn of War II (DOW) [6] is a real-time strategy game developed by Relic Entertainment in 2009. In campaign mode, players control squads to fight enemies and destroy buildings. In multiplayer, up to 8 players play matches on the same map to complete a mission, such as having specific positions.

Modern video encoders strive to achieve the highest quality video with the lowest bitrate by applying complex coding techniques. However, too complex coding techniques are not feasible for real-time videos given their long coding time. As such, empirically we studied the compensation between bit rate, video quality and frame complexity using x264. More specifically, we applied the coding parameters in real time, and exercised a wide spectrum of other coding parameters. Then they analyzed the quality of the resulting video and the coding time. According to the analyzes obtained in the study, the following x264 coding parameters are recommended:

- –profile main –preset faster –tune zerolatency
- –bitrate $r –ref 1 –media –merange 16
- –intra-refresh –keyint 48 –sliced-threads
- –slices 4 –threads 4 –input-res 1280 × 720,

where $r is the coding rate. The GamingAnywhere server is configured to use the encoding parameters mentioned above, and the coding bit rate is set to 3 Mbps. For a fair comparison, all games are transmitted at a resolution of 720 p. While we configure GamingAnywhere and OnLive to transmit at 50 fps, StreamMyGame only supports transmission at 25 fps. Experiments are designed to evaluate the three game systems from two critical aspects: response capacity and video quality. Experiments are also carried out to quantify the network loads incurred by different gaming systems in the cloud. The details of the experimental designs and the results are given below.

We define the response delay (RD) as the time difference between a user sending a command and the corresponding action in the game that appears on the screen. The RD is divided into three components:

- The processing delay (PD) is the time required for the server to receive and process the command of a player, and to encode and transmit the corresponding frame to that client.
- The reproduction delay (OD) is the time required for the client to receive, decode and represent a picture on the screen.

– The network delay (ND) is the time required for a round of data exchange between the server and the client. ND is also known as round trip time (RTT).

Therefore, we have RD = PD + OD + ND. Some remarks can be done regarding these parameters. First, the OD is small, $\leq 31$ ms, for all cloud-gaming systems and considered games. This reveals that all the decoders are efficient, and the decoding time of the different games does not fluctuate too much. Second, GamingAnywhere achieves a much smaller PD, at most 34 ms, than OnLive and SMG, which are observed up to 191 and 365 ms, respectively. This demonstrates the effectiveness of the proposed GamingAnywhere: OnLive and SMG PDs are 3+ and 10+ times longer than those of GamingAnywhere. Load, among the three systems, only GamingAnywhere achieves RD less than 100 ms, and can meet the strict delay requirements of network games [7].

The server and client of GamingAnywhere are well paired, in the sense that all the steps in the production line are quite efficient. Even for the video coding that consumes more time (on the server) and video playback (on the client), each frame ends at an average of 16 and 7 ms maximum. Such low delay contributes to the superior performance RD of GamingAnywhere, compared to the other known cloud gaming systems.

## 2.2 Video Quality

Video streaming quality directly affects the gaming experience, and network conditions are the keys to high-quality streaming. From this point of view, we use dummynet to control three network condition metrics: network delay (ND), packet loss rate and network bandwidth. We vary ND between 0–600 ms, packet loss rate between 0–10% and bandwidth 1–6 Mbps in the experiments. Also included are experiments with unlimited bandwidth. For OnLive, the ND on the Internet is already 130 ms and, therefore, you can not report the results from scratch ND. Two video quality metrics, PSNR and structural similarity (SSIM), are adopted. The average values of PSNR and SSIM of component Y are reported.

The proposed GamingAnywhere solution incurs much lower uplink traffic loads, compared to OnLive and SMG. The only exception is that, with Batman, SMG incurs a lower rate of uplink packets. However, SMG also produces a larger uplink payload size, which leads to a higher uplink bit rate than GamingAnywhere. The downlink bit rates of OnLive are between 3–5 Mbps, while those of SMG are between 10 and 13 Mbps. This finding indicates that the compression algorithm used by OnLive achieves a compression rate up to 4.33 times higher than that of SMG.

The values of PSNR and SSIM, respectively. We make four observations about these two figures. First, ND does not affect the quality of the video too much. Second, GamingAnywhere achieves much higher video quality than OnLive and SMG: up to 3 dB and 0.03, and 19 dB and 0.15 spaces, respectively. Third, GamingAnywhere suffers quality declines when the packet loss rate is not trivial. This can be attributed to the lack of fault tolerance mechanism in

GamingAnywhere. However, high packet loss rates are less common in modern networks. Finally, the video quality of GamingAnywhere drops suddenly when the bandwidth is less than the coding bit rate of 3 Mbps. A possible future work to address this is to add a speed adaptation heuristic to dynamically adjust the rate of coding bits, in order to use all the available bandwidth without overloading the networks.

# 3   Proposal #1

In this first proposal shown in Fig. 1 the generic architecture of the platform includes the use of GAW in the car, running in a Raspberri PI, and with GAW in the end of the client (player). In this possible solution the GAW would be embedded/connected to the FPGA in order to optimize the overall quality of the services. The LHE encoder would be embedded in the FPGA to optimize performance and minimize coding time. In this case, an ad-hoc software would have to be developed in the server as a videogame that forwards the commands produced by the player to the car.

## 3.1   Advantages

The main advantage of this architecture design of the platform is that its implementation is very simple.

## 3.2   Disadvantages

The main drawback in this scheme is that the frame is retained in GAW until it is complete, so time is lost in this task increasing the overall latency of the game.

# 4   Proposal #2

This solution shown in Fig. 2 would include the development of the following subsystems:

- development of an ad-hoc software at the end of Racingdrone to capture the video, compress it efficiently (fast and lightweight) and send it to the server.
- development of an ad-hoc software for the sending of the racingdrone control commands.
- development of ad-hoc software at the end of the client client.

## 4.1   Advantages

It is not required in this case to retain the frame in the car until it is complete, so that the overall latency is reduced.

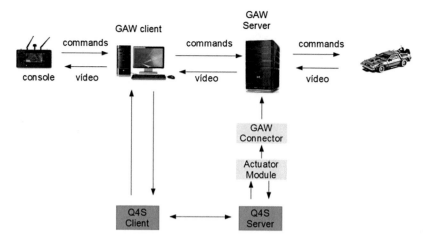

**Fig. 1.** Proposal #1 of system architecture

## 4.2 Disadvantages

In this solution proposal you have to program everything, since there would be no existing software that could carry out some of the necessary functionalities, software of the RPI, create a kind of proxy on the server side of video games, and also on the side client.

## 5 Proposal #3

In this possible solution shown in Fig. 3 the GAW is implemented in a server, but it does not carry out a transcoding, but it is used only for the handling of the control commands of the racingdrones and to forward the video flow to the authorized players.

In this possible solution it would be necessary to develop an ad-hoc software in RPI to capture the video. In addition, it would be necessary to develop ad-hoc software on the server as a videogame that forwards the commands to the car from the player interface.

### 5.1 Advantages

As advantages of this solution proposal are, first of all, that the frame is not retained on the side of the car, improving with this option the overall latency time of the system (taking into account all the factors that influence it) and, On the other hand, at the end of the client it is possible to make use of solutions that are already implemented, such as the case of the GAW Client, which although

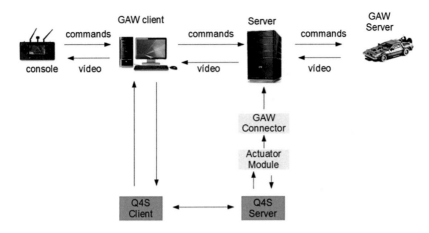

**Fig. 2.** Proposal #2 of system architecture

it is necessary to develop an integration work with the other components of the platform, allows to advance in the fastest development.

### 5.2   Disadvantages

The main drawback that appears in this solution is that the frame is retained in the server, although we have the intuition (and that its veracity will be analyzed) that it is not necessary to retain the frame in the server since it can leave the car in RTP, which it is the optimal solution that will be developed.

## 6   Results

Based on the analysis that has been carried out in Sects. 3, 4 and 5, the solution adopted, which guarantees the best quality of service and playability, is *Proposal #3*. In the decision of the choice of said solution have fundamentally influenced the following aspects: Latency: this solution is the one that provides the lowest overall latency time, as previously mentioned, because the frame is not retained at the end of the racingdrone, where the computing capacity is more compromised. In any case, it must be produced on the server. Solutions on the client side: there are solutions for the end of the client that are already developed and efficient, although it is necessary to carry out a rather intense work to be able to carry out the integration with the rest of the elements of the platform.

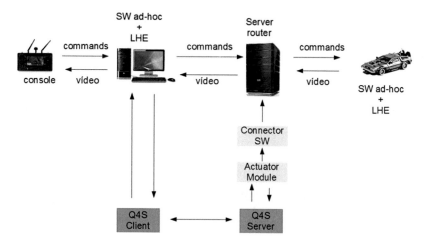

**Fig. 3.** Proposal #2 of system architecture

# 7   Conclusion

This research has the objective of analyzing the different options of platform implementation, highlighting the advantages and disadvantages of each of them in order to determine and finally propose the design specifications of the platform. We have analyzed three possible implementation schemes, each of them with particular implementation requirements that influence the total quality of service (QoS) offered to the user (player). In the decision of the solution adopted, various factors have influenced it, ultimately, impact on the quality of service offered to the user. Among these aspects is the latency time, which is influenced by aspects not only related to the quality of the communication link, but also others related to coding/decoding and in particular the influence of where this process is carried out, since It is very influential on the total latency. The use of the LHE will greatly improve the coding/decoding latency compared to other video encoders.

The use of an FPGA in which the LHE encoder is embedded will improve the coding latency and, therefore, the overall platform.

The impact of the latency on the manageability of the racingdrone has also been taken into account, so that the reaction times allow a real-time control of the vehicle that provides the user with a satisfactory experience in the game.

After carrying out the analysis of the different options taking into account the restrictions they impose, it has been found that the most optimal solution is *Proposal #3*, whose diagram was shown in Fig. 3 and based on this, the specifications of the platform were defined.

Depending on the system architecture and the aspects that affect the QoS, the limit values of the parameters that must be preserved have been determined so that the player has a good experience in the development of the game.

**Acknowledgments.** The authors wish to thank the Spanish Science and Innovation Ministry which funds this research through "INNPACTO" innovation program RTC-2016-4744-7.

# References

1. http://gaminganywhere.org/perf.html
2. http://gaminganywhere.org/perf.html#onlive
3. http://gaminganywhere.org/perf.html#smg
4. http://gaminganywhere.org/perf.html#batman
5. http://gaminganywhere.org/perf.html#fear
6. http://gaminganywhere.org/perf.html#dow
7. http://www.gaminganywhere.org/perf.html#CC06

# Optimised Scheduling Algorithms and Techniques in Grid Computing

Rajinder Vir[1]([envelope]) [ORCID], Rajiv Vasudeva[1] [ORCID], Vikrant Sharma[1] [ORCID],
and Sandeep[2] [ORCID]

[1] GNA University Phagwara, Dhak Khati, India
rvir75@gmail.com
[2] DAVIET, Jalandhar, India

**Abstract.** This paper provides an overview of grid computing, grid scheduling and grid scheduling algorithms. In this paper various study and comparison of Min-Min algorithm and IACO algorithm has been described. The problem to make full use of all types of resources in the grid tasks scheduling has been discussed in this paper. Ant Colony Algorithm in grid computing is described. The paper discusses the problem to make full use of all types of resources in the grid tasks scheduling. When a large number of tasks request the grid resources, according to the task type of the adoption of appropriate strategies, the different tasks are assigned to the appropriate resources nodes to run it, in order to achieve the optimum utilization of resources. As the heterogeneous and dynamic of grid environment, while the different requirements for applications, making the tasks scheduling become extremely complex, the algorithm will have a direct impact on task execution efficiency of the grid environment, as well as the success or failure.

**Keywords:** Grid computing · Scheduling · Ant colony optimization · Grid tasks and grid scheduling algorithms

## 1 Introduction

### 1.1 Grid Computing

Grid is defined as a type of parallel and distributed system that enables the sharing, selection, and aggregation of geographically distributed autonomous resources dynamically at runtime depending on their availability, capability, performance, cost, and users' quality-of-service requirements [1]. Grid technology is a growing information technology, where that the main purpose of grid is to build a kind of dynamic, distributed and heterogeneous computing environment and realize collaborative resource sharing and problem solving in dynamic and multiple virtual organizations [2]. In Fig. 1 the grid computing environment is shown.

Grid computing enables large-scale resource sharing and provides a promising platform for executing tasks efficiently. In order to get optimal resource utilization and avoid overload, task scheduling with load balancing in grid computing is of interest to both users and grid systems. Load balancing mechanism aims to distribute workload

© Springer Nature Switzerland AG 2019
C. Benavente-Peces et al. (Eds.): SEAHF 2019, SIST 150, pp. 231–244, 2019.
https://doi.org/10.1007/978-3-030-22964-1_24

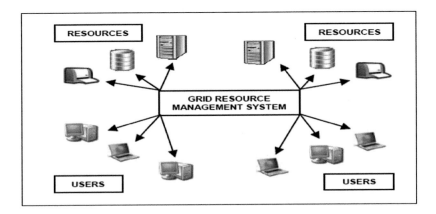

**Fig. 1.** Grid computing environment

evenly among computing nodes and minimize the total task execution time [3]. However, the scheduling problem with load balancing is usually computationally intractable. Grid computing is a technology that represents a kind of distributed computing. This approach is mainly used in the context of resource utilization and sharing across geographical boundaries. It helps in coordinating computing capabilities, data storage and network resources across dynamically dispersed organizational structures.

## 1.2  Grid Scheduling

Scheduling is a key concept in computer multitasking operating systems, multiprocessing operating system design and in real-time operating system design. In modern operating systems, there are typically many more processes running than there are CPUs available to run them. Scheduling refers to the way processes are assigned to run on the available CPUs [4]. This assignment is carried out by software known as a scheduler or is sometimes referred to as a dispatcher.

Grid Scheduling is the mapping of jobs to resources as per their requirements and properties. It is proposed to solve large complex problems. Grid scheduling is an intelligent algorithm capable of finding the optimal resource for processing a job [5]. The objectives of a grid scheduler are overcoming heterogeneousness of computing resources, maximizing overall system performance, such as high resource, utilization rate and supporting various computing intensive applications, such as batch jobs and parallel applications.

## 1.3  Grid Scheduling Algorithms

The resource scheduling in grid is a NP complete problem. The scheduling problem becomes more challenging because of some unique characteristics belonging to Grid computing. Various algorithms have been designed to schedule the jobs in computational girds. The most commonly used algorithms are OLB, MET, MCT, Min-Min, Max-Min, and MaxStd [6].

**Opportunistic Load Balancing (OLB)** Assigns each job in arbitrary order to the machine with the shortest schedule, irrespective of the ETC on that machine. OLB is intended to try to balance the machines, but because it does not take execution times into account it finds rather poor solutions.

**Minimum Execution Time (MET)** assigns each job in arbitrary order to the machine on which it is expected to be executed fastest, regardless of the current load on that machine. MET tries to find good job-machine pairings, but because it does not consider the current load on a machine it will often cause load imbalance between the machines.

**Minimum Completion Time (MCT)** assigns each job in arbitrary order to the machine with the minimum expected completion time for the job. The completion time of a job j on a machine m is simply the ETC of j on m added to m's current schedule length. This is a much more successful heuristic as both execution times and machine loads are considered.

**Min-min** establishes the minimum completion time for every unscheduled job (in the same way as MCT), and then assigns the job with the minimum completion time (hence Min-min) to the machine which offers it this time. Min-min uses the same intuition as MCT, but because it considers the minimum completion time for all jobs at each iteration it can schedule the job that will increase the overall makespan the least, which helps to balance the machines better than MCT.

**Max-min** is very similar to Min-min. The minimum completion time for each job is established, but the job with the maximum minimum completion time is assigned to the corresponding machine. Max-min is based on the intuition that it is good to schedule larger jobs earlier on so they won't "stick out" at the end causing a load imbalance. However experimentation shows that Max-min cannot beat Min-min on any of the test problems used.

**Maximum Standard deviation (MaxStd)** In MaxStd heuristic, the task having the highest standard deviation of its expected execution time is scheduled first. The task having low standard deviation of task execution has less variation in execution time on different machines and hence, its delayed assignment for scheduling will not affect overall makespan much. Moreover, the task with higher standard deviation of task execution time exhibits more variation in its execution time of different machines. A delayed assignment of such tasks might hinder their chances of occupying faster machines as some other tasks might occupy these machines earlier. It would increase the system makespan. The task having high standard deviation is assigned to the machine which has minimum completion time.

## 1.4 Ant Colony Optimization

Ants have inspired a number of methods and techniques among which the most studied and the most successful is the general purpose optimization technique known as ant colony optimization. Ant colony algorithm called ant colony optimization method (Ant Colony Optimization, ACO) [2], by the Italian scholar M. Dorigo, who first proposed, is a simulated evolutionary algorithm based on population.

The laying down of trails is shown in Fig. 2.

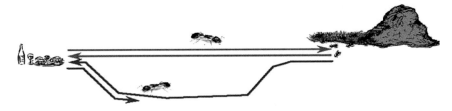

**Fig. 2.** Ants deposit pheromone on the ground

The ant colony optimization algorithm (ACO) is a probabilistic technique for solving computational problems which can be reduced to finding good paths through graphs [7]. The advantages of the algorithm are the use of the positive feedback mechanism, inner parallelism and extensible.

### 1.5    Ant Algorithm for Grid Scheduling Problem

The ACO algorithm uses a colony of artificial ants that behave as co-operative agents in a mathematical space were they are allowed to search and reinforce pathways (solutions) in order to find the optimal ones. Solution that satisfies the constraints is feasible [7]. After initialization of the pheromone trails, ants construct feasible solutions, starting from random nodes, then the pheromone trails are updated. At each step ants compute a set of feasible moves and select the best one (according to some probabilistic rules) to carry out the rest of the tour. The transition probability is based on the heuristic information and pheromone trail level of the move. The higher value of the pheromone and the heuristic information, the more profitable it is to select this move and resume the search. In the beginning, the initial pheromone level is set to a small positive constant value $\tau_0$ and then ants update this value after completing the construction stage. The ant algorithm for grid is [8]:

**procedure** ACO

 **begin**

   Initialize the pheromone

   **while** stopping criterion not satisfied **do**

    Position each ant in a starting node

    **repeat**

    **for** each ant **do**

     Chose next node by applying the state transition rate

    **end for**

   **until** every ant has build a solution

   Update the pheromone

  **end while**

 **end**

After the initialization of the pheromone trails and control parameters, a main loop is repeated until the stopping criteria are met. The stopping criteria can be a certain number of iterations or a given CPU time limit or time limit without improving the result. In the main loop the ants construct feasible solutions and then the pheromone trails are updated. More precisely, partial problem solutions are seen as states: each ant starts from random state and moves from state i to another state j of the partial solution. At each step, ant k computes a set of feasible solutions to its current state and moves to one of these expansions, according to a probability distribution specified as follows. For ant k the probability $p^k_{ij}$ to move from a state i to a state j depends on the combination of two values [8]:

$$p^k_{ij} = \begin{cases} \dfrac{\tau_{ij}\eta_{ij}}{\sum t \in \text{allowed}_k \ \tau_{ij}\eta_{ij}} & \text{if } j \in \text{allowed}_k \\ 0 & \text{otherwise} \end{cases}$$

where

- $\eta_{ij}$ is the attractiveness of the move as computed by some heuristic information indicating a prior desirability of that move;
- $\tau_{ij}$ is the pheromone trail level of the move, indicating how profitable it has been in the past to make that particular move (it represents therefore a posterior indication of the desirability of that move);
- $\text{allowed}_k$ is the set of remaining feasible states.

The higher the value of the pheromone and the heuristic information, the more profitable it is to include state j in the partial solution. In the beginning, the initial pheromone level is set to $\tau_0$, which is a small positive constant. In the nature there is not any pheromone on the ground at the beginning, or the initial pheromone in the nature is $\tau_0 = 0$. If in ACO algorithm the initial pheromone is zero, than the probability to chose next state will be $p^k_{ij} = 0$ and the search process will stop from the beginning. Thus it is important the initial pheromone to be positive value. The pheromone level of the elements of the solutions is changed by applying following updating rule [9]:

$$\tau_{ij} \leftarrow \rho.\tau_{ij} + \Delta\tau_{ij},$$

where the rule $0 < \rho < 1$ models evaporation and $\Delta\tau_{ij}$ is an additional pheromone and it is different for different ACO algorithms. The value of evaporation rate $\rho$ is adaptively changed and a minimum value $\rho_{min}$ is assigned. The evaporation rate $\rho$ is under control and will never be reduced to 0 [4].

## 1.6   Grid Scheduling Model

The grid scheduler's aim is to allocate the jobs to the available nodes. The grid scheduler finds out the better resource of a particular job and submits that job to the selected host. The best match must be found from the list of available jobs to the list of available resources. The selection is based on the prediction of the computing power of the resource. The grid scheduler must allocate the jobs to the resources efficiently. The efficiency depends upon two criteria; one is makespan and the other is flow time. These two criteria are very much important in the grid system. The makespan measures the throughput of the system and flow time measures its Quality of Service (QoS).

The following assumptions are made for the algorithm [10]. The collection of independent tasks with no data dependencies is called as a meta-task. Each machine executes a single task at a time. The meta-task size is one and the numbers of machines are 'm'. The selecting strategy can be based on the prediction of the computing power of the host. Some terms used in grid scheduling model are:

Expected execution time $ET_{ij}$ of task $t_i$ on machine $m_j$ is defined as the amount of time taken by $m_j$ to execute $t_i$ given that $m_j$ has no load when $t_i$ is assigned. The expected completion time $CT_{ij}$ of the task $t_i$ on machine $m_j$ is defined as the wall-clock time at which $m_j$ completes $t_i$ (after having finished any previously assigned tasks). Let M be the total number of the machines. Let S be the set containing the tasks. Let the beginning time of $t_i$ be $b_i$. From the above definitions, $CT_{ij} = b_i + ET_{ij}$. The makespan for the complete schedule is then defined as $\max_{t_i \in S}(CT_{ij})$. Makespan is a measure of the throughput of the heterogeneous computing system. The objective of the grid scheduling algorithm is to minimize the makespan [8].

The algorithm is for batch mode mapping. Let the number of the tasks in the set of tasks is greater than the number of machines in the grid. The result will be triples (task,machine, startingtime). The function free(j)—shows when the machine $m_j$ will be free. If the task $t_i$ is executed on the machine $m_j$ then the starting time of $t_i$ becomes $b_i$ = free(j) + 1 the new value of the function free(j) becomes free(j) = $b_i$ +$ET_{ij}$ = $CT_{ij}$.An important part of implementation of ACO algorithm is the graph of the problem [8, 10].

Let M = $\{m_1, m_2, \ldots, m_m\}$ is the set of the machines and t = $\{t_1, t_2, \ldots, t_s\}$ is the set of the tasks and s > m. Let $\{T_{ij}\}_{s \times m}$ is the set of the nodes of the graph and to machine $m_j \in M$ corresponds a set of nodes $\{T_{kj}\}_k^s$ = 1. The graph is fully connected. The problem is to choose s nodes of the graph thus to minimize the function F = max (free(j)), where $[b_i, CT_{ij}] \cap [b_k, CT_{kj}] = \emptyset$ for all i, j, k. Several ants are used and every ant starts from random node to create their solution. There is a tabu list corresponding to every ant. When a node $T_{ij}$ is chosen by the ant, the nodes $\{T_{ik}\}_k^m$ = 1 is included in tabu list. This prevents the possibility of the task $t_i$ to be executed more than ones. An ant adds new nodes in the solution till all nodes are in the tabu list as:

$$\eta_{ij} = \frac{1}{free(j)}.$$

At the end of every iteration we calculate the objective function $F_k$ = max(free(j)) over the solution constructed by ant k and the added pheromone by the ant k is:

$$\Delta\tau_{ij} = \frac{(1 - \rho)}{F_k}.$$

In the next iterations the elements of the solution with less value of the objective function will be more desirable. This ACO implementation is different from ACO implementation on traditional tasks machines scheduling problem. The new of the implementation is using of multiple node corresponding to one machine. It is possible because in grid scheduling problem every machine can execute any task. Two kind of sets of tasks are needed: set of scheduled tasks and set of arrived and unscheduled tasks. When the set of scheduled tasks becomes empty the scheduled algorithm is started over the tasks from the set of unscheduled tasks.

## 1.7  GridSim-Grid Modeling and Simulation Toolkit

Simulation has been used extensively for modeling and evaluation of real world systems, from business process and factory assembly lines to computer systems design. The notion of a metatask as a collection of independent jobs with no inter-job dependencies is defined to minimize the total execution time of the metatask [9]. As the scheduling is performed statically all necessary information about the jobs in the metatask and machines in the system is assumed to be available a priori. Essentially, the expected running time of each individual job on each machine must be known, and this information can be stored in an 'expected time to compute' (ETC) matrix [11].

GridSim is a toolkit for the modeling and simulation of distributed resource management and scheduling for Grid computing, which is developed by Dr. Rajkumar Buyya and his research group from Grid Computing and Distributed Systems Laboratory, the University of Melbourne, Australia [12]. The GridSim toolkit provides a comprehensive facility for simulation of different classes of heterogeneous resources, users, applications, resource brokers, and schedulers. Application schedulers in the Grid environment, called resource brokers, perform resource discovery, selection, and aggregation of a diverse set of distributed resources for an individual user. This means that each user has his or her own private resource broker and hence it can be targeted to optimize for the requirements and objectives of its owner.

## 2  Problem Formulation

The paper discusses the problem to make full use of all types of resources in the grid tasks scheduling. When a large number of tasks request the grid resources, according to the task type of the adoption of appropriate strategies, the different tasks are assigned to the appropriate resources nodes to run it, in order to achieve the optimum utilization of resources [11]. As the heterogeneous and dynamic of grid environment, while the different requirements for applications, making the tasks scheduling become extremely complex, the algorithm will have a direct impact on task execution efficiency of the grid environment, as well as the success or failure.

As the grid tasks scheduling faced with an NP complete problems [4]. At present, tasks scheduling algorithms are mainly based on Minimum Completion Time (MCT), User-Directed Assignment (UDA), and Minimum Execution Time (MET) [6]. This paper studies two kinds of algorithms, Min-min algorithm and ant colony algorithm, and introduces an improved ant colony algorithm IACO.

In a grid environment, tasks scheduling process involves finding suitable computing resources according to application requirements, and then from the computing resources choose a most appropriate resource according to the scheduling strategy. After obtained the right resources, the task runs on the resources, and subjects to local resource management mechanisms. The resources should be paid back to grid resource institutions when the tasks have been completed. Grid task management modules return the results and relevant information to the users.

Scheduling includes two aspects: the task mapping and task scheduling. Task mapping is the task of machine matching, match is a logical process, not to transmit; based on task mapping results, tasks scheduling is the process that sent the task to the execution queue of designated machine [8]. In the paper tasks scheduling refers to the tasks mapping.

The essence of the tasks scheduling is to schedule the task that needs m assigned to n available resources reasonably, to obtain the minimum total execution time. This paper introduces the scheduling model that consisted of a triple triples, namely $M = (J, H, f)$, where J said that the task collection that the user submitted, denoted by $J = (J_1, J_2, ..., J_m)$, $J_i$ said that the task i; H represented a resource node, denoted by $H = (H_1, H_2, ..., H_n)$; f is the objective function.

The tasks of the grid can be divided into compute-intensive, correspondent-intensive and compute-correspondent relatively balanced [3]. In order to describe the problem simple, the text referred to compute-intensive tasks in particular, that task of computing time is much larger than the data communication time.

Related definitions and assumptions are as follows:

Suppose 1: Between tasks independent of each other, there is no dependence.

Suppose 2: A task at the same time can only be assigned to a resource node, a resource node can only perform a task at the same time.

**Definition 1** Define R (j) storing ready waiting time of node $H_j$.

**Definition 2** The definition of E (i, j), said the expected execution time that $J_i$ at the $H_j$, the matrix E (i, j) stores the execution time that the first i-task on the first j node.

**Definition 3** Define C (i, j), said $J_i$ at the $H_j$ 's earliest completion time, and C (i, j) = R (j) + E (i, j), matrix C (i, j) stores the earliest completion time that the first i-task on the first j node.

**Definition 4** Makespan, namely, the completion time that all of the tasks are assigned to the corresponding node by scheduling program.

**Definition 5** f = min (Makespan).

There are two criteria to evaluate scheduling algorithms good or bad in this paper:

(1) Makespan: Makespan is smaller, indicating the performance of scheduling algorithm is better and the execution is more efficient.
(2) Resource Load: refers to the system resources load equilibrium condition. If a scheduling algorithm cannot fully utilize all kinds of resources in a grid environment and assign tasks to machines of good performance, the load on these machines will be very high. When the load exceeds a certain threshold value, the machine may crash. Therefore, whether the resource load balanced is also one of the evaluations.

## 2.1    Description of Algorithms

### 2.1.1    Min-Min Algorithm

Min-min algorithm [6] completes heuristic choice of resources through the calculation of the minimum two. Its purpose is to assign a large number of tasks to computing resources that complete them earliest and execute them fastest to enable the minimum task completion time.

For each task, compute the earliest completion time C (i, j) that the task allocate to the n resources, find the smallest C(i, j), signed $C_{min}$ (i,j); and then find the smallest $C_{min}$ (i,j) from all of the $C_{min}$ (i, j), namely the smallest task, schedule the smallest task $J_i$ on the resource of $H_j$ and update R (j); finally remove the task i from the tasks waiting for scheduling collection, adjust the earliest completion time matrix C (i, j), to find the smallest tasks again. Repeat, until the task waiting for scheduling collection is empty.

Min-min algorithm is able to quickly determine the computing resource of the smallest tasks, but likely to cause a result that schedules small task to the computing nodes of better performance excessively. It will cause that computing nodes of poor performance are idle. It is unable to balance the resource load, so the efficiency of execution is not high.

### 2.1.2    The Basic Ant Colony Algorithm

Ant colony algorithm called ant colony optimization method (Ant Colony Optimization, ACO) [7], by the Italian scholar M. Dorigo, who first proposed, is a simulated evolutionary algorithm based on population. The algorithm imitates the behavior of ants when foraging through the pheromone to adjust the selection of solution, and gradually converges to the global optimal solution of the problem.

The advantages of the algorithm are the use of the positive feedback mechanism, inner parallelism and extensible. The disadvantages are: (1) initial pheromones shortage, slow.

(2) the stagnation phenomenon, or searching for to a certain extent, all individuals found the same solution exactly, cannot further search for the solution space, make the algorithm converge to local optimal solution [8].

### 2.1.3    Improved Ant Colony Algorithm (IACO)

For the shortcomings of Ant colony algorithm, this paper introduces an improved ant colony algorithm (IACO). First initialize the pheromone with Min-min algorithm to make up for the lack of initial pheromone, and then improve the parameters of the ant colony algorithm to avoid falling into the part of local optimum. Results show IACO algorithm has a smaller Makespan, resource load balancing.

The procedure of IACO is described as follow:

Step 1    Obtain the earliest completion time by Min-Min algorithm, conduct a pre-allocation of tasks to find the resources to complete its first, record the resource node matching probability, denoted by pro (j), the times that each node is matched are $M_j$, the total times are m. The formula is given by (1).

$$pro(j) = M_i/m \tag{1}$$

Step 2    Initialize pheromone of each resource node $\tau_j(0) = a * L + b * T + \text{pro}(j)$, in which, L and T express separately computing ability and communication ability of the resource, a, b, respectively is computing ability pheromone and communication ability pheromone, and $0 < a < 1$, $0 < b < 1$, $a + b = 1$.

Step 3    Create an ant for each task. When the task requests resources, the probability of resource j being assigned to the task is determined by the pheromone density of resource j, according to the formula (2), then record the visited resource.

$$P_j(t) = \begin{cases} \dfrac{[\tau_j(t)]^\alpha \times [\eta_j]^\beta}{\sum\limits_U \{[\tau_U(t)]^\alpha \times [\eta_U]^\beta\}} & U \in resource \text{ left} \\ 0 & others \end{cases} \tag{2}$$

where $\tau_j(t)$ is the pheromone of resource j at time t; $\eta_j$ is the intrinsic property of resource j which equals to $\tau_j(0)$; $\alpha$ is the information inspiration factor, indicating importance of pheromone, and $\alpha = 0.5$; $\beta$ is the hope inspiration factor, indicating the importance of the inherent properties of resources, and $\beta = 1$.

Step 4    When each ant assigned the task to a resource, the pheromone of resource will change accordingly, that is, local pheromone updates. At the same time get a current solution. Suppose the first i ant selected resource node j, then the resource node j according to formula (3) updates pheromone:

$$\tau_j(t+1) = (1 - \rho)\tau_j(t) + \tau_j(t) \tag{3}$$

Step 5    If the ants were traversing each of the available resources, continue to step 6; otherwise, jump back to step 3.

Step 6    Calculate the objective function and get the current optimal solution. Restore the initial value of pheromone, then update the global pheromone of the resource node on the current optimal match according to the formula (4). The global pheromone updating means that when get the current optimal solution update global pheromone before the next resource selection.

$$\tau_j(t+1) = (1 - \rho)\tau_j(t) + \rho \cdot \Delta\tau_j^{gb}(t)$$
$$\Delta\tau_j^{gb}(t) = \frac{1}{S_{gb}} \tag{4}$$

where $\rho$ is pheromone evaporation parameter, $0 < \rho < 1$, in this paper $\rho = 0.7$, $S_{gb}$ is the current optimal solution.

Step 7    Limit the amount of the pheromone in a certain range. To prevent "premature" phenomenon, avoiding the pheromone is too high or too low, this article made reference to the MMAS [4] in the limited way to constrain the pheromone concentration. The amount of pheromone is limited to the sector [τmin, τmax], to avoid the difference between the pheromones is too large in the process of the algorithm. For all the pheromones, if $\tau_j(t) > \tau_{max}$, then $\tau_j(t) = \tau_{max}$; if $\tau_j(t) < \tau_{min}$, then $\tau_j(t) = \tau_{min}$. Set $\tau_{min} = 0.01$, $\tau_{max} = 100$.

Step 8    NC = NC + 1.

Step 9    Up to satisfy the terminal condition, output the global optimal solution; otherwise, repeat step 3–8.

## 2.2   Simulation Results

This paper uses GridSim simulator to carry out the algorithm simulation experiments. GridSim is a toolkit for the modeling and simulation of distributed resource management and scheduling for Grid computing, which is developed by Dr. Rajkumar Buyya and his research group from Grid Computing and Distributed Systems Laboratory, the University of Melbourne, Australia [12]. Simulation system makes the simulation work simple by avoiding a huge overhead of the resource scheduling process in the real system, and this is very conducive to algorithm research.

GridSim is extended on the basis of the SimJava to simulate the heterogeneous grid resources and application tasks. GridSim can simulate the resources of time-sharing and space-sharing; can define load, performance and even time zones and other parameters, quite robust. GridSim did not specifically limit the use of any allocation strategy or model, can add it according to their own specific circumstances.

In this paper two experiments were conducted to simulate the efficiency of the improved ant algorithm for independent tasks scheduling.

**Experiment 1**: Design a grid environment, which contains 7 resources and the number of tasks at 5 to 95, the number of each task is from 500 to 1,000, respectively use Min-Min algorithm and IACO algorithm to schedule. Compare the algorithms makespan under the same number of resources, Fig. 3.

**Results of experiment 1**: The results show that with the increase in the number of tasks, task completion time will increase, and IACO algorithm completion time than Min-Min algorithm is shorter.

**Experiment 2**: Design a grid environment, which contains 7 resources and the number of tasks is 455, the number of each task is from 500 to 1,000, respectively use Min-Min algorithm and IACO algorithm to schedule. Compare the resources load under the same total number of tasks, Fig. 4.

**Results of experiment 2**: The results show that using Min-Min algorithm for tasks scheduling, resources load are imbalance in serious condition, and some resources, overloading, while some resources load are too light. IACO algorithm can achieve a balanced utilization of resources, will not present the resources load to be excessively light or overweight.

Experimental results show that compared with Min-Min algorithm and IACO algorithm, IACO algorithm not only can get better results, but will balance the load between resources. With the increase in the number of tasks, IACO algorithm is more obvious advantages in completion time.

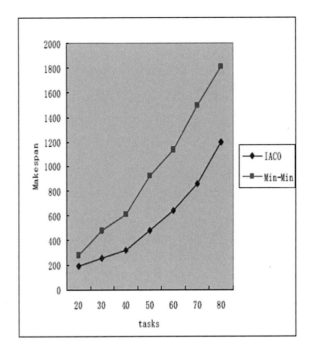

**Fig. 3.** Comparison of the makespan of IACO and Min-Min

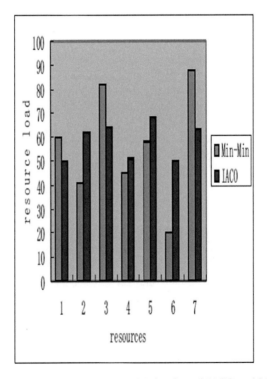

**Fig. 4.** Comparison of resource load balancing of IACO and Min-Min

# 3  Conclusions and Future Work

Grid technology is a growing information technology, where the main purpose of grid is to build a kind of dynamic, distributed and heterogeneous computing environment and realize collaborative resource sharing and problem solving in dynamic and multiple virtual organizations. Grid Scheduling is the mapping of jobs to resources as per their requirements and properties. It is proposed to solve large complex problems. Grid scheduling is an intelligent algorithm capable of finding the optimal resource for processing a job.

Task scheduling is an integrated part of parallel and distributed computing. Intensive research has been done in this area and many results have been widely accepted. With the emergence of the computational grid, new scheduling algorithms are in demand for addressing new concerns arising in the grid environment. In this environment the scheduling problem is to schedule a stream of applications from different users to a set of computing resources to maximize system utilization. This scheduling involves matching of applications needs with resource availability. There are three main phases of scheduling on a grid. Phase one is resource discovery, which generates a list of potential resources. Phase two involves gathering information about those resources and choosing the best set to match the application requirements. In the phase three the job is executed, which includes file staging and cleanup. In the second phase the choice of the best pairs of jobs and resources is NP-complete problem.

The scheduling problem becomes more challenging because of some unique characteristics belonging to Grid computing. Various algorithms have been designed to schedule the jobs in computational girds. The most commonly used algorithms are OLB, MET, MCT, Min-Min, Max-Min, and MaxStd. For the complexity of tasks scheduling in grid many algorithms have been developed. The improved ant colony algorithm (IACO) solves the problems of ant colony algorithm that the deprivation of initial pheromone and easily trapping into local optimization. Experiments with this algorithm show that the algorithm can increase scheduling efficiency, and has improved overall performance of grid tasks scheduling. In the future work the algorithm can be applied to scheduling problem of dependent tasks and testing its performance.

# References

1. Foster, I., Kesselman, C.: The anatomy of the grid. Int. J. Supercomput. Appl. 1–25 (2001)
2. Zhu, Y., Wei, Q.: An Improved Ant Colony Algorithm for Independent Tasks Scheduling of Grid, vol. 2, pp. 566–569. IEEE (2010)
3. Chang, R.S., Chang, J.S., Lin, P.S.: Balanced job assignment based on ant algorithm for computing grids. In: Asia-Pacific Services Computing Conference, pp. 291–295 (2007)
4. Chen, M.: Toward adaptive ant colony algorithm. In: International Conference on Measuring Technology and Mechatronics Automation, pp. 1035–1038 (2010)
5. Lorpunmanee, S., Sap, M.N., Abdullah, A.H., Inwai, C.C.: An ant colony optimization for dynamic job scheduling in grid environment. World Academy of Science Engineering and Technology, pp. 314–321

6. Braun, T.D., Siegel, H. J., Beck, N., Boloni, L.L., Maheswaran, M., Reuther, A.I., Robertson, J.P., Theys, M.D., Yao, B., Hensgen, D., Freund, R.F.: A comparison of eleven static heuristics for mapping a class of independent tasks onto heterogeneous distributed computing systems. J. Parallel Distrib. Comput. **61**(6), 810–837 (2001)

7. Dorigo, M., Stutzle, T.: The ant colony optimization metaheuristic: Algorithms, applications and advances. In: International Series in Operations Research and Management Science, vol. 57, pp. 251–285 (2002)

8. Fidanova, S., Durchova, M.: Ant Algorithm for Grid Scheduling Problem, pp. 405–412. Springer (2006)

9. Ritchie, G., Levine, J:. A hybrid ant algorithm for scheduling independent jobs in heterogeneous environments. J. Am. Associat. Artif. Intell. 178–184 (2004)

10. Kousalya, K., Balasubramanie, P.: To improve ant algorithm for grid scheduling using local search. Int. J. Comput. Cogn. **7**(4), 47–57

11. Yan, H., Shen, X.Q., Li, X., Wu, M.H.: An improved ant algorithm for job scheduling in grid computing. In: Proceedings of the Fourth International Conference on Machine Learning and Cybernetics, pp. 2957–2961 (2005)

12. Buyya, R., Murshed, M.: GridSim: a toolkit for the modeling, and simulation of distributed resource management, and scheduling for grid computing. Concurr. Computat. Pract. Exper. **14**, 1175–1220 (2002)

13. Bai, L., Hu, Y., Lao, S., Zhang, W.: Task scheduling with load balancing using multiple ant colonies optimization in grid computing. In: Sixth International Conference on Natural Computation (ICNC), pp. 2715–2719 (2010)

14. Tang, B., Yin, Y., Liu, Q., Zhou, Z.: Research on the application of ant colony algorithm in grid resource scheduling. Wirel. Commun. Netw. Mobile Comput. 1–4 (2008)

15. Author, F.: Contribution title. In: 9th International Proceedings on Proceedings, pp. 1–2. Publisher, Location (2010)

16. LNCS Homepage. http://www.springer.com/lncs. Last accessed 21 Nov 2016

# Deep Learning Algorithm for Predictive Maintenance of Rotating Machines Through the Analysis of the Orbits Shape of the Rotor Shaft

R. Caponetto[1] , F. Rizzo[2], L. Russotti[2], and M. G. Xibilia[3(✉)]

[1] University of Catania, DIEEI, Catania, Italy
[2] TEORESI S.p.A., Catania, Italy
[3] Department of Engineering, University of Messina, Messina, Italy
mxibilia@unime.it

**Abstract.** The prediction of failures in rotating machines is an important issue in industries to improve safety, to reduce the cost of maintenance and to prevent accidents. In this paper a predictive maintenance algorithm, based on the analysis of the orbits shape of the rotor shaft is proposed. It is based on an autonomous image pattern recognition algorithm, implemented by using a Convolutional Neural Network (CNN). The CNN has been designed, by using a suitable database, to recognize the orbits shape, allowing both fault detection and classification.

**Keywords:** Maintenance · Deep learning · Convolution neural network · Fault detection

## 1 Predictive Maintenance of Rotating Machines

An effective predictive maintenance program is measured by how well it works to predict imminent failures, identify their causes and avoid expensive recovery costs.

Electrical testing of motors and rotating machines is a critical component of predictive maintenance. Static and dynamic analysis, along with data acquisition and analysis, provide the information needed to take good decisions regarding the maintenance of each motor.

Many approaches have been proposed in literature. In [1], the frequency analysis is the suggested approach, while in [2] soft computing techniques are introduced and in [3] the use of wireless sensors is proposed for application in hazardous area. Recently, image processing approaches are attracting more and more interest [4].

To accomplish predictive maintenance, suitable sensors must be added to the system, to monitor and collect data about its operations. In the approach proposed in this paper the acquired data are images representing the orbits of the motor axis. The goal of predictive maintenance is to predict at the time $t$, using the data up to that time, whether the equipment will fail in the near future.

© Springer Nature Switzerland AG 2019
C. Benavente-Peces et al. (Eds.): SEAHF 2019, SIST 150, pp. 245–250, 2019.
https://doi.org/10.1007/978-3-030-22964-1_25

Predictive maintenance can be formulated by using two different approaches:

- Classification approaches: predict whether there is a possibility of failure in next n-steps.
- Regression approaches: predict how much time is left before the next failure, in other words the Remaining Useful Life (RUL).

The above-mentioned methods need the collection of data directly on the real system but is possible also to build predictive maintenance algorithms on the basis of a dataset obtained from the model of the system or of the fault.

In this paper, faults occurring in a rotating machine are considered. The orbit shapes of the rotor shaft are used, by means of a CNN, to detect and classify the faults. Mathematical models of the faulty orbits are used to create a database of orbit shapes on which a CNN can be trained.

## 2    Deep Learning Algorithm for Predictive Maintenance

Machine learning methods [5], can create category (or class) of patterns from raw data, building auto-cognitive systems for a given task.

A variety of machine learning algorithms have been used to diagnose faults in rotating machines. They can be used to extract fault symptoms from a suitable dataset, which can be extracted automatically from the learning algorithm or using empirical knowledge which came from direct experience of engineers.

Representation learning is a set of methods that allow a machine to be fed with raw data and automatically discover the representation needed for detection or classification. Among these, deep learning methods are representation learning methods with multiple levels of abstraction, obtained by composing simple, but non-linear, modules, each transforming the representation of the lower level (starting with the raw input) into a representation at a higher, more abstract level. With the composition of such transformations, very complex functions can be learned, and good feature can be automatically extracted using general-purpose learning procedure. This is key advantage of deep learning. Among the possible deep learning structures, Convolutional Neural Networks are used in this paper [5].

## 3    Orbits Shape Classification and Modelling on MATLAB

In rotating machines, different causes of faults and symptoms can be considered. The most common kind of faults are unbalance, shaft misalignment and oil whirl in a rotor shaft. For each type of fault there are a corresponding orbit shape of the shaft, which, in ideal condition is a circle. Possible orbits due to the different kind of fault are an ellipse, a heart, a tornado and an eight, as represented in Fig. 1.

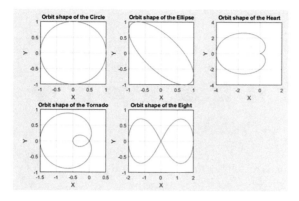

**Fig. 1.** Orbits shape

So, to our purpose there are 5 classes of orbits shape:

(1) Circle.
(2) Ellipse.
(3) Heart.
(4) Tornado.
(5) Eight.

The circle is related to normal condition of the rotor shaft, the ellipse is due to an unbalance of the rotor, while the heart, the tornado and the eight are associated to a misalignment on the rotor shaft.

To realize the different orbits shape, that will be used as a learning dataset for a CNN, their mathematical models have been implemented in the MATLAB environment. By changing suitable parameters, different orbits shapes, belonging to the same class have been generated, as described in the next Section.

## 4    Implementation, Training and Test of the CNN

The five steps that have been followed for the implementation of the CNN are:

1. Load and explore image data.
2. Define the network architecture.
3. Specify training options.
4. Train the network.
5. Predict the labels of new data and calculate the classification accuracy.

On the basis of the 5 classes of possible orbit shapes, 5000 images (1000 for each class), collected in a dataset for the CNN training have been created. An example is shown in Fig. 2.

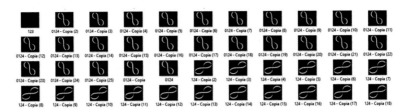

**Fig. 2.** Images for class of orbits shape

After a trial and error phase, the CNN structure, implemented by using the Neural Toolbox of MATLAB has been fixed as follows:

1. Image Input Layer
2. Convolutional Layer
3. Batch Normalization Layer
4. ReLU Layer
5. Max Pooling Layer
6. Convolutional Layer
7. Normalization Layer
8. ReLU Layer
9. Max Pooling Layer
10. Convolutional Layer
11. Batch Normalization Layer
12. ReLU Layer
13. Fully Connected Layer
14. Softmax Layer
15. Classification Layer

The Convolutional Layer has the filter size 3-by-3. The number of filters is 8 in the 2nd layer, is 16 in the 6th layer and 32 in the 10th layer, moreover by using the 'Padding' name-value pair is possible to add the padding to the input feature map. For a filter size of 3, a padding of 1 ensures that the spatial output size is the same as the input size.

Different number of epochs are used during the learning. The results obtained in the training process after 5 epochs are reported in Fig. 3. An accuracy of the 0.9960 is obtained, i.e. the 99.6% of the images given to the network are correctly classified.

Best results have been obtained after 15 epochs and an accuracy of the 0.9992 was obtained, as reported in Fig. 4.

The training time is around 4 min with a single CPU of a commercial PC. It can be reduced using GPU based machines.

After the CNN training, it is necessary to validate its performance on a new dataset. The validation dataset contains 250 new orbits shapes that are translated/rotated with respect to the centre, to obtain a robust validation dataset.

In test phase the CNN was able to recognize the new input images in the correct category of fault with an accuracy of the 99.9%. The cases of incorrect clustering can

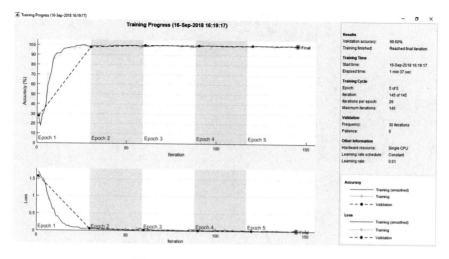

**Fig. 3.** Training process for 5 epochs

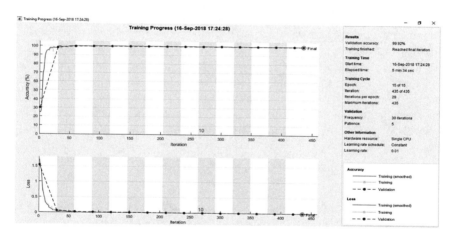

**Fig. 4.** Training process for 15 epochs

be eliminated adding more orbits shape in the Dataset. The obtained results are reported in Fig. 5.

| Results of the test with a Datatest of 250 images | | | | | |
|---|---|---|---|---|---|
| Orbits shape | Circle | Ellipse | Heart | Tornado | Eight |
| Circle | 46 | 0 | 0 | 0 | 0 |
| Ellipse | 4 | 46 | 0 | 4 | 0 |
| Heart | 0 | 0 | 48 | 0 | 0 |
| Tornado | 0 | 4 | 2 | 46 | 0 |
| Eight | 0 | 0 | 0 | 0 | 50 |

**Fig. 5.** Table of the image pattern recognition during the validation

## 5   Conclusion

The proposed algorithm for the predictive maintenance of the rotating machines through the analysis of the orbits shape of the rotors shaft with a CNN, is able to recognize the orbits shape and therefore the faults of the rotating machines with an accuracy of the 99.9%. To further validate the proposed method, an experimental test bench of a rotating machine, in which some proximity sensors detects in real time the orbits shape of the rotor shaft will be developed. To avoid the use of images of the motor axis, accelerometers can also be used. Three accelerometers can be displaced on the motor with a shift of 120° among them, in order to measure the radial acceleration along the three axes. By integrating and filtering these data, the 120° shift displacements of the motor axis will be obtained. From this displacement the axis orbit, useful for the CNN application, will be reconstructed.

## References

1. Graney, B., Starry, K.: Rolling element bearing analysis. Mater. Eval. **70**(1), 78–85 (2012)
2. Unal, M., Onat, M., Demetgul, M., Kucuk, H.: Fault diagnosis of rolling bearings using a genetic algorithm optimized neural network. Meas. J. Int. Meas. Confed. **58**(1), 187–196 (2014)
3. Hashemian, H.: Wireless sensors for predictive maintenance of rotating equipment in research reactors. Ann. Nucl. Energy **38**(2–2), 665–680 (2011)
4. Jeong, H., Park, S., Woo, S., Lee, S.: Rotating machinery diagnostics using deep learning on orbit plot images. In: Procedia Manufacturing (44th Proceedings of the North American Manufacturing Research Institution of SME, vol. 5, pp. 1107–1118 (2016). http://www.sme.org/namrc
5. Goodfellow, I., Bengio, Y., Courville, A.: Deep Learning. Mit Press (2016)

# False Data Injection in Smart Grid in the Presence of Missing Data

Rehan Nawaz$^{(\boxtimes)}$, Muhammad Awais Shahid, and Muhammad Habib Mahmood

Air University, Islamabad, Pakistan
rehan.nawaz@mail.au.edu.pk

**Abstract.** The cyber-attacks in smart grid can be so vulnerable and has the potential to damage the complete power network as well as the utilities of the consumers that are connected to that power network. Blind false data is injected into the communication network with the help of measurement matrix. However, some of the entries of the measurement matrix might get lost during the collection of measurement vectors. In this paper, we propose an algorithm to complete the measurement matrix with missing data. The simulations are done and results are shown quantitative in comparison with the complete data. To establish the effectiveness of our technique, we calculate the accuracy by using Bad Data Detection.

**Keywords:** Blind false data injection · Regression · Incomplete measurement matrix

## 1 Introduction

The robust false data injection requires some sensitive information of smart grid like susceptances and connectivity e.t.c.. Recent false data injection techniques requires only the measurement matrix to inject the false data [3]. The measurement matrix is just the measurement vectors collected over time. During the collection of measurement vectors, some of the entries might get corrupted or for some time instants attacker may not have access to all the the measured data of all the sensors.

The measurement matrix can still be constructed by estimating the missing entries of the measurement vectors [5]. There are a lot of imputation techniques to estimate the missing data like mean imputation, hot deck imputation, cold deck imputation, regression imputation, multiple imputation. our focus is on the multiple imputation, in which single value is predicted multiple times and then final value is calculated through mean or polling selected.

In this paper, we used multiple imputation to predict each missing entries and then the missing value is replaced with the weighted mean of all the estimated values. The recovered measurement vectors are tested on conventional

© Springer Nature Switzerland AG 2019
C. Benavente-Peces et al. (Eds.): SEAHF 2019, SIST 150, pp. 251–257, 2019.
https://doi.org/10.1007/978-3-030-22964-1_26

false data detection technique to show the accuracy. Secure measurement vectors follow the dependencies or some patterns. However, in Bad data due to the presence of some incorrect measurements, measurement vector does not follow the existing correlations. BDD detection is a technique that is used to check that either estimated measurement vector follows the dependency among all the measurements or not. That is why we used BDD to show the robustness of our technique.

## 2  Literature Review

The significance and the challenges of using power networks with communication networks are described in [1]. Most of the communication protocols are based on internet. The communication through the internet is not very secure and through cyber-attacks these communication networks can be hijacked. Stealthy attack is the most robust false data injection technique in literature [7]. Formation of stealthy attack requires the information of the jacobian matrix. According to M. Esmalifalak [2], M. Rahman [4] and A. Anwar [9] Jacobian matrix can be constructed by using the partial measurement matrix or local area information. If it is to be assumed, that partial information of measurement matrix that is useful, has not any missing value and Jacobian matrix is successfully constructed. Then stealth attack that is constructed by using that jacobian matrix can be detected by using False Data detection classifiers like SVM, Extended Nearest Neighbor, Nave Bayes, Random Forest, Adaboost, JRipper, OneR, NNge, Perceptron, k-Nearest Neighbour and Sparse Logistic Regression. ICA-based stealthy false attack can be injected into the communication network without the knowledge of the Jacobian matrix [6] but through the measurement matrix [8]. In recent past machine learning based supervised learning algorithms are used to inject false data in smart grids that has the ability to bypass those false data detection classifiers. So, to inject false data in smart grid robustly without having the complete secret information of smart grid, complete measurement matrix is needed. If $Q$ is the original low rank measurement matrix, $D$ is the matrix of missing values that is sparse and $G$ is the incomplete measurement matrix.

$$G = M + D \tag{1}$$

This sparse convex optimization matrix recovery problem is solved it through the augmented Lagrange multiplier (ALM) method in [9]. By using this technique, they estimated the measurement matrix, constructed Jacobian and injected stealthy false data. That stealthy attack successfully bypassed BDD as obtained residue was less then threshold. However, they just added 1% of missing values and they performed their experiments by just taking three missing measurements. Lanchao Liu in [10] addressed this problem using nuclear norm minimization and low rank matrix factorization. Missing data was varied from 0% to 10%. Then at their worst possible case when they have just 90% of complete data, they obtained true positive rate of 87% and 81% for the low rank matrix factorization and nuclear norm minimization respectively. In this paper, we used regression based multiple imputation to complete the measurement matrix for the false data injection that can handle the more percentage of missing data with more accuracy then the previous methods.

## 3   Proposed Methodology

Incomplete measurement matrix have missing values that are lost during the collection of real time measurement vectors or due to some other reason. If $Z$ is the incomplete measurement matrix that has zeros at the missing locations, $E$ is the Imputation matrix that is sparse and has non-zero entries only at the missing locations of the measurement matrix. $M$ is the complete measurement measurement matrix that has all the nonzero values of $Z$ and $E$. $Z$ is collected from the communication network of the power grid over time. Our task is to estimate $E$.

$$M = Z + E \qquad (2)$$

Accuracy of learning is based on the the correlation between the data. The correlations among the sensors are maximum at the same instants of different days as data follows daily demand curve periodically. So, linear regression is applied by taking one sensor as input and all the other sensors as output one after another for the same instant of the different days. Taking using one column of $Z^t$ as an independent variable and all other columns as output variables separately, linear regression is applied and weights are calculated. The indexes where both the dependent and independent variables are present, taken for the learning purpose.

$$J(\alpha_{ij}^t, \beta_{ij}^t) = \min_{\alpha_{ij}^t, \beta_{ij}^t} \frac{1}{2n_{ij}} \sum_{u=1}^{n_{ij}} ((\alpha_{ij}^t + \beta_{ij}^t Z_{(ind,j)}^{t(u)}) - Z_{(ind,i)}^{t(u)})^2 \quad \forall j, i, \text{ and } t \quad (3)$$

Where $i = 1, 2, 3, \ldots, m$, $t = t_1, t_2, t_3, \ldots, t_d$, $m$ is the total number of sensors, $t_d$ is the last day when measurements are received, $n_{ij}$ is the total number of training examples where both $Z_i^t$ and $Z_j^t$ are non missing. $ind$ are the indexes of training examples where both $Z_i^t$ and $Z_j^t$ are non missing. $Z_j^t$. $Z^t$ is an incomplete measurement matrix containing all the measurement vectors received at time instant $t$. $Z_{(ind,i)}^t$ are the non missing values of sensor$-i$ when sensor$-j$ is also non-missing at time instant-t. After the successful learning of $\alpha$ and $\beta$ the indexes where the values of dependent variable are present but the values of independent variable are missing are taken and by using the calculated linear regression parameters the missing values of independent variable are calculated.

$$Q_i(j)^t = \alpha_{ij}^t + \beta_{ij}^t Z_{(index,i)}^t \qquad \forall j, i \text{ and } t \qquad (4)$$

$Q_i^t$ is the sparse imputation matrix that is calculated by using $i_{th}$ sensor as input for time instant-t. $Q_{ij}^t$ is the vector for the $j_{th}$ sensor that is calculated by using $i_{th}$ sensor as input(independent variable) for time instant-t. $Z_{(index,i)}$ are the non missing measurements of sensor-i when measurements of sensor-j are missing.

Each sensor gives a one imputation matrix for each time instant-t. $n$ number of imputation matrices are calculated for each instant. Final prediction matrix $E^t$ is calculated by taking the weighted mean of all the calculated imputation matrices. $E^t$ is sparse and has nonzero enteritis only at the missing entries of $Z^t$. The weighted mean is calculated because all the sensors have different type of decency on each other. Power network is an interconnected infrastructure and all the measurements are dependent on each other with different dependency that is based on the connectivity. Sensors of the directly connected buses or transmission lines are more dependent on each other as compare to the sensors that are installed far away from each other.

$$E^t = w_1^t \odot Q_1^t + w_2^t \odot Q_2^t + w_3^t \odot Q_3^t \ \ldots \ w_n^t \odot Q_n^t \quad \forall t \tag{5}$$

$\odot$ is the dot multiplication. $w_1^t, w_2^t, w_3^t, \ldots, w_n^t$ are the weight matrices. $w_i^t$ are initialized by:

$$w_i^t(r_1, c_1) = \begin{cases} \frac{1}{\sum_i Q_i^t(r_1, c_1)} & Q_i^t(r_1, c_1) \in \mathbb{R}_{\neq 0} \forall i \\ 0 & \text{otherwise} \end{cases} \tag{6}$$

$(r_1, c_1)$ correspond to the single index.

$$M^t = Z^t + E^t \tag{7}$$

Jacobian matrix is calculated by applying PCA on the non-missing training examples, threshold ($\tau$) is also calculated by applying BDD on the complete measurement vectors that does not have any missing value. Then states(phase angles of voltage buses) are estimated for every estimated measurement vector. For the measurement vectors that has failed to bypass BDD, weights are readjusted randomly by using two constraints.

$$w_i^t(r_1, c_1)\epsilon[0, 1] \quad \text{and} \quad \sum_i w_i^t(r_1, c_1) = 1 \tag{8}$$

Again residue is calculated. The process of random selection of weights goes on until and unless residue becomes less then $\tau$ or some maximum limit of iteration is reached.

$$\exists w_i^t(r, c) \forall i | M^t(r, c) \text{Bypasssed BDD} \tag{9}$$

pseudo code of our proposed algorithm is shown in Appendix ??.

## 4    Simulations

The simulation are done on the MATLAB. MATPOWER is used to generate the data for 2-years. All the simulations are done on the modified IEEE case-5 bus system, IEEE case-9 bus system, IEEE case-14 bus system and IEEE case-30 bus system. By considering the demand pattern of complete day and load variance of 90–110%, data is generated periodically with the step of 1 h. Then missing values

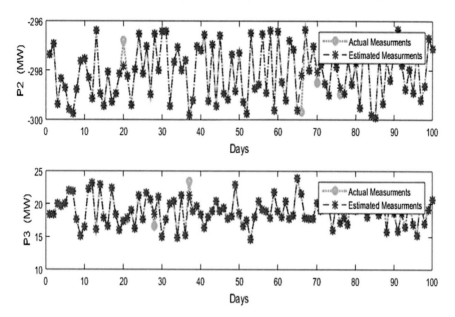

**Fig. 1.** Actual measurements and the estimated measurements

of 5–50% are added into the dataset randomly. Then proposed algorithm shown in Appendix ?? is applied. Actual measurements and the predicted measurements of sensor-2 and sensor-3 are shown in Fig. 1. Estimated results are very close to the actual measurements. To see the accuracy of our proposed algorithm BDD is applied on all the estimated measurement vectors. Threshold for the BDD is 4.89. Total measurement vectors received in two years with one hour step 17520. Total number of measurements vectors received for an individual time instant are 730. With 5% missing data for all the sensors on the random indexes through Algorithm ?? we obtained 7680 training example that has at least one missing value. After estimating the complete measurement matrix we applied BDD on the training examples that had at least one missing value from 7680 training examples 7428 training examples has successfully bypassed BDD. If $T_E$ is the total number of estimated measurement vectors that has bypassed BDD and $T_{nan}$ is the total number of measurement vectors that has at-least one missing value. Then,

$$Accuracy = \frac{T_E}{T_{nan}} \tag{10}$$

Then Accuracy can be calculated by using Eq. 10. Accuracy of 94.4% is obtained for the 5% missing data. Accuracy is calculated for the 5–50% missing data and results are shown in Fig. 2. with 50% missing data still 72% of accuracy is obtained. This accuracy shows the confidence level of training of supervised learning algorithm to inject false data.

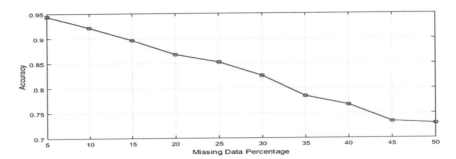

**Fig. 2.** Accuracy of proposed technique is shown above

## 5 Conclusion

In latest grids, communication system is not so much secure, cyber-attacks can be done to mislead the control actions. Blind false data injection based algorithms don't have susceptances and other inside information of power network they just collect measurement vectors over time and make false data attacks. In case of incomplete measurement vectors to inject false data measurement matrix is to be completed. we showed that our proposed algorithm can complete the incomplete measurement matrix efficiently with accuracy of 94.4–72.9% for the missing data of 5–10%. Therefore supervised machine learning algorithms can be trained on that data for the formation of false attack vectors. As in [10] they obtained accuracy of 87% with their best approach at 10%. By our proposed technique we obtained accuracy of 92.2% with same percentage of missing data.

## References

1. Gungor, V.C., Lambert, F.C.: A survey on communication networks for electric system automation. In: International Journal of Computer and Telecommunications Network, vol. 50, pp. 877–897 (2006)
2. Esmalifalak, M., Liu, L., Nguyen, N., Zheng, R., Han, Z.: Detecting stealthy false data injection using machine learning in smart grid. In: IEEE Systems Journal (2014)
3. Kim, J., Tong, L., Thomas, R.: Subspace methods for data attack on state estimation a data driven approach. In: IEEE Transactions on Signal Processing, pp. 1102–1114 (2015)
4. Rahman, M., Mohsenian-Rad, H.: False data injection attacks with incomplete information against smart power grids. In: IEEE Global Communications Conference (2012)
5. Srivastava, A., Morris, T., Ernster, T., Vellaithurai, C., Pan, S., Adhikari, U.: Modeling cyber-physical vulnerability of the smart grid with incomplete information. In: IEEE Transactions on Smart Grid, vol. 4, no. 1, pp. 235–244 (2013)
6. Liu, X., Li, Z.: Local load redistribution attacks in power systems with incomplete network information. In: IEEE Transactions on Smart Grid, vol. 5, pp. 1665–1676 (2014)

7. Tan, S., Song, W.Z., Stewart, M., Yang, J., Tong, L.: Online data integrity attacks against real-time electrical market in smart grid. In: IEEE Transactions on Smart Grid (2018)
8. Yu, Z.H., Chin, W.L.: Blind false data injection attack using PCA approximation method in smart grid. In: IEEE Transactions on Smart Grid (2015)
9. Anwar, A., Mahmood, A., Pickering, M.: Data-driven stealthy injection attacks on smart grid with incomplete measurements. In: Pacific-Asia Workshop on Intelligence and Security Informatics (2016)
10. Liu, L., Esmalifalak, M., Ding, Q., Emesih, V.A., Han, Z.: Detecting false data injection attacks on power grid by sparse optimization. In: IEEE Transactions on Smart Grid (2014)

# A Predictive Real-Time Energy Management Control for a Hybrid PEMFC Electric System Using Battery/Ultracapacitor

Nasri Sihem[1](✉), Ben Slama Sami[2], Bassam Zafar[2],
and Cherif Adnane[1]

[1] Department of Physics, Analysis and Processing of Electrical and Energy Systems Unit, Faculty of Sciences of Tunis El Manar, 2092 Belvedere, PB, Tunisia
nasri_sihem@live.fr, adnane.cher@fst.rnu.tn
[2] Information System Department, King Abdulaziz University, Jeddah, Saudi Arabia
benslama.sami@gmail.com, bzafar@kau.edu.sa

**Abstract.** The optimization of energy consumption in transport applications, especially dual-energy vehicles (hybrid electric System), was selected as an interesting problem for future transportation applications due to its high efficiency, reduced emissions and the required consumption optimization. To solve this embedding problem, a Hybrid Electrical System (HES) powered by a proton exchange membrane fuel cell (PEMFC) was proposed. The hybrid electric system has some main drawbacks derived from the PEMFC due to its inability to meet the energy demand. Battery (BT) and Ultracapacitor (UC) devices are added in this aspect, wherein each power source comprises a DC/DC converter. These devices are used as a potential energy storage that deals to minimize the transitional response associated with the PEMFC. So, the HES is evaluated and demonstrated through an accurate Multi-Input Single-Output (MISO) state space model. To optimize the energy demand, Road-Speed-profiles was chosen. The selecting profiles are systematically compared in terms of hydrogen fuel consumption and tank state of charge prediction. To maintain the required energy through an effective cooperation between the PEMFC, BT and UC, a predictive real-time Energy Management Unit (EMU) is developed. The obtained results reflect the system performance and the proposed EMU effectiveness using Matlab/Simulink environment.

**Keywords:** Prediction · Battery · PEMFC · Ultracapacitor · Multi-Input/Single-Output model · Real-time energy management

© Springer Nature Switzerland AG 2019
C. Benavente-Peces et al. (Eds.): SEAHF 2019, SIST 150, pp. 258–265, 2019.
https://doi.org/10.1007/978-3-030-22964-1_27

# 1 Introduction

Nowadays, PEMFC-Vehicles (FCVs) can provide uninterrupted power [1, 2]. Consequently, hydrogen powered PEMFC becomes an interest energy carrier to replace the traditional fuels. Indeed, PEMFC emerges as one of the most promising candidates for transport applications [3, 4]. However, PEMFC is still unable to provide the energy required continuously for the electrical load demand due to their lower power and lower starting density respectively, their slower response power; their inability to regenerate the needed power [5–7].

Nevertheless, the PEMFC drawbacks can be solved by adding a secondary energy sources like batteries or Ultracapacitor or a both combination [3]. Generally, BTs devices are characterized by its specific higher energy compared to the UCs. Indeed, BTs can provide a long time additional power [8]. However, compared to BTs, UCs are integrated to control transient power due to its high power, higher efficiency and longer charging/discharging cycles [9].

The cooperation between sources and the required average power optimization are considered the main concerns. For this reason, the Energy Management Unit is included as an interest solution to control the demand for average power required. However, the main Energy Management Unit concerns are the operation effectiveness of the control method. Energy management strategies have been discussed and developed in various works to provide knowledge and information to the community as a whole. For example, the authors in [10] proposed an Energy Management Unit for a hybrid (PEMFC/BT) configuration. The proposed system deals to supply an Electrical vehicle using energy storage to compensate the power fluctuations. While, the authors in [6], proposed a hybrid (PEMFC/BT/UC) for electrical vehicles that tends to cooperate between the sources. The authors proposed an accurate Energy Management Unit for a hybrid (PEMFC/BT) configuration. The presented EMU aims to provide and to optimize the required power for the electrical load [11, 12]. Authors, in [3] proposed an Energy Management Unit for hybrid (PEMFC/BT/UC) system to optimize and to control the required average power demand.

This article proposed an efficient hybrid electric system configuration which combines FC, BT and UC. Indeed, the system under study uses PEMFC as the main source to meet the electrical load requirements. A battery and Ultracapacitor are used as energy backup components that are used to compensate the power deficit. The developed design, integrating, PEMFC, BT and UC devoted to transport application are evaluated and investigated using a given real load profile. A Multi-Input/Single-Output model is proposed and discussed.

The article is organized as follows. Section 2 presents the proposed HES system configuration. The energy management unit (EMU) devoted to the proposed system is developed and detailed in Sect. 3. Simulation results are shown in Sect. 4, in which a real-time simulation test is evaluated and discussed. Finally, the conclusions are drawn in Sect. 5.

## 2  Design of HES

The proposed design is illustrated in Fig. 1 which is developed and processed using MATLAB, Simulink. As shown, the system consists of a PEMFC, BT and UC. The PEMFC is used as the primary power source to deliver the required power and UC and BT are used as a backup energy. The PEMFC, BT and the UC exchange power through a DC bus. A unidirectional boost converter is included to connect the PEMFC to the EMU. Although a bidirectional boost converter is necessary for the UC to ensure the needed power when the PEMFC is switched off. The included UC is suitable for correcting and controlling instantaneously the transient peak power demands.

BT issued especially to increase the power energy. Furthermore, the BT can provide the required power during the permanent phases like hydrogen fuel lack and the energy braking. To regulate the distribution power between PEMFC, BT and UC, two operating modes are proposed which are PEMFC-UC mode and UC-BT mode. To achieve high efficiency and to optimize electrical power demand, a new control algorithm is developed for EMU. Indeed, this algorithm tends to manage and supply respectively the load demand through UC state of charge and the vehicle velocity.

To model the HES, a Multi-Input/Single-Output state space (MISO) is presented as follows:

$$\begin{cases} \dot{X} = AX + BU \\ y = CX + DU \end{cases} \tag{1}$$

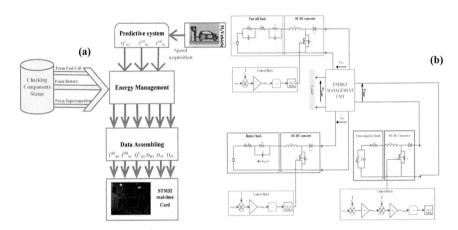

**Fig. 1.**  HES Model: (a) Design of HES; (b) Electrical model of HES

$$X = \begin{bmatrix} U_a & U_c & I_{FC} & U_{UC} & I_{UC} & U_{CBT} & I_{BT} & U_{Load} \end{bmatrix}^T$$

$$A = \begin{bmatrix}
-\frac{\varepsilon_a}{R_a} & 0 & \varepsilon_a & 0 & 0 & 0 & 0 & 0 \\
0 & -\frac{\varepsilon_c}{R_c} & \varepsilon_c & 0 & 0 & 0 & 0 & 0 \\
-\alpha & -\alpha & -\alpha R_{ohm} & 0 & 0 & 0 & 0 & -\alpha(1-\alpha_{BFC}) \\
0 & 0 & 0 & 0 & -\frac{1}{C_{UC}} & 0 & 0 & 0 \\
0 & 0 & 0 & \frac{1}{L_{LUC}} & \frac{-R_{UC}}{L_{LUC}} & 0 & 0 & \frac{(1-\alpha_{BUC})}{L_{LUC}} \\
0 & 0 & 0 & 0 & 0 & \frac{1}{C_{BT}R_{BTp}} & \frac{1}{C_{BT}} & 0 \\
0 & 0 & 0 & 0 & 0 & \frac{-1}{L_{LBT}} & \frac{-2R_{BT}}{L_{LBT}} & \frac{-(1-\alpha_{BBT})}{L_{LBT}} \\
0 & 0 & \frac{(1-\alpha_{BFC})}{C_L} & 0 & \frac{(1-\alpha_{BUC})}{C_L} & 0 & \frac{(1-\alpha_{BBT})}{C_L} & 0
\end{bmatrix}$$

$$B = \begin{bmatrix}
0 & 0 & \alpha U_{Load} & 0 & 0 & 0 & 0 & \frac{I_{FC}}{C_L} \\
0 & 0 & 0 & 0 & \frac{U_{Load}}{L_{UC}} & 0 & 0 & \frac{I_{UC}}{C_L} \\
0 & 0 & 0 & 0 & 0 & 0 & \frac{U_{Load}}{L_{LBT}} & \frac{I_{BT}}{C_L}
\end{bmatrix}$$

$$U = \begin{bmatrix} \alpha_{BFC} \\ \alpha_{BUC} \\ \alpha_{BBT} \end{bmatrix}$$

Where:

$$C = \begin{bmatrix} 0 & 0 & 0 & 0 & 0 & 0 & 0 & \kappa \end{bmatrix} \, || \, \kappa = \frac{1}{R_L}$$

(2)

## 3 Energy Management Unit

The proposed hybrid electric vehicle scheme provides a flexible and a new management system which is controlled through various parameters like the load demand and the BT-UC state of charge (SOC). The proposed EMU aims to maintain the proper functioning of the system by appropriately distributing the energy flows between the components of the system. To do this, the work proposed by this control strategy acts to achieve a continuous load under various conditions, according to two operation modes which are PEMFC-UC and BT-UC modes. In this context, the PEMFC and BT have both selective sources to ensure the load demand. In fact, the required energy can be fully met from the energy emitted by at least one source. Thus, the EMU chooses the appropriate component to provide the load demand, according to its constraints during the fluctuation of the load based on a decision-making. Hence, the system is based on the interaction between all the constitutive elements by message exchanging for checking status and activation order. Figure 2 describes the principle of the EMU with the different operation mode according to the control algorithm (Table 1).

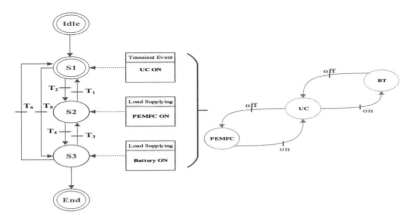

**Fig. 2.** EMU State Diagram

**Table 1.** Control principle devoted for FC/BT/UC system configuration

| State | Transition | Transition description | Previous state | Description |
|---|---|---|---|---|
| Idle | No transition | | | System Startup |
| $S_1$ | No transition | Transition event detection | Idle | Activation of UC |
| | $T_1$ | | $S_1$ | |
| | $T_6$ | | $S_2$ | |
| $S_2$ | $T_2$ | Sufficient hydrogen fuel | $S_1$ | Activation of PEMFC |
| | $T_3$ | Insufficient battery power | $S_3$ | |
| $S_3$ | $T_4$ | Insufficient hydrogen fuel | $S_2$ | Activation of BT |
| | $T_5$ | Sufficient battery power and insufficient hydrogen fuel | $S_1$ | |

## 4    Findings and Results

This section is devoted to the performance evaluation of the proposed hybrid electrical system configuration (PEMFC/BT/UC). Indeed, numerous simulation tests are made in Matlab/Simulink environment to obtain the desired results which are, then, analyzed and discussed in details. In this work, each simulation test lasts 254 min. During its operation, the HES refers to an experimental vehicle data profiles as speed, acceleration and power consumption.

Figures 3 and 4 depict the results for generated system currents ($I_{FC}$, $I_{BT}$ and $I_{UC}$) and the BT and UC state of charge variations among three recognized roads. As shown, the higher energy demand is observed on the road "A" characteristics caused by its higher registered speed as summarized by Table 2.

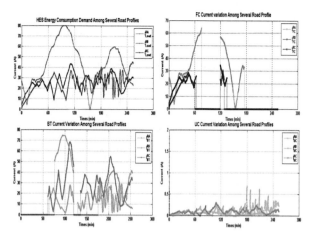

**Fig. 3.** HES characteristics among several road profiles

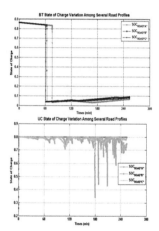

**Fig. 4.** HES state of charge control

**Table 2.** Roads performance indices and consumptions

| Road number | Distance (km) | Duration (min) | Average speed (km/h) | Fuel consumption (mol) |
|---|---|---|---|---|
| Road "A" | 2.059 | 2:48 | 49.62 | 0.015 |
| Road "B" | 2.606 | 3:44 | 45.41 | 0.01 |
| Road "C" | 3.600 | 5:12 | 42.11 | 0.0095 |

As shown, the higher energy consumption is observed in road "A" followed by road "B" while road "C" is in the rear. So, on the basis of these results, if we aim to reduce the energy consumption, we can notice that the road "C" presents the best way to follow according to its characteristics.

The best performances have been recorded for road "A" compared to the other ways (see Table 3) despite it possesses the highest consumption rate. Therefore, it can be concluded that the road "A" seems to be the most suitable forward way since it offers the best features with a constraint of more fuel consumption compared to the performances of the other roads "B" and "C".

**Table 3.** HES efficiencies average values

| Road | Efficiency (%) | | | |
|---|---|---|---|---|
| | HES | FC | BT | UC |
| Road "A" | 68 | 68 | 78 | 55 |
| Road "B" | 63 | 63 | 98.7 | 81.5 |
| Road "C" | 65 | 63 | 98.8 | 82.5 |

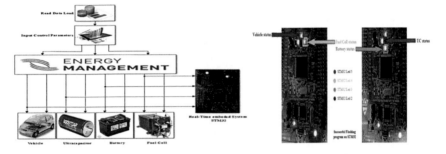

**Fig. 5.** Real-time simulation model of FC/BT/UC configuration

The real-time simulation model is used to control and to test the system behavior through referring to the Road characteristics. There by, Fig. 5 presents Simulink design of the embedded hybrid electric system configured with STM32 discovery F4 board. The decision keys ($D_{HES}$, $D_{FC}$, $D_{BT}$, $D_{UC}$) have been selected as main system outputs. These outputs are visualized through the card's LEDs: HES $\rightarrow$ LED 2, BT $\rightarrow$ LED 3, FC $\rightarrow$ LED 4 and UC $\rightarrow$ LED 5.

## 5   Conclusion

In this paper, we are concerned with a dynamic Hybrid Electrical System comprises by PEMFC, BT and UC. The proposed system is treated and evaluated according two modes which are PEMFC/UC mode and BT/UC mode. The (PEMFC/UC) mode proved the synchronization between the PEMFC and UC to meet efficiently the load requirements. Depending on our outcomes, we have shown the great function of the UC during the off-peak-period times and peak power transition. So, the UC can offer an efficient power with a fast dynamic response, which makes it well suitable for the load. While, during the (BT/UC) mode, the PEMFC is kept off owing to the hydrogen fuel lack. In this case, the UC participates with the battery (BT) to provide the load requirements. Finally, the enhancements of HES system performances are proven through numerous simulation tests using an experimental power profile.

Future work is aimed at implementing our proposed approach in an embedded system using a Raspberry Pi with RT Preempt card.

# References

1. Sami, B.S., Sihem, N., Zafar, B., Adnane, C.: Performance study and efficiency improvement of hybrid electric system dedicated to transport application. Int. J. Hydrog. Energy (2017)
2. Bernard, J., et al.: Fuel cell/battery passive hybrid power source for electric powertrains. J. Power Sources **196**(14), 5867–5872 (2011)
3. Thounthong, P., Raël, S., Davat, B.: Energy management of fuel cell/battery/supercapacitor hybrid power source for vehicle applications. J. Power Sources **193**(1), 376–385 (2009)
4. Thounthong, P., Raël, S., Davat, B.: Control strategy of fuel cell/supercapacitors hybrid power sources for electric vehicle. J. Power Sources **158**(1), 806–814 (2006)
5. Henao, N., Kelouwani, S., Agbossou, K., Dubé, Y.: Proton exchange membrane fuel cells cold startup global strategy for fuel cell plug-in hybrid electric vehicle. J. Power Sources **220**, 31–41 (2012)
6. Azib, T., Hemsas, K.E., Larouci, C.: Energy management and control strategy of hybrid energy storage system for fuel cell power sources. Int. Rev. Model. Simul. (IREMOS) **7**(6), 935 (2014)
7. Nasri, S., Sami, B.S., Cherif, A.: Power management strategy for hybrid autonomous power system using hydrogen storage. Int. J. Hydrog. Energy **41**(2), 857–865 (2016)
8. Huang, M., Li, J.-Q.: The shortest path problems in battery-electric vehicle dispatching with battery renewal. Sustainability **8**(7), 607 (2016)
9. Thounthong, P., Rael, S., Davat, B.: Analysis of supercapacitor as Second source based on fuel cell power generation. IEEE Trans. Energy Convers. **24**(1), 247–255 (2009)
10. Odeim, F., Roes, J., Heinzel, A.: Power management optimization of an experimental fuel cell/battery/supercapacitor hybrid system. Energies **8**(7), 6302–6327 (2015)
11. Jung, H., Wang, H., Hu, T.: Control design for robust tracking and smooth transition in power systems with battery/supercapacitor hybrid energy storage devices. J. Power Sources **267**, 566–575 (2014)
12. Kim, Y., Koh, J., Xie, Q., Wang, Y., Chang, N., Pedram, M.: A scalable and flexible hybrid energy storage system design and implementation. J. Power Sources **255**, 410–422 (2014)

# Social Networkopia: Metamorphosing Minds with Hypnotic Halo

Neetu Vaid[✉] and Disha Khanna

Department of FLA, GNA University, Phagwara, Punjab, India
Sharmaneetu2009@gmail.com,
Dean.fla@gnauniversity.edu.in

**Abstract.** The globe today has enormous earmarks of social media imprinted on its sensitive skin. Everyone is in the grip of gadgets scratching their seconds on social ladder. Social media has emerged as panacea for present generation education industry. The Himalayan help that is vouchsafed by social media learning tools is undoubtedly incredible. Gifting a rare gadget to the contemporary world for leaning, presenting plethora of information on multiple topics, bestowing the best research bounties for intellect strata, laying upon an ocean of knowledgeable treasure for students as well as teachers- Social media has sprung up as a panama of perennial possessions cupped inside our palms.

## 1 Social Media- An Oasis of Mattered Much Material

The resourceful learning tools including YouTube, Face book, Google+ , Instagram, Blackboard, Skype, Lab Roots, Research Gate, Teacher Tube and many more reach everyone effectively. The exchange of motivational and informative videos on such podiums radiates youthful minds. Furthermore, there emerges an edifying environment for online discussions, healthy video notes and other fruitful practices among pupils. Videos currently are so happeningly inspiring and best way to share your emotion to millions just with the help of social media leaning mechanism. Erudite experts deliver highly weighty discourse. The online leaning courses offer a peculiarly compassionate niche to a student whose classroom shyness casts away. The original voice of a student receives up-to-the-minute apparatus in hand.

Social media is a roaring tool in the hands of leaning buds. Leaning programs not only open a vast gate of knowledge for the students but also pocket precious notes to preserve inside their memory guards. Under one unimpaired canopy of crazy technology is procurable.

## 2 Social Media – A Hypnotic Independent Halo

Individuals of all age groups are basking in the social media sunshine. The students envision an extremely independence with the help of social media learning tool kits. They surf, observe, explore and in the end find themselves enriched in their respective

C. Benavente-Peces et al. (Eds.): SEAHF 2019, SIST 150, pp. 266–269, 2019.
https://doi.org/10.1007/978-3-030-22964-1_28

areas of learning and education. Their ability to access has sharpened and it is upscale according to the smart age wherein we all are dwelling.

## 3 Social Media- A Rich Reservoir for Researchers

Scholars throughout the world are granted with an enormous ration of research related write ups. Academia.Edu is one of the many search engines where researchers get vast data about both researches- already existing and ongoing ones. The real motif of these social media tools of learning are to make global communication happen and then boost the brainy scientists as well intellectuals swell their research process. An affluent ocean to extract information meets researcher's eye.

## 4 Social Media – A Global Classroom

The world has stepped into the globally infrastructure social media classroom. Walls are adorned with latest news, happening all across the world. The smart boards display learning to make it luring as well as long lasting for students with technological markers. The students' minds are jeweled with the forthcoming far-flung facts.

## 5 Social Media – An Acquisition Tool

It's a distinguished fact that internet based life is led by millions of people. Web based life hits strikingly how we live, how we work, and now like never before, how we learn. As indicated by an ongoing report, an ever increasing number of instructors and educators are joining web based life into their classrooms to draw in understudies and bolster their instructive improvement, regardless of whether on the web or face to face. In short – web based life is forming and impacting how understudies learn and collaborate today.

## 6 Importunes of Social Media as Learning Tool

The ascent of internet based life in the classroom isn't about what number of individuals "like" your posts. The synergistic condition and open gathering that web based life energizes, alongside the fast pace of data sharing that it encourages, implies that understudies can quicken the improvement of their inventive, basic reasoning, and correspondence forms in certain ways when they utilize it.

Online life advances self-coordinated realizing, which plans understudies to scan for answers and settle on choices autonomously. At the point when fortified in a classroom setting, these internet based life aptitudes can be guided and refined to deliver better learning results and basic mindfulness. Web based life additionally enables understudies more opportunity to associate and team up past the physical classroom, which implies understudies anyplace can begin to encounter the comprehensively associated world some time before they enter the workforce.

Social learning is dynamic realizing, which implies that understudies take part specifically in their own adapting instead of inactively engrossing data they will undoubtedly overlook once the exam is finished. Online networking shapes and introduces data in a way that bodes well to and energizes understudies more than customary devices do, regardless of whether it's through a common article with remark usefulness, a live stream of an essential occasion, a study identified with course materials, or an inquiry presented to the more extensive network.

Besides, offering presents and data on different understudies, instead of just submitting assignments to the instructor, advances further commitment and better execution from all understudies. On the off chance that understudies know from the begin that they and their associates will collaborate with course materials and each other on different internet based life stages, they may invest more exertion to both their work and online nearness.

## 7    Plentiful Platter of Social Media Tools

Educators may utilize any of the accompanying online life stages as learning instruments:

- Blogs with remark usefulness to share and talk about data;
- Twitter and course hashtags to empower open discussion and discussion;
- Skype or Google+ Hangouts to connect all the more profoundly with the material and one another;
- Google Docs, Wikis and other community oriented archive devices to store and refine information;
- Project Management Apps to encourage and streamline joint effort;
- LinkedIn and other informal organizations to fabricate association;
- YouTube to make both course and understudy introductions;
- And more!

Not exclusively can internet based life in the classroom make more communitarian and innovative work from understudies, it can likewise enable understudies to recognize true applications for the online networking apparatuses they as of now utilize. Today like never before, understudies, working experts, and even organizations should be clever via web-based networking media to succeed. Particularly when precisely created and coordinated amid school, online networking aptitudes can subsequently enable understudies to discover employments in a commercial center that undeniably depends on computerized devices for systems administration and data sharing.

Whenever utilized with aim, internet based life can decidedly impact the manner in which every individual learns and retains data in the classroom. Joining online life into a more conventional learning condition can extend understudies' imaginative opportunity and urge them to work harder and connect more. Obviously, as the internet based life scene changes, classrooms will likewise need to adjust – yet with online networking as of now affecting the manner in which we learn and associate outside of the classroom, applications inside the classroom will probably just increment.

# Reference

1. www.ashford.edu/blog/online-learning/using-social-media-as-a-learning-toolwww.lcibs.co.uk/the-role-of-social-media-in-education/Vaid, Neetu (Self Citation)

# Design of the Furniture and Fixture Smart Automatic Batch Simulation Mobile Application

Jung-Sook Kim[1(✉)], Taek-Soo Heo[2], and Tae-Sub Chung[3]

[1] Kimpo University, Gimpo-Si, Gyeonggi-do, Korea
kimjs@kimpo.ac.kr
[2] Swithchspace Co, Seoul, Korea
tsheo@switchspace.co.kr
[3] Yonsei University, Gangnam Severance Hospital, Seoul, Korea
tschung@yuhs.ac

**Abstract.** In this paper, we design a smart mobile application which is a furniture and fixture smart automatic batch simulation. It is an Android-based mobile application system that, anticipating previously the furniture and fixtures as the same as those in the actual measurement according to the plane structure, if providing for a disaster and risk to the characteristics of space and presenting the ecological equipment element smartly, the user can choose it easily, and that can presents several furniture and fixtures for the efficient and quick batch of repeated apparatus in a prescribed space considering the safety and greening to the maximum, and that can provide several layouts of template representing the shape and size of space.

**Keywords:** Mobile application · Furniture and fixture smart automatic batch simulation · Arrangement layout · Natural disaster · Smart mobile application

## 1 Introduction

In planning the arrangement of office areas and service areas such as company, bank, public institution, school, hospital, and in case of arranging the apparatus manually with the existing blueprint, a bad arrangement is frequently made with a wrong transcription, and it results in the repeated process of drafting and relocation to infinity. Also, with the element to maximize the efficiency of work according to the characteristics of each space its safety and ecological element should be considered, and a system development to generate it automatically in real time is required [1–4]. This paper designed technology provides the automatic method of drafting by arranging at the rate based on the real size the furniture and fixtures in the same virtual plane structure as in the actual measurement through the mobile terminal. Specially, the furniture and fixtures should be arranged according to smart technology that can consider first the safety against the natural disaster (ex. earthquake) and disaster (ex. fire) and that can consider the ecological elements such as fine dust, and it would be more desirable if considering them. Accordingly, in this research development, smart technology development function is

© Springer Nature Switzerland AG 2019
C. Benavente-Peces et al. (Eds.): SEAHF 2019, SIST 150, pp. 270–273, 2019.
https://doi.org/10.1007/978-3-030-22964-1_29

important characteristics that, grasping the characteristics of space and maximizing the efficiency of work, can recognize the natural disaster and calamity as soon as possible and that can arrange the furniture and fixtures for the most efficient element of ecologic equipment to be made with its surrounding environment.

The composition of this paper is as follows. In Sect. 2, we describe the related works. In Sect. 3, we design the technique that explain the smart automatic batch simulation system of the furniture and fixture. Lastly, in Sect. 4, we conclude our presentation and take a look at the direction of future research.

## 2  Related Works

For an efficient arrangement of space, the method of generating and editing the interior space and of arranging the props in the virtual space and of producing previously the three dimensional design space has been already preceded. Of the drawing production programs, 'Auto CAD' and 'Sketch up' of 'Auto Desk'company is generally used and calculates an exact drawing, but it is optimized to PC and has a weakness that common people cannot use it without any professional knowledge and technique. 'Magic Plan' can produce a drawing in a mobile device and has several libraries but with its interface and function unrelated to practical affairs, cannot be used by expert group, while the space simulation app. provided by 'IKEA' was developed to promote the furniture, with its library configuration made only of this company's products, which is difficult to be used in a real spot. The existing space arrangement simulation service uses the interior design DB information whose user terminal is provided by the service provision server through internet net to choose the interior space desired by user and choose props in props library, and used the information saved in the 3D drawing information file DB to initiate the virtual 3D interior space design configuration, which has much time and effort needed because the user should arrange and confirm each furniture.

## 3  Furniture and Fixture Smart Automatic Batch Simulation Introduction

The desirable functional arrangement of office is minimize the unnecessary movement in a pleasant environment, make much use of space, and arrange the facility to achieve effectively its corporate activity destined with less efforts in less time. The objective of arranging the facility such as company, bank, public institution, school, hospital is to optimize the arrangement of facility to increase the efficiency of office system, and therefore in the arrangement of facility the business effect is direct and indirectly searched, and such an effect should include the minimization of movement, high utilization of space, flexibility of change, satisfaction of physical environment and the convenience of client. With the automatic resizing of icon, it has the same immediacy as arranging the equipment in a real space, so it is possible to produce efficiently several layout drawings in less time. It can shorten the time of planning and confirming the arrangement plan, and reduce the inefficiency originated from the drafting task, and anyone or beginner without specialized knowledge on the drawing can produce the drawing, which has its range of use is wide. This automatic arrangement layout function allows a group or a person to estimate simple and quickly the number or location of desk expected when leasing the space, so the group or the person can lay his plan of space lease more planned. In case that the arrangement of office furniture should be flexible such as the appearance of the worldwide 'WeWork' company and the advent of office sharing era, it can present the optimal number of automatic arrangement cases in real time to be more easy for business, and the furniture company can discuss on the layout with the client immediately in the spot, and provide a service of high satisfaction to the client. This technology allows you to arrange the furniture more efficiently in the statistics through the technology with which can process the big data saved for every design and propose a new situation of furniture arrangement to the user through smart algorithm development and technology. It is a smart system that can present the location of optimal installation of ecological elements, namely the equipment to eliminate the fine dust and that also can consider the other ecological elements. The following Fig. 1 is an example of realizing the mobile application and shows the screen that imports the objective floor plan through this company's app of actual measurement in development.

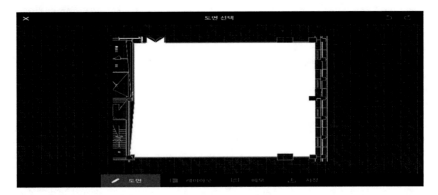

**Fig. 1.** Example of realizing application

The following Fig. 2 is an example of realizing the mobile application and shows the screen that chooses the category and designates the item to arrange.

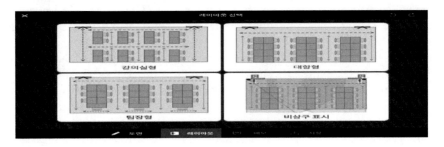

**Fig. 2.** Floor planning system for a catastrophic event

# 4   Conclusion and Future Research

In this paper, we design the smart mobile application which is for the automatic batch simulation of the furniture and fixture in the school, hospital, office and, etc. Also, this paper design the way of presenting the location of sensors installed that can recognize quickly the natural disaster and calamity being in harmony with the furniture and equipment. It is a smart system that can present the location of optimal installation of ecological elements, namely the equipment to eliminate the fine dust and that also can consider the other ecological elements. In the future, we will implement the smart automatic batch simulation of furniture and fixture. And we will apply the mobile application to user in field site.

**Acknowledgement.** The work (Grants No. C0541497) was supported by the Business for Cooperative R&D between Industry, Academy, and Research Institute Funded Ministry of Korea Small and Medium Business industry in 2017.

# References

1. Magic Plan: http://www.magic-plan.com. Last accessed 1 Dec 2017
2. Planner 5D: https://planner5d.com/. Last accessed 1 Dec 2017
3. Je-Hun, Yu., Ahn, Seong-In, Lee, Sung-Won, Sim, Kwee-Bo: Mobile robot control using smart phone for internet of things. J. Korean Inst. Intell. Syst. **26**(5), 396–401 (2016)
4. Kim, J.S., Jeong, J.H.: Development of the smart drawing editor for the interior design. In: iFUY2018: 2018 International Conference on Fuzzy Theory and Its Applications Proceedings, Korea (2018)

# Performance Evaluation of an Appraisal Autonomous System with Hydrogen Energy Storage Devoted for Tunisian Remote Housing

Ben Slama Sami[1], Nasri Sihem[2(✉)], Bassam Zafar[1], Cherif Adnane[2], and A. Elngar Ahmed[3]

[1] Information System Department, FCIT, King Abdulaziz University, Jeddah, Saudi Arabia
benslama.sami@gmail.com, bzafar@kau.edu.sa
[2] Analysis and Processing of Electrical and Energy Systems Unit, Faculty of Sciences of Tunis El Manar, PB 2092, Belvedere, Tunisia
nasri_sihem@live.fr, adnane.cher@fst.rnu.tn
[3] Faculty of Computers and Information, Beni-Suef University, Salah Salem Str., Beni Suef 6251, Egypt
elngar_7@yahoo.co.uk

**Abstract.** Autonomous hybrid energy systems are a completely green system and are considered as an attractive research problems that utilize all power requirements using energy storage. Deploying green systems is seen as an option to improve power security. For this reason, the main objective is to ensure the efficient production of electricity without interruption. To achieve this objective, we have proposed a precise simulation model that combines solar energy with an energy recovery component (fuel cell). A long-term energy storage component includes a water electrolyzer which is regarded as a primary storage and an ultracapacitor storage component deployed as a short-term storage of energy. To reach the correct system operation, an accurate schema approach for energy management unit (EMU) is developed and discussed according to an excess and deficit modes. To prove the reliability and effectiveness of the applied control strategy and its impact on the system operation, the proposed design is simulated using the Matlab/Simulink environment by referring to an experimental data profile extracted from the Tunisian meteorological database with anticipated conditions in a typical mode Weekly working period.

**Keywords:** Solar energy · Fuel cell · Ultracapacitor · Hydrogen · Energy management unit · Simulation

## 1 Introduction

Nowadays, interest in the economic homes in the world using green energy sources continues to increase and attracts the solution to replace traditional energy. The renewable energy sources like tide, wind, solar are an alternative energy source [1]. Among these renewable green sources, photovoltaic sources have been widely used in

© Springer Nature Switzerland AG 2019
C. Benavente-Peces et al. (Eds.): SEAHF 2019, SIST 150, pp. 274–281, 2019.
https://doi.org/10.1007/978-3-030-22964-1_30

low power industrial applications. Actually, solar energy is chosen as a promising candidate for the research and industry development due to its performance and reliability [2]. The development of the renewable stand-alone using solar source can be custom designed for variety industrial applications like the telecommunication areas, the electrification, and the Remote areas [3]. Thus, the solar system incorporates several alternative energy sources - storage systems (such as fuel cells, ultracapacitor, electrolyte, Hydrogen Storage Tank) to overcome its problems [4]. Several industries have benefited from the use of solar energy to automatically ensure the necessary of the power demand to the green home [5]. However, stand-alone hybrid power becomes an efficient way to minimize the energy production and to decrease natural resources consumption [6]. Various configurations were developed and discussed to deploy an autonomous hybrid power for different load requirements [7].

For example, in [8], the authors proposed a hybrid system using Fuel Cell as a backup source to supply the electrical load based on energy management approach. Whereas in [9], the authors proposed a predictive configuration control model using a solar system and Fuel cell. The proposed topology has used hydrogen as energy storage for a long period. Consequently, hydrogen inventory, obtained through a surplus of energy delivered by the photovoltaic array, is considered as a promising and an efficient energy storage. When the energy produced became insufficient, with the respect to the load demand, the stored hydrogen was fed to a PEMFC in order to serve the load [10, 11]. Also, other various works have been considered the hydrogen as an appreciating solution for several applications. For example, the hydrogen inventory is providing a clean energy with Zero-Emissions of the pollutants and can offer and greenhouse gases instead of the grid power [12, 13]. Indeed, the Hydrogen was considered as a promising power source for numerous Hybrid electrical vehicles by replacing the chemical batteries with the PEMFC [14, 15].

## 2 Autonomous Hybrid Energy System: System Design and Modeling

To provide the required power in a remote area, an accurate autonomous hybrid system design is developed and controlled. The proposed design seems to be an effective solution to resolve the remote area electrification problems (see Fig. 1). As seen, the Solar Energy system was chosen as the main power source providing the required power. Sometimes the generated power exceeds the requirements. Hence, to keep the optimum system functioning, all the exceeded energy must be stored by the most appropriate storage unit. As seen, two energy storages were included and specified by Energy Storage Component (ESC) and Ultracapacitor Storage Component (USC). Due to its intermittent Complaining character, the SEC does provide the opportunity for the desired and necessary energy. For this reason, the system includes a secondary power source specified the energy recovery Component (ERC) characterized by the use of a fuel cell. To optimize the operation of the system, an accurate scheme Energy Management Unit (EMU) and was developed and discussed through two operation modes. The proposed EMU deals to organize the participation of each system element through a various decision-making.

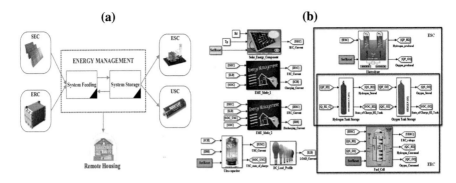

**Fig. 1.** System physical configuration of the proposed autonomous hybrid system. **(a)** Descriptive scheme of the autonomous hybrid system; **(b)** Model matlab/simulink of the autonomous hybrid system

## 3    Energy Management Analysis

The EMU is proposed at this stage to improve the system performance. Indeed, the EMU role lies in the energy production management by protecting storage elements against all overloads occurred in the excess power event; Energy consumption management through protecting the Fuel cell source against the massive hydrogen consumption and the deep discharge and the energy demands. The descriptive scheme of the proposed system modeled using two operations modes is presented by Fig. 2a. The developed algorithm dedicated to the autonomous hybrid system is based on the selection operation mode to meet all system requirements and constraints. Indeed, the proposed algorithm deals to supervise the flow energy distribution in order to select the appropriate mode (excess/deficit).

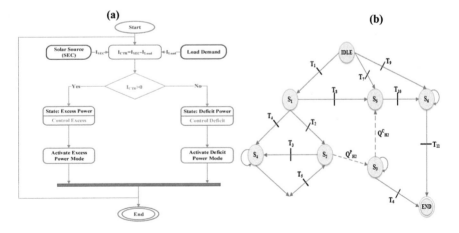

**Fig. 2.** Descriptive scheme of the energy management unit: **(a)** System control; **(b)** System behavior

Indeed, the control current, defined as the difference between the SEC current and the electrical load current ($I_{CTR} = I_{SEC} - I_{Load}$), is used to determine the state of the system. Therefore, when the $I_{CTR} > 0$, the system provides surplus power ($I_{EX}$) which must be controlled and stored. Otherwise, when $I_{CTR} < 0$, the system rectifies the identified power by the deficit current $I_{DEF}$ when the $I_{CTR} < 0$, the system corrects the identified power by the deficit $I_{DEF}$ current. In the next control step, the excess ($I_{EX}$) and deficit ($I_{DEF}$) currents are used as the basic decision input parameters. These currents aim to active or to deactivate the system components as required. The proposed EMU is characterized by several states {S1 … S6} (see Fig. 2b).

# 4  Findings and Discussion

In this section, the simulation results of the proposed autonomous system block diagram performed under Matlab/Simulink are presented and discussed. Thus, an experimental data profile extracted from the Tunisian meteorological database has been introduced and exploited (temperature and solar radiation). Moreover, a variable load profile takes into account the energy consumption of a remote housing is also presented. The simulation time considers the power consumption during seven consecutive days (168 h).

During the HPS system operation, the SEC is used as a primary source for ensuring the energy demand. However, this feeding source cannot provide continuously the required energy consumption. For that, the system sustains a behavior change based on the decisions taken by the EMU to withstand any occurred fluctuations. The EMU refers to a control current ($I_{CTR}$) to check the system state (Excess/Deficit) to select the appropriate operation mode (energy storage/energy recovery). Thus, the given Fig. 3 presents the power balancing between the energy source SEC and the load demand to separate the mode intervals. Figure 4 illustrates the excess power time periods in which the out-coming energy from SEC exceeds the system requirements.

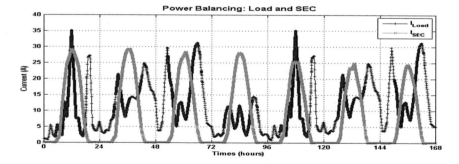

**Fig. 3.** HPS system control current

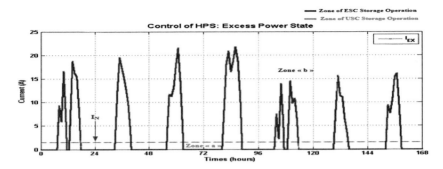

**Fig. 4.** Excess power state

As seen, the system behavior is controlled through the excess current ($I_{EX}$). So, Two zone of system operation are selected:

- When $I_{EX} > I_N$, the ESC is operated: **zone "b"**. Indeed, the specified component is turned on to ensure the storage process ($D_{ESC} = 1$) by producing hydrogen.
- When $I_{EX} < I_N$, the USC is running to accomplish the storage process ($D_{USC} = 1$): **zone "a"**. So, the USC is being charged while ESC is turned off ($D_{ESC} = 0$).

Figure 5 illustrates the state operation of each storage component during the excess power running mode also the energy flow production (hydrogen, USC charging).

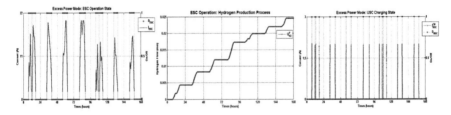

**Fig. 5.** Energy storage process

The deficit power problem is generally caused by the disability of the SEC to sustain the load demand.

So, the system, in this state, is controlled through the deficit current ($I_{DEF}$) that is considered as the main parameter responsible for the decisions taken by the EMU (see Fig. 6).

- When $I_{DEF} > I_N$, the ERC is turned on ($D_{ERC} = 1$) to cover the energy scarcity while the USC is being off ($D_{USC} = 0$). So, the energy recovery is ensured by hydrogen consumption process.
- When $I_{EX} < I_N$, ERC is switched off ($D_{ERC} = 0$) and the USC is intervening ($D_{USC} = 1$) through its discharge to ensure the remaining energy.

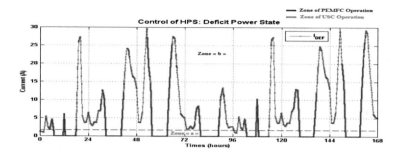

**Fig. 6.** Deficit power state

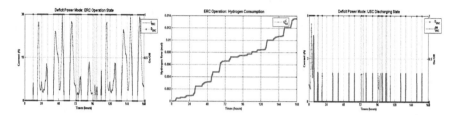

**Fig. 7.** Energy recovery process

Figure 7 highlights the recovery components behavior during the deficit power running mode also it describes the energy consumption state control (hydrogen and USC discharging). In addition, the operation periods of both ERC and USC is given by the table below.

Referring to these states, the EMU chooses the appropriate component that complies all the system necessities. In this stage, the system changes its behavior according to any occurred fluctuation in $SOC_{H2}$ or $SOC_{USU}$. So, the Fig. 8 illustrates the overall system control states and the entire USC efficiency registered during the simulation test. Indeed, the hydrogen state value $SOC_{H2}$ fluctuates between 0% and 41% while the USU state toggles in range [80%–18%]. Hence, it is shown that the system is not exposed to any constraint during its operation.

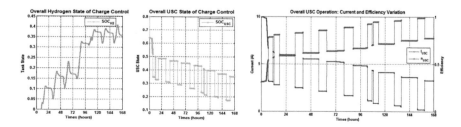

**Fig. 8.** Overall system behavior

The overall system efficiency is calculated by identifying the operation mode (excess/deficit). Indeed, the overall system performance value ($\eta_{SYS}$) depends on the partial efficiencies per mode ($\eta_{M1}$ and $\eta_{M2}$) (see Fig. 9). As observed, the overall HPS efficiency reaches at maximum 35%. Moreover, the higher registered efficiency value is about 42% in excess power mode versus 85% in deficit power mode. Relying on these obtained results, the proposed HPS system proves its reliability and effectiveness to be one of the most effective legacy systems.

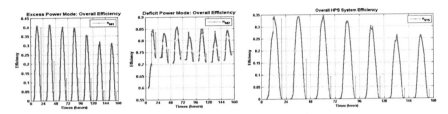

**Fig. 9.** Overall system performance

## 5   Conclusion

This article aims with autonomous hybrid energy system (Solar energy/Fuel cell) using hydrogen and ultracapacitor energy storages for long and short times components which are observed to have better potential for controlling energy demand. Moreover, an efficient scheme energy management approach using a deficit and excess modes was proposed to control the required energy (production/consumption). The excess and deficit operation modes were included as a challenging task for designing the EMU strategy that is made to compensate dynamically the delivery power on the autonomous system under several scenarios. The simulation results showed the performance of the presented EMU as well as the reliability of the proposed autonomous system to provide high-quality power to the load by being against undesirable peaks.

Future work aimed at implementing an intelligent energy management (Using multi-agent systems) and implements it through an embedded system where the delay will be a very critical factor.

## References

1. Mathews, J.: Seven steps to curb global warming. Energy Policy **35**, 4247–4259 (2007)
2. Agbossou, K., Chahine, R., Hamelin, J., Laurencelle, F., Anouar, A., St-Arnaud, J., Bose, T.: Renewable energy systems based on hydrogen for remote applications. J. Power Sources **96**, 168–172 (2001)
3. Banat, F., Qiblawey, H., Al-Nasser, Q.: Economic evaluation of a small RO unit powered by PV installed in the village of Hartha, Jordan. J. Desalin. Water Treat. **3**,169–174 (2009)
4. Mainwaring, A., Culler, D., Polastre, J., Szewczyk, R., Anderson, J.: Wireless sensor networks for habitat monitoring. In: Proceedings of the 1st ACM International Workshop on Wireless Sensor Networks and Applications, pp. 88–97 (2002)

5. Kanase-Patil, A., Saini, R., Sharma, M.: Integrated renewable energy systems for off grid rural electrification of remote area. J. Renew. Energy **35**, 1342–1351 (2010)
6. Lee, K., Shin, D., Lee, J., Kim, T., Kim, H.: Experimental investigation on the hybrid smart green ship. In: Conference Intelligent Robotics and Applications, pp. 338–344. Springer (2013)
7. Abadlia, I., Bahi, T., Bouzeria, H.: Energy management strategy based on fuzzy logic for compound RES/ESS used in stand-alone application. Int. J. Hydrog. Energy **41**, 16705–16717 (2016)
8. Fazelpour, F., Soltani, N., Rosen, M.A.: Economic analysis of standalone hybrid energy systems for application in Tehran, Iran. Int. J. Hydrog. Energy **41**, 7732–7743 (2016)
9. Wang, F.-C., Peng, C.-H.: The development of an exchangeable PEMFC power module for electric vehicles. Int. J. Hydrog. Energy **39**, 3855–3867 (2014)
10. Kim, B.H., Kwon, O.J., Song, J.S., Cheon, S.H., Oh, B.S.: The characteristics of regenerative energy for PEMFC hybrid system with additional generator. Int. J. Hydrog. Energy **39**, 10208–10215 (2014)
11. Nfah, E.M., Ngundam, J.M., Tchinda, R.: Modelling of solar/diesel/ battery hybrid power systems for far-north Cameroon. Renew. Energy **32**, 832–844 (2007)
12. Zhang, X., Mi, C.C., Masrur, A., Daniszewski, D.: Wavelet-transform based power management of hybrid vehicles with multiple on-board energy sources, including fuel cell, battery and ultracapacitor. J. Power Sources **185**, 1533–1543 (2008)
13. Gao, W.: Performance comparison of a fuel cell battery hybrid power train and a fuel cell-ultra capacitor hybrid power train. IEEE Trans. Veh. Technol. **54**, 846–855 (2005)
14. Payman, A., Pierfederici, S., Meibody-Tabar, F.: Energy control of supercapacitor/fuel cell hybrid power source. Energy Convers. Manag. **49**, 1637–1644 (2008)
15. Miller, J.R., Simon, P.: Fundamentals of electrochemical capacitor design and operation. Electrochem. Soc. Interface (2008, spring)

# A Technology Centric Strategic Approach as Decision Support System During Flood Rescue for a Better Evacuation and Rehabilitation Plan

Mohammad Nasim[1]([envelope]) [ORCID] and G. V. Ramaraju[2] [ORCID]

[1] Department of Computer Science Engineering, Lingayas University,
Faridabad, India
mnasimsiddiqui@gmail.com
[2] Former Pro Vice Chancellor, Lingayas University, India and former
Scientist G, Ministry of Electronics & Information Technology,
Government of India, Faridabad, India
gvramaraju54@gmail.com

**Abstract.** This study explains, how a strategically defined software solution can play an important role in the response and recovery phase of a flood disaster. It explains a better use of the available technology like Hadoop, Twitter, Tableau, Python, Kafka to create a managed solution to make operations easy at the time of recue. It is a decision support system which takes data feed from Twitter batch, Twitter streaming, and Movie data, stores it in big data and then perform various useful analytical solutions and helps in performing actions on the ground. Disaster Management team, Local Response Team, Visitors, Volunteers, NGO's, Doctors, Citizens are the key contributor in the support system. There is a continuous monitoring of flood factors and their current situation like rivers, mountains, valleys, rain fall, sea and keep on updating relevant info in Decision Management system. This system helps in allocation and monitor resources in terms of Shelters, Hospital, Buildings, transport and Adhoc Assets. It provides an important role in communication between two modules of the systems. It also explains about the standards and communication policy which is very much required to make an effective system [1] on the ground reality.

**Keywords:** Flood · Rescue · Twitter · Disaster response · Kafka · Hadoop

## 1 Introduction

Disaster can be defined as a sudden accident or a natural catastrophe that causes great damage or loss of life. Below pic (Fig. 1) is the picture of Kerala (India) Flood 2018. No one imagined about this huge loss to society. When we talk about the disaster, only one thing came into our mind that we as an educated human being, are destroying our nature very fast which in return resulting into these kind of disasters. We cry a lot at the time of disaster about the actions, standards and policy but after somedays again everybody forgets and become busy on other jobs. Nature reminds us time to time but

© Springer Nature Switzerland AG 2019
C. Benavente-Peces et al. (Eds.): SEAHF 2019, SIST 150, pp. 282–288, 2019.
https://doi.org/10.1007/978-3-030-22964-1_31

we don't hear it or follow the nature rules. In this study, we shall talk about the response and recovery phases specific to flood. Also how software technology can support us to make the system in a better way during response. This study also provides information for developing an effective decision support system.

**Fig. 1.** Kerala flood 2018

## 2    Problem Statement

To define the problem statement, taking few recent examples of flood in India that is heavy rains in Assam in July–Aug 2016 resulted in floods affecting 1.8 million people and flooding the Kaziranga National Park killing around 200 wild animals. Following heavy rain in Gujarat in July 2017, Gujarat state of India was affected by the severe flood resulting in more than 200 deaths. Following high rain in Kerala in late July 2018 and heavy Monsoon rainfall from august 8, 2018, severe flooding affected the Indian state of Kerala resulting over 445 deaths. Several lacs people are forced to live in camps across the world as everything being destroyed for them. They have to start the life again from zero with all struggles. In-spite of several plans implemented by central govt, state govt. and concerned disaster authorities, we are not able to manage these disasters effectively [2, 3]. These counts could be increased to thousands but thanks to the people who came in for support specially army, ngo's and local boat-mans/bodies who collectively worked and made this death counts to few hundreds only. In case of flood disaster, generally there is not any sudden increase of water level. Nature gives us little time to evacuate us. Also, it is not like that everybody was sleeping in the affected area and they become dead. So here our management planning partially fails instead of investing so much money.

## 3    Research Goals and Objectives

- With better practices and strategies, we can have a better planning and hence could save more people and assets.
- Although death counts could also be made very less if there would be a proper management and proper responsibility and seriousness individual/department (Local/Central) would take and handle.

- Concerned authorities in the affected area can develop mechanism about the ways to find out the flood zones levels and can categorize based upon severity and accordingly make an evacuation plan.
- From the social media front: Twitter can become a better medium at ground level to face this unholy events by providing its best possible support.

## 4  Purpose of the Study

To provide a better approach during rescue operations and rehabilitation process below are some key points/questions to be rolled out at ground reality.

### 4.1  How to Prepare for a Crisis on Social Media?

The only reason the saying "hope for the best but prepare for the worst". First of all, as per the water level, we can tag flood zones number to each area of any particular state/county.

### 4.2  Create a Social Media Crisis Management Plan

Having a nation-wide plan in place will empower you to act quickly and effectively when a crisis begins. Instead of wasting time debating how to handle things on social media, you'll be empowered to take action and prevent the crisis from growing out of control.

### 4.3  Run Social Media Crisis Simulations

The middle of a real crisis is the worst time to realize your plans don't hold up. Identify a type of crisis that would have a big impact and practice running through every step outlined in your plan. It will give everyone a better sense of how long it really takes to execute the plan, and help to identify any gaps or weak spots that require more attention.

### 4.4  How to Identify a Potential Crisis on Twitter?

**Listen carefully**

As a twitter management team member, with at least one eye on the internet at all times, you'll often be the first person to notice a potential threat starting to emerge on twitter. Setting up streams in Twitter Management tool to monitor specific keywords or hashtags can help you stay proactive in identifying these potential crises.

**Communicate internally and externally**

As important as it is to quickly send out external messaging about the crisis, communicating internally with concerned staffs/authority is also crucial to prevent misinformation and the spread of rumors.

## 4.5 After the Disaster—Next Steps?

Basically there are two sections in this DSS (Decision support system). During response phase, government actions has to be taken based on the processed twitter data. During rehabilitation/recovery phase, government need to prepare a transparent system in web portal about the affected families.

# 5 Research Design and Methodology

This section includes the conceptual design and technical architecture of the system by keeping the objective and purpose in mind. Anything is possible if there is awareness and willingness of responsibility and transparency in the society. This is only possible when we make our people as a contributor to the system and also appraise too.

## 5.1 Conceptual Architecture

Please find the below conceptual architecture for a better understanding from the work-flow perspectives (Fig. 2.).

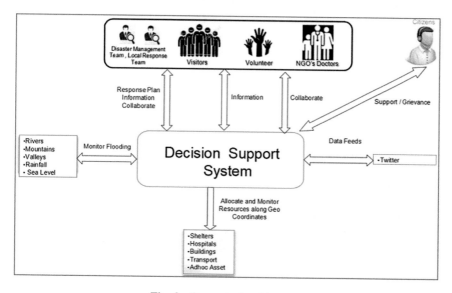

**Fig. 2.** Conceptual architecture

## 5.2    Technical Architecture

Please find the technical architecture of the system below (Fig. 3):

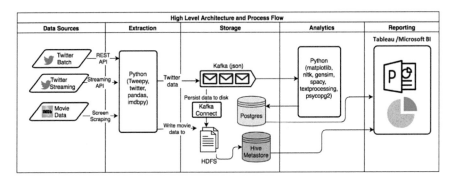

**Fig. 3.**  Technical architecture of system

Twitter Batch, Twitter Streaming and Movie Data are the important data sources [4]. Through Python scripting this data can be extracted and stored into JSON and HDFS formats. After running analytics on the JSON data, the processed (analytical) data gets stored in PostgreSQL. Tableau or Microsoft BI or any other analytical software can be used to display the dashboard and reporting sections of the system.

# 6    System Development

## 6.1    Key Files

**Twitter Movie Data:**
　　**extract_tweets_restful.py**
　　**Kafka_Consumer.py**
　　**create_tweets_table.sql**
　　Consumer preprocessing.py
　　**textfeatures.py.** – Textfeatures program combines functions from pandas, gensim and sklearn to create vectorization, feature extraction and sentiment scoring capabilities
　　**utils.py.** – Utils function offers a logger
　　**app.py.** – app.py program extracts movie names from hive tables, transforms names into hashtags and passes that onto batch and streaming producer scripts.

## 6.2    Pre-requisites

PYTHON, TWITTER, ZOOKEEPER, KAFKA Installation.

### 6.3    Implementation Notes

Kafka is designed for scale-out distributed messaging system for log. Kafka cluster can store large dataset for longer period [5]. Allow clients to read in different orders.

Internal zookeeper allows producers and consumers to work at different speeds Handle data bursts and fault tolerance.

### 6.4    Tableau Analytics Dashboard

**Time Series Analysis**

The time series analysis provides visualization of the number of tweets per candidates as a function of time. Simply select a starting and ending date, as well as a period to see the related date.

Below on the left panel is the time series visualization (Fig. 4).

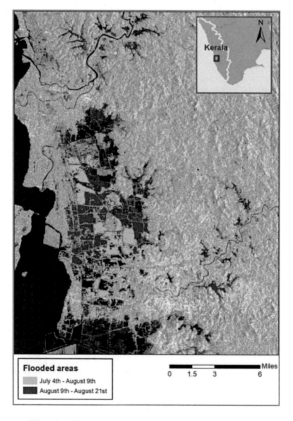

**Fig. 4.** Time series visualization of Kerala flood

# 7  Technology/Software Used

1. Kafka version 2.11
2. Hadoop 2.6
3. Anaconda Python 2.7
4. Postgres 9.6
5. Apache Hive 2.1
6. Tableau Desktop/Microsoft BI 10.0.2.

# 8  Conclusion and Benefits

The flood crisis management plan should include:

- Guidelines for identifying the type and magnitude of a crisis.
- Roles and responsibilities for every concerned government department.
- An internal and external communication plan for internal updates.
- Up-to-date contact information for critical people stuck in flood.
- Approval processes for messaging posted on twitter specific to flood.
- Any pre-approved external messaging, images, or information.
- A law on the nation-wide flood twitter policy. Need to build twitter standards during flood and awareness to people.
- There should be only one government financial contribution account.
- Location coordinate and geo-tagged images plays an important role to identify the help-spots. Government should provide awareness program for the same.

# References

1. Gunawong, P., Butakhieo, N.: Social media in local administration: an empirical study of twitter use in flood managemen. In: 2016 Conference for E-Democracy and Open Government (CeDEM), pp. 77–83 (2016)
2. Funayama, T., Yamamoto, Y., Tomita, M., Uchida, O., Kajita, Y.: Disaster mitigation support system using Twitter and GIS. In: 2014 Twelfth International Conference on ICT and Knowledge Engineering, pp. 18–23 (2014)
3. Ogie, R.I., Forehead, H.: Investigating the accuracy of georeferenced social media data for flood mapping: The PetaJakarta.org case study. In: 2017 4th International Conference on Information and Communication Technologies for Disaster Management (ICT-DM), pp. 1–6 (2017)
4. Layek, A.K., Pal, A., Saha, R., Mandal, S.: DETSApp: an app for disaster event tweets summarization using images posted on twitter. In: 2018 Fifth International Conference on Emerging Applications of Information Technology (EAIT), pp. 1–4 (2018)
5. Anbalagan, B., Valliyammai, C.: # ChennaiFloods: leveraging human and machine learning for crisis mapping during disasters using social media. In: 2016 IEEE 23rd International Conference on High Performance Computing Workshops (HiPCW), p. 5 (2016)

# Specification of the Data Warehouse for the Decision-Making Dimension of the Bid Process Information System

Manel Zekri[1]([✉]), Sahbi Zahaf[2], and Faiez Gargouri[2]

[1] Department of Computer Science, Faculty of Sciences of Tunis, Tunis, Tunisia
manel.zekri@planet.tn
[2] MIRACL Laboratory, Higher Institute of Computer and Multimedia, Sakiet Ezzit Technopole, Sfax 242-3021, Tunisia
sahbi@zahaf.net, faiez.gargouri@isimsf.rnu.tn

**Abstract.** Bid process is a key business process which influences the enterprise's survival and strategic orientations. Therefore, the Bid Process Information System (BPIS) that supports this process must be characterized by integrity, flexibility and interoperability. Nevertheless, the specification of this system, has to deal with "three fit" problems. Four dimensions have been identified to cope with such failures: operational, organizational, decision-making, and cooperative dimensions. In this paper we are particularly interested to organize the decision-making dimension of the BPIS. Thus, we propose an approach for representing data warehouse schema based on an ontology that captures the multidimensional bid-knowledge.

**Keywords:** Bid process · Information system · Data warehouse · Conceptual data model · Multidimensional design · Ontology

## 1 Introduction

The bid process is a particular environment for the exploitation of business process [7]. Indeed, this process corresponds to the conceptual phase of the lifecycle of a product. A bid process interacts upstream and involves other processes being design process. It aims to examine feasibility of the bid before negotiating any contract with any owner following a pre-study carried out before a project launch.

The IS (Information System) [4] that allows to exploit the bid process (Bid Process Information System or BPIS) must be [7]: integrated, flexible and interoperable. Nevertheless, the enterprise architecture approach [4], on which we rely to implement this system, has to deal with "three fit" problems: vertical, horizontal and transversal fit (Fig. 1). Such problems handicap the exploitation of these three criteria [7].

The "vertical fit" represents the problems of integrity from a business infrastructure, which is abstract, to a technical infrastructure, which represents implementations. The "horizontal fit" translates not only the software's problems of identification (induced by the "vertical fit" problems) that can cover the entire infrastructure of the company's business, but also the intra-applicative communications problems (internal

C. Benavente-Peces et al. (Eds.): SEAHF 2019, SIST 150, pp. 289–295, 2019.
https://doi.org/10.1007/978-3-030-22964-1_32

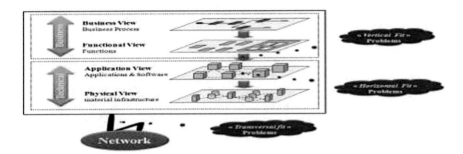

**Fig. 1.** Enterprise Architecture I.S reference model [4]: "three fit" problems [7]

interoperability) to ensure the interactions between software's of the same technical infrastructure in the company. The "transversal fit" translates the inter-applicative communications problems (external interoperability carried out dynamically through a network). In this context, we have identified an IS that must be support four dimensions to deal with such failures: operational, organizational, decision-making, and cooperative dimensions [7].

Afterwards we are particularly interested to organize the decision-making dimension of the BPIS. In this context, data warehouse is mainly used for making better decisions that can improve the business of organizations and optimize bid process [1]. Hence, building a data warehouse is a complex task that aims to satisfy the needs of decision makers. One of the key points to the success of a bid data warehousing project is the design of the multidimensional schema [5]. Many researches showed that the use of ontology in IS design, especially in BPIS, becomes more and more promising [3]. In this paper, we propose a method for multidimensional design of data warehouses, from an operational data source, using ontologies. Furthermore, we introduce the concept of multidimensional ontology as a tool for the specification of multidimensional bid knowledge. In addition, we present an ontology-based method for data modeling schema that eventually covers different phases of the data warehouse lifecycle, and takes into account the users by considering their personalized needs as well as their bid knowledge.

This work is organized as follows. The second section describes the resolution of the "three fit" problems. The third section shows the related work: extending Multidimensional Ontology. The fourth section describes our design approach. We end this work by the conclusion and with the prospects of our future works.

## 2 Resolution of the "Three Fit" Problems: BPIS Integrated, Flexible and Interoperable

We have identified four dimensions to deal with "three fit" problems [7]: (i) the operational dimension that serves to specify the bid exploitation process by undertaking a specific project; (ii) the organizational dimension which allow to organize the set of skills and knowledge, that the company acquired during the previous bid in which it participated: the objective is a possible reutilization of this patrimony in future bid projects; (iii) the decision-making dimension which aims at optimizing and making the right decisions that concerns the market offers and that takes place during the company's eventual participation in bid processes; and (iv) the cooperative dimension which aims at ensuring communication intra-enterprise (internal interoperability) and at planning the inter-enterprise communication on demand, in order to realize a common goal (dynamic interoperability).

We suggest filling these dimensions while relying on the following hypothesis (Fig. 2) [7]: "ERP (Enterprise Resources Planning) allows us to build the techno-economic proposal of an offer (cover the operational dimension), as we rely on the organizational memory (to cover the organizational dimension). The set of solutions that make this proposal realistic, are going to be evaluated by a data warehouse (to cover the decision-making dimension)".

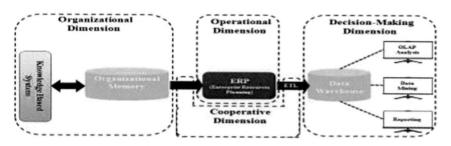

**Fig. 2.** Our urbanized bid process information system (BPIS) [7]

# 3 Related Work: Extending Multidimensional Ontology

In data warehouse design, different modeling techniques are used to represent the multidimensional concepts extracted from data sources, as well as the sources themselves. It can be ER diagram, UML diagram or graphs [8], etc. Unlike ontologies, which are ready for computing, these techniques are conceptual formalizations intended to graphically represent the domain and not thought for querying and reasoning. This work is a continuation of a previous research [8, 9] and it aims at integrating ontologies in the data warehouse design process. So our starting point was our meta-model of data warehouse scheme [9].

The multidimensional ontology is a representation of knowledge dedicated to decisional systems. It specifies the multidimensional concepts and their semantic and multidimensional relations. Its use covers different steps of the data warehouse life-cycle. During these steps, it assists the designer to solve problems of data sources heterogeneity. In the OLAP requirement specification step, the decisional ontology helps to validate the multidimensional concepts (fact, measure, dimension, etc.) and relations between these concepts. Also, it prevents associations between concepts not semantically associable (e.g. associating fact-to-fact, dimension-to-dimension, hierarchy-to-fact, etc.). Various approaches were proposed to guide creating ontologies [6, 5], we mainly based our multidimensional ontology construction process on the approach proposed in [2]. While other techniques (mentioned above) for representing multidimensional knowledge are an appropriate choice for their respective approaches, their use ends when the task of designing the data warehouse is accomplished. On the other hand, the multidimensional ontology can still be useful. It can cover the various phases of the life cycle of a decisional IS, that is to say, from the requirements specification, the design of data warehouse and data marts, until the exploitation and evolution phases. This is convenient because the recent DBMSs allow storing ontologies alongside data in the same database structure [1]. Such database is called OBDB (ontology-based database). Ontologies are scalable and extendable, and they showed their effectiveness for IS and requirements specification [3]. More concretely and in the same way, as ontologies are used for clarifying the semantics of data sources, they are used to identify and manage semantic conflicts between concepts. This allows us to add as many extensions as needed to the multidimensional ontology to cover the various phases of the life cycle of a decisional IS. To demonstrate this aspect, we have proposed an extension that represents the operational data source conceptual schema, in this case, an ER diagram [9].

## 4  Our Data Warehouse Lifecycle: Ontology-Based Method for Data Modeling Schema

In this section, we will present an overview of the first phase of our approach. Our approach is progressive and iterative. The assistance of the designer throughout this construction is optional. The approach can be executed autonomously, but the intervening of a designer during validation steps is advised and will result in better output. The steps are represented by Fig. 3.

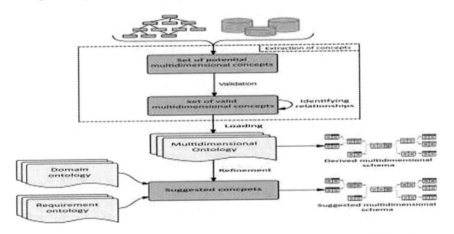

**Fig. 3.** Our data warehouse lifecycle: ontology-based method for data modeling schema

In order to ensure the availability of data, we consider first the data sources that are represented in a conceptual data model of the production base. The next step is to extract the multidimensional concepts. It is divided into three stages that are repeated for each multidimensional concept. First we determine a set of potential multidimensional, using extraction rules.

## 4.1    Fact Extraction

Facts describe the daily activities of the bid companies. These activities result in transactions and produce transaction objects. A transaction object is an entity registering the details of an event such as the payment of the bid proposition, etc. These entities are the most interesting for the data warehouse and are the basis for the construction of the fact tables. However, they are not all important for decision making, so we must choose those that have an interest in our decisional IS.

Usually a transaction object is a complex object containing multiples information's. Therefore its modeling requires its decomposition into associated sub-objects. In the ER model, a transaction object may be represented in one of two forms:

- An entity connected to an association;
- Two entities linked by an association.

In order to determine "Fp" (set of potential facts), we define the following heuristic: HF: All transaction objects are potential facts.

For each identified transaction object identified, we associate a more descriptive name, which will be the name of the fact. These facts are necessarily all pertinent, thus a validation phase where the designer may intervene is essential to retain a subset of valid facts "(Fv)".

## 4.2    Measure Extraction

As we previously stated, a transaction object is the result of the bid companies' activities. Accordingly, the attributes may be measurements of a fact are encapsulated in the transaction object.

The following heuristics determine the potential measures:

Hm1: Mp (fv) contains the non-key numeric attributes belonging to the transaction object representing "fv".

Hm2: If the attribute is a Boolean, we add to Mp (fv) the number of instances corresponding to the values "True" and "False" of this attribute.

- Fv: a valid fact from the previous step;
- Mp (fv): the set of potential measures of "fv";
- Mv (fv): the set of measures of "fv" approved by the designer, which is a subset of "Mp (fv)".

The extraction of measures is also followed by a validation step by the designer in order to determine "Mv (fv)" whose elements satisfy the following assertion:

$$\forall fv \in Fv, \forall m \in M v\,(fv) \rightarrow Is_{\text{meaure}}(m,\, fv)$$

## 4.3 Decision Extraction

Extraction of dimensions is based on a second type of object called base object. A base object determines the details of an event by answering the questions "who", "what", "when", "where" and "how" related to a transaction object. For example, the "Bid Project" event is defined by several components such as (Owner: who bought) and (Bid Proposition: we sold what). A base object completes the meaning of the event represented by a transaction object thus providing additional details.

Each object corresponding to one of these questions, and directly or indirectly linked to a transaction object is a potential dimension for the fact representing the transaction object. The extraction of dimensions consists of determining the name, ID and hierarchy(s) through these heuristics:

Hd1: Any base object directly or indirectly connected to the transaction object of "fv" is a potential dimension of "fv".

Hd2: All IDs of a base object obtained by Hd1 is an "id" of "d".

- Dp (fv): the set of potential dimensions of "fv";
- Dv (fv): the set of valid dimensions of "fv"; which is a subset of "Dp (fv)";
- d: a dimension;
- idd: the "id" of a dimension "d",

The validation step produces two subsets Dv (fv) and IDDv (fv), satisfying the following assertion:

$$\forall fv \in Fv, \forall d \in Dv(fv), \exists idd \rightarrow Is_{dimension}(d, fv) \wedge Is_{dimension_{ID}}(idd, d)$$

## 4.4 Attributes Extraction

We define the following heuristics to determine the potential attributes:

Ha1: Any attribute belonging to the base object containing "idd" is a potential attribute of "d".

Ha2: Any attributes belonging to a base object that generated a valid dimension "dv" is a potential attribute of "dv".

The validation step produces a set of valid attributes Av, satisfying the following assertion:

$$\forall av \in Av(dv) \rightarrow Is_{attribut}(av, dv)$$

Each extracted element becomes an individual (i.e. instance) of the concept representing its role (the concepts are detailed

# 5   Conclusion and Perspectives

In this work, we have interested to define the characteristics of the decision-making dimension of the BPIS. Thus, we have presented an approach for representing data warehouse schema based on an ontology that captures the multidimensional knowledge. We discussed our vision for extending the multidimensional ontology to eventually cover different phases of the data warehouse life cycle. We focused on the design phase, and showed how the use of the multidimensional ontology combined with an extension can be beneficial. In the future we intend to continue to exploit the extensibility of ontologies by considering bid ontologies as extensions to try to improve the resulting data warehouse schema.

# References

1. Bellatreche, L., Valduriez, P., Morzy, T.: Advances in databases and information systems. Inf. Syst. Front. **20**(1), 1–6 (2018)
2. Bentayeb, F., Maïz, N., Mahboubi, H., Favre, C, Loudcher, S., Harbi, N., Boussaïd, O., Darmont, J.: Innovative Approaches for efficiently Warehousing Complex Data from the Web (2017). CoRR https://arxiv.org/abs/1701.08643
3. Calero, C., Ruiz, F., Piattini, M.: Ontologies for Software Engineering and Software Technology. Springer, Berlin (2006)
4. Fournier-Morel, X., Grojean, P., Plouin, G., Rognon, C.: SOA le Guide de l'Architecture du SI. Dunod, Paris (2008)
5. Gallinucci, E., Golfarelli, M., Rizzi, S.: Schema profiling of document-oriented data-bases. Inf. Syst. **75**, 13–25 (2018)
6. Guarino, N.: Formal Ontology and Information Systems. In: Guarino, N. (ed). IOS Press, Amsterdam (1998)
7. Zahaf, S., Gargouri, F.: Business and technical characteristics of the bid-process information system (BPIS). In: 17th International Conference on Information and Knowledge Engineering, pp. 52–60, 17–20 Juillet, 2017. CSREA Press, Las Vegas, Nevada, USA (2017). ISBN: 1-60132-463-4
8. Zekri, M., Marsit, I., Abdellatif, A.: A new data warehouse approach using graph theory. In: IEEE International Conference on eBusiness Engineering (ICEBE 2011) Octobre 19–21, China (2011)
9. Zekri, M., Marsit, I., Abdellatif, A.: Query history based approach for data warehouse design. In: 1st International Conference on Model and Data Engineering (MEDI'2011) Septembre 28–30, Obidos, Portugal (2011)

# Insights in Machine Learning
# for Cyber-Security Assessment

César Benavente-Peces[(✉)] and David Bartolini

Universidad Politécnica de Madrid, ETSIS Telecomunicacin, Campus Sur,
Calle de Nikola Tesla sn, 28031 Madrid, Spain
cesar.benavente@upm.es
http://www.upm.es

**Abstract.** Information transactions and sharing is a common activity
in daily life, at work, private and leisure activities. Currently, companies perform most of information delivery, storing and sharing electron-
ically through internal and external connections. External connections
are source of cyber-risk as unauthorized connections could provoke enormous damages. In this framework, cyber-risk assessment is a must for
companies as their incomings are in risk, but its implementation is cum-
bersome. Financial issues produced by breaches in information infrastructures and procedures security are covered by the so called *cyber-
insurance*. Cyber-security management requires a lot of resources both
economical and human, involving all the departments and people of the
company. Among other activities, alerting of potential risks and damages before any attack is produced, or providing the appropriate means,
after a cyber-attack, to prevent a new one and restore, when possible,
the damages, are responsibilities of decision-makers. In this scenario,
*insurability* assessment is a must for insurance companies, and risk management engineers are the staff responsible to analyse and report about
cyber insurance acceptance test. In order to optimize and automatize the
procedure of insurability assessment, artificial intelligence and machine
learning techniques constitutes an appropriate tool. In this paper the
authors focus the attention on the analysis of the collected data used to
take decisions in order to determine dependencies, correlations and take
further steps to predict insurability, which in next investigations will be
optimized through machine learning techniques.

**Keywords:** Cyber-security · Cyber-risk management ·
Information security · Data protection · Cyber-insurability prediction

## 1 Introduction

The availability of a global cyber-space and the integrity, authenticity, reliability and confidentiality of data in cyber-space have become key issues of the
globalization in the 21st century. Ensuring cyber-security and avoiding vulner-

C. Benavente-Peces et al. (Eds.): SEAHF 2019, SIST 150, pp. 296–305, 2019.
https://doi.org/10.1007/978-3-030-22964-1_33

abilities has thus turned into a main challenge to be faced by governments, public services, business and society both on national and international context. The cyber-security strategy is intended to assess and improve the framework conditions in this field.

Cyber-insurance represents a new dynamic segment and a market with considerable potential growth for insurers. Companies estimate that there is a premium potential of at least 700 million € in Germany by the end of 2019. Many companies, especially small and medium-sized (SME) ones, continue to underestimate the risks associated with using the Internet. In large companies, safety management is in general better trained than in medium-sized companies. However, further challenges for companies are the regulatory challenges in the context of the General Data Protection Regulation (GDPR) and the requirements of the IT (Information Technologies) security law, among others for operators of critical infrastructures. The global network creates problems that have gained significance under the term cyber-risks. Any company connected to the Internet is vulnerable to intrusions. Attacks on Sony, Google, Amazon and the German Bundestag are few examples which show the dimensions. IT security experts point out that it has become impossible to prevent data breaches. An additional protection is thus a Cyber Police. The focus of this paper is to present a risk-related approach in customer analysis, which helps to assess the question of insurability. The conclusion obtained in [1] served to jointly develop the insurance-relevant customer risk. According to the name, a dialogue cannot represent a risk assessment and should also be conducted openly and serve as an exchange between clients and insurers. However, to subsequently implement the insurability check, the findings must be recorded in a structured manner. To guarantee this, an own-used question board is used, which is sorted in the respective question categories.

The cyber-risk assessment in a company is traditionally developed using a questionnaire. In general, such a questionnaire is structured considering the various domains and different cyber-security maturity levels of the addressed customer. Each domain and maturity level have a number of properties which are classified in accordance to the factors which impact the result. Engineers specialized in cyber-risk assessment collect the data and systematically introduce them in the assessment questionnaire [2–4, 7–10, 12–16].

Depending on the Cyber-incidents can impact remarkably on corporate turnover. Costs added by the incident may include forensic investigations, Public Relation campaigns, legal fees and court fees, consumer credit monitoring, technology changes and comprehensive recovery measures [3,5] Cybersecurity therefore needs to be integrated across the enterprise as part of corporate governance processes, information security, business continuity and third-party risk management. Cybersecurity roles and processes referred to in the assessment may be separate roles within the security group (or outsourced) or may be part of broader roles within the institution.

Among other factors, during cyber-security risk assessment potential impacts must be reliably estimated. In order to get a reliable estimation of the entire organization, including all existing departments and a deep knowledge of the incident impact to the lowest level considering any potential ramification is required [18].

First and foremost, companies are affected by the regulations of the IT GDPR, which are counted among the critical infrastructures (CRITIS) and therefore have a special significance for the common good of the state and its citizens.

The remaining part of this paper is structured as follows: In Sect. 2 the cyber-risk related data collection composing the reference database is introduced. Additionally, the parameters are defined to give an idea of their meaning. In Sect. 3 the dependencies among the various parameters are analyzed in order to identify correlation between them and determine the model which better allows predict results from given inputs. The results obtained along this investigation are discussed in Sect. 4. Finally, some concluding remarks are provided in Sect. 5.

## 2    Data Collection

In this section, the various parameters considered in the cyber-risk analysis of a company are described and a real example is considered to conduct a realistic evaluation. Table 1 shows the parameters involved in cyber-risk assessment in numbers.

**Table 1.** Cyber-risk parameters in numbers.

| count | $1.296 \cdot 10^6$ | $1.296 \cdot 10^6$ | $1.296 \cdot 10^6$ | $1.296 \cdot 10^6$ | $1.296 \cdot 10^6$ | $1,30 \cdot 10^9$ | $1.296 \cdot 10^6$ |
|---|---|---|---|---|---|---|---|
| mean | 39.366.512 | 505.401 | 30.666.917 | 2.887.114 | 202.160 | $2,74 \cdot 10^{12}$ | 585.648 |
| std | 14.165.184 | 500.164 | 6.077.783 | 382.690 | 401.766 | $7,21 \cdot 10^{12}$ | 492.800 |
| min | 18.000.000 | 0 | 15.960.000 | 2.000.000 | 0 | $1,12 \cdot 10^9$ | 0 |
| 25% | 27.000.000 | 0 | 26.305.000 | 3.000.000 | 0 | $6,3 \cdot 10^9$ | 0 |
| 50% | 39.000.000 | 1.000.000 | 30.447.500 | 3.000.000 | 0 | $1,31 \cdot 10^{10}$ | $10^6$ |
| 75% | 51.000.000 | 1.000.000 | 34.770.000 | 3.000.000 | 0 | $4,74 \cdot 10^{10}$ | $10^6$ |
| max | 99.000.000 | 1.000.000 | 52.580.000 | 3.900.000 | 1.000.000 | $4,82 \cdot 10^{13}$ | $10^6$ |

As shown in the Table 1, several statistics are obtained and provided for each parameter, including the cumulative value, the mean, the standard deviation, the minimum value, some values at a given percentage and the maximum value. These are the individual features of each group of data collection. In the following subsection the various parameters are described in order to understand their properties individual properties.

### 2.1    Definitions

In the next paragraphs the definition of the various parameters concerning the cyber-risk and cyber-analysis evaluation are defined.

- **Turnover** of the company (given in Millions of Euros). The greater the company's turnover, the larger its size and thus its IT infrastructure. In addition, it is well-known that a big and widely known company attracts the attention and it is more exposed to cyber-attacks than a small one.
- **Other IT insurances.** This parameter refers to the fact whether the company has already contracted other IT insurances to the customer (e.g., electronics insurance, etc.). At the same time, in the context for cyber-risk and aimed at getting a well-deployed risk-based approach, this parameter is a relevant aspect if the customer has more IT insurances besides the cyber-insurance and whether this additional risk transfer has a positive correlation. This parameter takes value 1 for additional IT insurance, or 0 for non-additional IT insurance.
- **CC_PII** refers to how much personal data regarding credit card data is processed (in thousands). Data breaches have dramatic results on the privacy of personal data as far as the card-holder data are stolen by criminals through cyber-security breaches. In such a case, several claims would be addressed to the company which will be forced to pay fines for the produced damages, given the legal context (e.g. General Data Protection Regulation) and/or in the regulatory context (e.g. PCI DSS).
- **Critical infrastructures (KRITIS).** The customer is one of the critical infrastructures (if yes the value is 1 and if not teh value is 0). As critical infrastructures, they are an attractive target for criminals. Hence, these companies needs to have a higher maturity level in order to prevent cyber-risks.
- Investments on IT/Cyber Programs (**Cyber_Invest**). It is another economic parameter which is worthy to be studied through correlation analysis. In general, it has been observed that a strong development in this area will have a positive impact on the cyber-risk factor.
- (**Insurance Claim**: Already happened 1 or not happened 0) In the case a cyber-attack succeeds, a cyber-damage is produced and finally, this factor impacts the economy of the company if already a damage has occurred. Then, it is also a very important aspect within the customer portfolio.
- The result of the technical cyber-risk assessment (The cyber-risk questionnaire 37 questions within are the 11 show-stoppers). As described in [1], a company can be technically insured if the result of the cyber-risk assessment reaches a minimum score of 2,00. Therefore, these criteria is also interesting for this approach.

First of all, before studying the inter-dependencies in order to introduce prediction techniques in machine learning, an individualized analysis of the different parameters will be carried out; specifically the analysis focuses on the parameters histogram, that is to say, how the values of the given parameters are distributed on its range observing whether they mainly concentrate around certain values or they spread in a wide range.

Figure 1 shows the histogram of the *CC_PII*. The analysis of the picture shows that this histogram is not evenly distributed but the values are mainly concentrated at the middle of the scale.

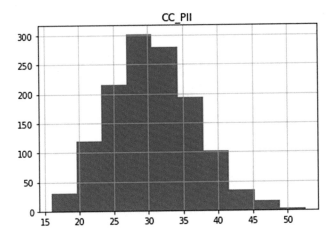

**Fig. 1.** Histogram of the *CC/PII*

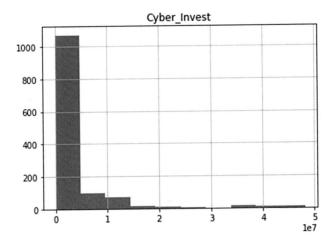

**Fig. 2.** Histogram of the *Cyber-investment*

Figure 2 depicts the histogram referring to the investment in cyber-security of the company. Observing the picture we realize that within the range of analysis of cyber-investment most of the cyber-security company investments concentrate at low values, with few investments at high values.

Figure 3 shows the histogram regarding the company *Turnover*. It can be observed that there is a greater concentration of values in the lowest middle part of the range of analysis, i.e., the distribution is more or less uniform, while in the highest middle part the values can be almost neglected.

The analysis of these histograms gives an idea of the behaviour of the different companies regarding the investment in cyber-security and the preoccupation about cyber-attacks.

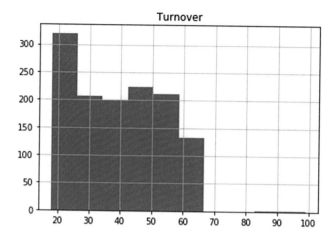

**Fig. 3.** Histogram of the *Turnover*

# 3  Parameters Dependencies Analysis

In this section we'll analyze the different inter-dependencies between the parameters that concern the cyber-risk assessment. Next, an analysis of the interdependence between the different variables will be carried out among the different parameters in order to see how each of them influences the insurability of a given company company. A linear relationship between a couple of parameters would mean that they are strongly correlated and hence machine learning based prediction techniques can be easily used.

One likes investing in certain aspects as, in consequence, cybersecurity increases in the returns of the investment linearly in proportion to that investment. Knowing this interdependence can be a crucial and impacting parameter to finally determine which is the best strategy of investment in cybersecurity of the company. From the point of view of the possibility of assuring the risks of cyber-attacks from companies, insurance companies must analyse the possibility that they have to give compensation in the event that the company suffers a cyber-attack and there are third parties or third parties Companies affected by these cyber-attacks and sue requesting a solution to your problem. When a company wishes to take out insurance related to cybersecurity, the insurance company must carefully analyse the risks that this company may face in the face of cyber-attacks, and for this purpose it must carefully analyse the different parameters mentioned above. The analysis of these parameters is done nowadays in a non-automated way so the design of appropriate strategies based on artificial intelligence machine learning and Big data are the most optimal ways to obtain a fast and efficient analysis of the risk to cyber-attacks of the companies . Therefore, the analysis of the parameters that affect cybersecurity and the cyber-risk of attacks that a company may suffer are of crucial importance when determining which algorithms reliably allow us to decide on the convenience or

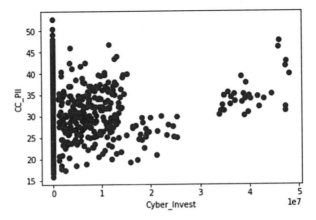

**Fig. 4.** Relation between the *CC/PII* and the *Cyber Investment*

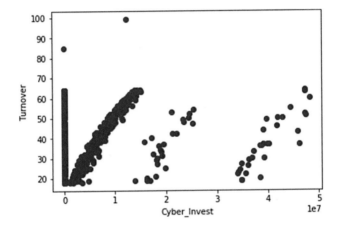

**Fig. 5.** Dependence between the *Cyber Investment* and the *Turnover*

not of hiring a company. safe and in addition to being able to determine the fundamental risk factors that affect the company. Figure 4 represents the relation between the CC/PII and the Cyber. The careful analysis of the figure concludes that these parameters are not strongly correlated given the spatial dispersion of the points which concentrate at low values of the *Cyber_invest*. Figure 5 represents the company turnover with respect the cyber-investment. As depicted in the figure, there is a strong linear relation between both variables, i.e., they are quite correlated. Figure 6 depicts the relation between the *KRITIS* and the *insurance claim*. In this case, the analysis of the points spatial location reveals that these parameters are not strongly correlated given the spatial distribution which seems to be like a discrete function where the points concentrate at four corners of a square.

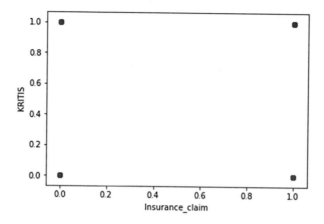

**Fig. 6.** Relation between the *KRITIS* and the *Insurance claim*

## 3.1 Multiple Regression

In reality, there are multiple variables that influence the prediction of the company insurability. When more than one independent variables (parameters) are involved in the process, then it is usually called a multiple-linear-regression. An example for this case is predicting insurability using the company *turnover*, *cyber-investment* and existence of *other insurance*. The positive fact here is that the multiple-linear-regression is the extension of the simple-linear-regression model. The multiple regression coefficients result in:

$$[[1.73077633e - 07 - 1.22304544e + 002.32521460e - 01]] \qquad (1)$$

In order to obtain the coefficients of the corresponding model appropriate Python libraries are used. As mentioned before, *coefficient* and *intercept*, are the parameters of the fit line. Given that the model we are using is a multiple linear regression, with 3 parameters, and knowing that the parameters are the intercept and coefficients of hyperplane, sklearn can estimate them from our data. Scikit-learn uses plain Ordinary Least Squares method to solve this problem.

The ordinary Least Squares (OLS) is a popular method to estimate the unknown parameters in a linear regression model. OLS chooses the parameters of a linear function of a set of explanatory variables by minimizing the sum of the squares of the differences between the target dependent variable and those predicted by the linear function. In other words, it tries to minimize the sum of squared errors (SSE), or mean squared error (MSE), between the targeted variable (y) and the predicted output $(\bar{y} - \hat{y})$ over the whole samples available in the data-set.

## 4    Results

To determine the convenience of using these techniques, it is important to analyze and clarify the results. Given the estimated target output denoted as $\bar{y} - \hat{y}$, the corresponding (correct or actual) targeted output $y$, and the Variance $Var$, i.e., the square of the standard deviation, then the *explained variance* (*expVar*) is estimated as follow:

$$expVar(y, \hat{y}) = 1 - \frac{Var(y - \hat{y})}{Var(y)} \qquad (2)$$

The best possible score of the explained variance is 1.0, and getting lower values are considered as worse results. In the example concerning companies insurability we are focusing on the resulting values are:

- Residual sum of squares: 183.47
- Variance score: 0.02

## 5    Conclusion

Cyber-risk assessment is a hard and tedious tasks which companies should perform in order to avoid potential cyber-attacks. Based on the conclusions provided by the evaluation results insurance companies decide whether a company is insurable or not, as a cyber-attack can produce costly damages in the information system. Furthermore, personal data could be under risk and in the case a leak is produce due to security breaches the insurance company must compensate the damages produces at a really high cost. Furthermore, new GDPR imposes stronger conditions to personal information holders in order to guarantee the inviolability of individuals rights. Hence, appropriate methodologies must be applied in order to implement the cyber-risk assessment in an efficient and reliable way, taking under consideration all those variables which affect the decision of whether a company is insurable or not (insurance company balance between incoming and potential costs).

In this scenario, machine learning techniques seem to be a promising and appropriate way to assess cyber-risks in companies and to drive insurability decisions for insurance companies. This paper has analyzed the various parameters related with the cyber-risk assessment. When we consider the evaluation of a concrete model o compare different ones, the overall reason is the need to point out that one which provides the most precise result.

In this paper, the results have demonstrated the strong correlation between some of the parameters which are considered in the problem because their influence in the result.

After analysing the result obtained in our research using the linear regression and multi-linear regression approaches we conclude that machine learning techniques can be satisfactorily applied in the cyber-risk assessment of companies and, at the end, deciding about its insurability.

The promising results obtained from this insight in machine learning demonstrate the potential of these techniques to improve the reliability, trustfulness and efficiency of the cyber-risk and insurability assessments.

# References

1. Bartolini, D.N., Benavente-Peces, C., Ahrens, A.: Risk assessment and verification of insurability. In: Proceedings of the 7th International Joint Conference on Pervasive and Embedded Computing and Communication Systems (PECCS 2017), Madrid, Spain, 24–26 July 2017, pp. 105–108 (2017). https://doi.org/10.5220/0006485101050108
2. BSI: Bundesamt fr sicherheit in der informationstechnik (BSI 2017). leitfaden zur basis-absicherung nach IT-Grundschutz. https://www.bsi.bund.de (2017)
3. Corporate, A.G., (AGCS), S.: Allianz risk barometer. http://www.agcs.allianz.com/insights/white-papers-and-case-studies/allianz-risk-barometer-2016 (2016)
4. COSO: The committee of sponsoring organizations of the treadway commission. https://www.coso.org/Pages/ermintegratedframework.aspx (1992)
5. Eckert, C.: Concepts, procedures and protocols. De Gruyter Oldenbourg (2014)
6. Harris, S., Maymi, F.: Certified Information System Security Professional. McGraw Hill Education, New York (2016)
7. ISACA: Cobit 5 framework. https://www.isaca.org/COBIT/Pages/COBIT-5-Framework-product-page.aspx (2012)
8. ISO: ISO/IEC 20000-1:2011. Information technology service management. https://www.iso.org/standard/51986.html (2011)
9. ISO: ISO/IEC 27001:2013. Information technology - security techniques - information security management systems requirements. https://www.iso.org/standard/54534.html (2013)
10. ISO: ISO/IEC 27017:2015. Information technology security techniques code of practice for information security controls based on ISO/IEC 27002 for cloud security services. https://www.iso.org/standard/43757.html (2015)
11. Kushner, D.: The real story of stuxnet. IEEE Spectr. **50**(3), 48–53 (2013). https://doi.org/10.1109/MSPEC.2013.6471059
12. NIST: NIST 800-45: Guideline on electronic mail security. https://nvlpubs.nist.gov/nistpubs/legacy/sp/nistspecialpublication800-45ver2.pdf (2007)
13. NIST: NIST 800-123: Guide to general server security. https://nvlpubs.nist.gov/nistpubs/legacy/sp/nistspecialpublication800-123.pdf (2008)
14. NIST: NIST 500-291: NIST cloud computing standards roadmap. https://www.nist.gov/publications/nist-sp-500-291-nist-cloud-computing-standards-roadmap (2013)
15. NIST: NIST 800-40: Guide to enterprise patch management technologies. https://csrc.nist.gov/publications/detail/sp/800-40/rev-3/nal (2013)
16. NIST: NIST 800-53: Security and privacy controls for federal information systems and organizations. https://nvd.nist.gov/800-53 (2013)
17. OWASP: Open web application security project. http://www.owasp.org (2017)
18. Warren, C., Bayuk, J., Schutzer, D.: Enterprise Information Security and Privacy. Artech House Inc., London (2009)

# Application of 3D Symbols as Tangible Input Devices

Hirofumi Shishido◉ and Rentaro Yoshioka(✉)◉

University of Aizu, Tsuruga, Ikki-Machi, Aizu-Wakamatsu, Fukushima 965-8580, Japan
maikel.3751.coro@gmail.com, rentaro@u-aizu.ac.jp

**Abstract.** A new type of tangible input devices employing three dimensional (3D) symbols for smart input is considered. A 3D symbol is an object derived from a 2D symbol that can be physically touched and manipulated. In this paper, application of 3D symbols as general-purpose tangible input devices to control software is presented. A systematic method to design effective and useful 3D symbols is briefly proposed. An overview of reassuring results from studies with 3D symbols designed with the method is reported. Such 3D symbols are anticipated to take full advantage of human abilities to improve human-computer interaction.

**Keywords:** 3D symbols · Smart input · Tangible input devices

## 1 Introduction

In entertainment, a wide variety of controllers are used to allow players to control software through more direct means, such as, body motions and hand gestures. The use of 3D objects as input is a subject frequently considered in augmented reality (AR) and virtual reality (VR). AR markers have been widely used to integrate the real world and virtual worlds [1, 2]. While AR markers are virtual, the effect of producing physical feedback through direct interactions such as throwing and hitting motions have been studied [3]. These specialized input devices incorporate natural actions, like shaking and waving, that people feel comfortable in performing. In general, somesthetic sense (tactile sense) is useful in building a feeling that something "certainly exists" [4]. Successful attempts have been made to realize natural interactions with virtual world objects by adding sense of reality through somesthetic sense [5]. New input devices that would make greater use of peoples' capabilities to control software is a promising direction for future human-computer interaction.

As a step in this direction, research on producing tangible and versatile input devices with 3D symbols is being pursued to show the necessary direction of innovation. The concept of 3D symbols is based on "3D Kanji", a unique 3D form created from 2D characters and symbols [6]. Figure 1a shows an example of a 3D Kanji object created from a Japanese Kanji character "跳" which has the meaning "jump". In the research of this paper, the concept 3D Kanji is applied to designing 3D symbols for which user's intentions can be encoded and effectively used to control software. This is

© Springer Nature Switzerland AG 2019
C. Benavente-Peces et al. (Eds.): SEAHF 2019, SIST 150, pp. 306–309, 2019.
https://doi.org/10.1007/978-3-030-22964-1_34

different from specialized input devices in the point that the symbols can be designed for general purpose usage, in other words, to be applicable in different applications. In this paper, a systematic method for designing 3D symbols as tangible input devices is proposed and an overview of findings from two pilot studies are briefly reported.

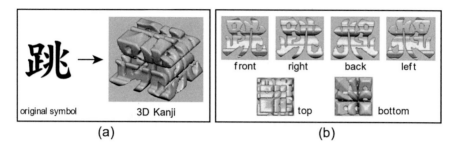

**Fig. 1.** Example of 3D Kanji "jump" and its structure viewed from six directions

## 2  3D Symbol Design

A 3D symbol is an object designed so that its features, mainly shape and construction, represent certain meanings. To manipulate software with such symbols, functions are also mapped to actions. The objective of the proposed design method is to provide a guideline for systematic assignment of semantics and actions.

**Shape and composition.** 3D symbols are based on 3D Kanji so that its shape and composition are inherited from the original character. As depicted in Fig. 1b, the resulting object has six sides (front, right, back, left, top, and bottom) among which front and right views are identical, back and left are mirrored. When this 3D Kanji is used as a 3D symbol, it is possible to attach one meaning to the object as a whole or attach different meanings to each of the different views. 3D symbols can also be created from symbols other than characters, such as the "person" shape depicted in Fig. 2a.

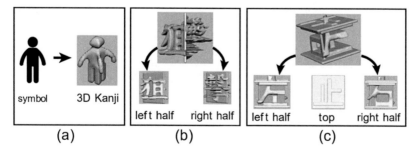

**Fig. 2.** 3D Kanji of a graphical symbol and two composed 3D symbols

To input multiple meanings, the composition idea proposed in the original 3D Kanji research [6] is adopted. Figure 2b is an example of combining two characters "撃 (shoot)" and "狙(aim)" back-to-back diagonally. The user may input either "shoot" or "aim" by turning the 3D symbol, so that the corresponding view is recognized by the computer. Figure 2c is an example of combining three characters "左(left)", "右 (right)", and "止(stop)". These are general design methods and can be applied to any combination of symbols (characters) depending on the application.

**Functional mapping**. To perform input with the 3D symbols, a user will manipulate them according to a set of predefined actions. A manipulation is defined by location, orientation and motion of the 3D symbols within a "workspace". The three manipulations are depicted in Fig. 3. The example in the figure assumes that a camera is used to recognize the 3D symbols. The combination of location, orientation, and motion with specific features of a symbol shape constitute a command.

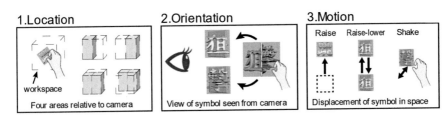

**Fig. 3.** Types of possible manipulations

## 3   Application Design and Preliminary Studies

Two pilot studies were conducted to explore how to utilize 3D symbols as a versatile new input device. The first study focused on a set of 3D symbols with simple construction and compared Kanji characters and graphical symbols. The second study involved a set of 3D symbols with composition of two and three symbols. In both studies, image recognition software, trained by machine learning, was used to recognize the 3D symbols with a camera.

In the first study, a simple 2D (horizontal scrolling) game was played with 3D symbols. The player can control a character, that runs left or right and jump low or high. The design and functional mapping of the 3D symbols used are described in Fig. 4a. There were 23 participants. From a questionnaire performed after playing the game, all participants confirmed that the symbols and manipulations were intuitive and that they enjoyed the interaction. In addition, the same game was played by three participants using 3D symbols based on the graphical symbol. The design and functional mapping of the 3D symbols used are described in Fig. 4b. Compared with the Kanji version, participants responded that the manipulations were more intuitive.

In the second study, a set of 3D symbols was designed to control a slightly more sophisticated game in which a tank can move left or right along a ridge, tilt a cannon to aim and shoot at a target. The design and functional mapping of the 3D symbols used are described in Fig. 4c. This study was carried out with six participants followed by informal interviews. All participants expressed positive feedback on the use of selecting different functions based on symbol's orientation and simultaneous usage of two 3D symbols using both hands.

Overall, participants of the studies expressed positive feedback on the meaning and physical feedback provided by 3D symbols and comfort in operating the game.

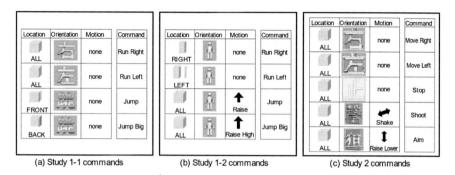

(a) Study 1-1 commands          (b) Study 1-2 commands          (c) Study 2 commands

**Fig. 4.** 3D symbols and functional mapping in first study and second study

## 4   Conclusion

A feasible design method and functional mapping for 3D symbols as tangible input devices has been presented. Methods to design 3D symbols to reflect semantics of multiple symbols for arbitrary applications have been devised. Moreover, a basic set of manipulations with 3D symbols to map user actions with semantics was introduced. From results of two pilot studies, the 3D symbol design and functional mappings were reasonably effective in providing both semantic and physical feedback on the software operations. The obtained results will be basis for exploring effective applications of the proposed 3D symbols.

## References

1. Starner, T., et al., A.: Augmented reality through wearable computing. Presence **6**, 386–398
2. Want, R., Fishkin, K. P., Gujar, A., Harrison, B.L.: Bridging physical and virtual worlds with electronic tags. In: Proceedings of CHI '99, pp. 370–377. ACM
3. Tobita, H.: Aero-marker: blimp-based augmented reality marker for objects with virtual and tangible features. In: VACAI'13 Proceedings of the 12th ACM SIGGRAPH International Conference on Virtual-Reality Continuum and Its Applications in Industry, pp. 223–230
4. SEKAI, http://toshin-sekai.com/interview/16/, Last accessed 13 Dec 2018
5. Nomura, R., Unuma, Y., Komuro, T., Yamamoto, S., Tsumura, N.: Mobile augmented reality for providing perception of materials. In: MUM'17 Proceedings of the 16th International Conference on Mobile and Ubiquitous Multimedia, pp. 501–506
6. Yoshioka, R., Mirenkov, N., Sekine, H., Noda, K.: 3D Kanji: A new paradigm of 3D objects to exploit additional dimensions of human sense for enhancing expression. J. Vis. Lang. Comput. **28**, 250–272 (2015)

# Portable Braille Reading Device with Electromagnetic Actuators

Daniel Aparicio$^{(\boxtimes)}$, Andhers Piña, and Alexis Bustamante

Instituto Politécnico Nacional, Unidad Profesional Interdisciplinaria en
Ingeniería y Tecnologías Avanzadas, Mexico City, Mexico
driano@live.com.mx

**Abstract.** Nowadays, human beings are experiencing more problems related to vision. These problems are caused by the deterioration and degeneration of the eyes, resulting from diseases, accidents, aging or inadequate use of the eyes in risky environments. It is estimated that worldwide there are about 253 million people suffering from visual impairment [1]. The Braille writing and reading system has been developed so that people with visual disabilities have access to written information. This system is made up of units of space, called Braille cells, which consist of six points where the positions of these represent letters, numbers and characters. In order to develop technology designed for people with visual impairments, Braille tactile displays have been created to translate a text from a computer into a readable system of Braille characters. In the present work, the proposal is made to carry out the design, construction and implementation of an updated Braille visualization system so that people with visual disabilities have greater access to the information that is in electronic format. The development of this proposal arises from the need to provide an alternative that complements existing technologies designed for people with visual disabilities.

**Keywords:** Braille cell · Braille display · Visual impairment

## 1 Introduction

The World Health Organization estimates worldwide that 253 million people suffer from visual impairment, of which 36 million are totally blind and 217 million are moderately or severely visually impaired [1]. In Mexico, the results of the 2014 National Survey of Demographic Dynamics (ENADID) indicated that about 3.5 million people have visual disabilities. This study also points out that 42.4% of the population with visual impairment from 3 to 29 years of age attends school, and compared with different disabilities, people with visual impairments are most attend school [2] On the other hand, in the Diagnosis on the situation of people with disabilities in Mexico, it is noticed that, from the population with visual disability, 63.5% do not use any type of technical assistance, 12.2% use a walking stick as a guide, 4.6% use the Braille system, 1.6% use an audio computer and 18.1% use another type of technical help [3].

© Springer Nature Switzerland AG 2019
C. Benavente-Peces et al. (Eds.): SEAHF 2019, SIST 150, pp. 310–315, 2019.
https://doi.org/10.1007/978-3-030-22964-1_35

The American Foundation for the Blind exposes in its magazine [4] various technologies that have been designed for people with visual disabilities. These technologies cost between $ 2,300 and $ 4,000 USD. Braille displays are the most important technology; however, their prices are high due to the number of braille cells they contain and the type of actuator they use to raise the points of the Braille cells. These devices are made up of between 20 and 40 braille cells. This situation within the country shows us the need to create spaces for inclusive education, in order to incorporate the visually impaired population into the teaching mechanisms to consolidate their permanence in schools.

## 2  Theoretical Framework

### 2.1  Braille System

Braille is a system of elevated points that can be read with the fingers of the people who are blind or have partial vision. The Braille system is not a language, it is a code with which many languages can be written or read, this system is used by thousands of people around the world. The symbols of Braille are made up of units of space known as Braille cells. A Braille cell consists of six raised points arranged in two parallel columns with each column having three points. The positions of the points are identified by numbers ranging from one to six. A single cell can be used to represent a letter of the alphabet, a number, a score or even a whole word (see Fig. 1) [5].

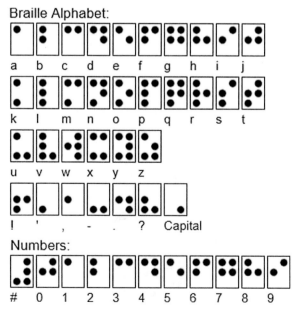

**Fig. 1.** Braille alphabet

## 2.2  Braille Display

A tactile visualizer provides information to the user through the sense of touch. This information can be represented using characters, symbols, signals or physical forces. In the field of touch technology, research has focused on devices for people with disabilities and medical instruments. A refreshable Braille display contains touch devices for people with partial or total visual impairment, which translates a text from a computer into a readable system of characters. A Braille display consists of several Braille cells, containing six or eight points organized in two columns. Conventional refreshable Braille cells use metal pins that are independently lifted. Today, the demand for updatable Braille displays is increasing among those who are partially or totally visually impaired who wish to access modern information systems (see Fig. 2) [6].

**Fig. 2.**  Braille display

## 2.3  Methods of Braille Cell Action

The fundamental part of the project is to implement a mechanism for the operation of the refreshable Braille display. Permanent magnets have been widely used in the construction of various actuators in recent years. The new designs of the permanent magnet actuators are used for different purposes. One of the main applications is the facilitation of the perception of images by people with visual impairment using Braille screens [7]. The main design of an electromagnetic actuator is shown in Fig. 3. The moving part is the axial magnetized cylindrical permanent magnet. The two coils are connected in series in such a way that they create the magnetic flux of opposite directions in the region of the permanent magnet. In this way, depending on the polarity of the power supply, the permanent magnet will move up or down.

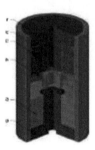

**Fig. 3.**  Electromagnetic actuator

# 3   Proposal

Due to the accelerated growth of new technologies and the existing technological gap for people with visual disabilities, it has been proposed the development of a portable Braille visualization prototype for people with visual disabilities to have access to electronic media information. The proposal will seek the translation of electronic files to the Braille system through a portable device. The first part of the process will consist in carrying out the programming algorithm for the translation of the types of documents that the system will interpret. Then, this information will be processed to control the actuators that make up the Braille display mechanism. The device will have buttons for the user to interact with the system functions. In Fig. 4 the general block diagram of the device is shown, it shows the most important elements as well as the flow of information that exists between each one.

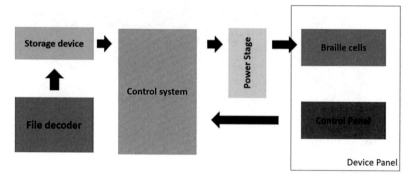

**Fig. 4.** Device block diagram

For a better understanding of the proposal, the following functional diagram will be used. This diagram shows the most important tasks that the project should cover, as well as the flow that exists between each one. This diagram can be seen in Fig. 5.

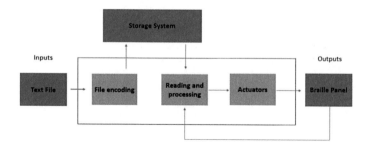

**Fig. 5.** Functional diagram

In Fig. 6 it is shown the physical model of the portable Braille visualization prototype. This model was designed in a computer drawing program considering the international standards of the Spanish language for Braille writing [8]. This model is the first design proposed for the project, it will be refined as the prototype develops.

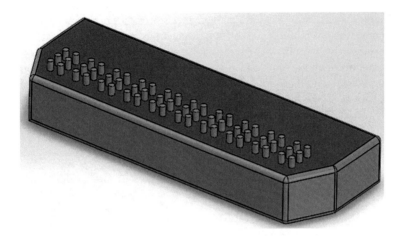

**Fig. 6.** First model of the proposed Braille visualizer

## 4  Conclusion

The Braille system is the largest communication tool, in addition to speech, with which people with visual disabilities have, therefore, it is a tool that should be potentiated its expansion and teaching. With the help of this device, people with visual disabilities will have the opportunity to take advantage of the great advantages that the digital age has brought with it: access to the information. This project has many possibilities for expansion, for example, the develop a variant of the prototype in which the user can learn to write and read in the Braille system, which would boost the communication skills and abilities of people with visual disabilities. In addition, the development of new national technologies, which lack support and reliability, would be promoted, two fundamental factors when talking about a globalized world. The main advantage of these national technologies is that they know specifically the problems that our country has as a society, so that they can offer a solution more in line with the real needs and possibilities of our societies.

## References

1. Bourne, R.: Magnitude, temporal trends, and projections of the global prevalence of blindness and distance and near vision impairment: a systematic review and meta-analysis. Lancet Glob. Health, e888–e897 (2017)

2. INEGI.: Estadísticas a propósito del día internacional de las personas con discapacidad. Datos Nac. 1–17 (2015)
3. SEDESOL.: Diagnóstico sobre la situación de las personas con discapacidad en México. Government, 49–50 (2016)
4. AccessWorld Magazine. Product Evaluations and Guides. http://www.afb.org/afbpress/pub-new.asp?DocID=aw180405, 30 Mar 2018
5. Braille: American foundation for the blind. http://www.afb.org/info/living-with-vision-loss/braille/12, 4 Mar 2018
6. Lee, J.S.: A micromachined refreshable braille cell. J. Microelectron. Syst. 2–4 (2005)
7. Karastoyanov, D.: Development of a Braille tactile device driven by linear Magnet Actuators. Int. J. Eng. Innov. Technol. (IJEIT) 1–4 (2014)
8. Documento Técnico B1 de la Comisión Braille Española: Parámetros Dimensionales del Braille. Comisión Braille Española. http://www.once.es/new/servicios-especializados-en-discapacidad-vi-sual/braille/documen-tos/DOCUMENTO%20TECNICO%20B1%20PARAMETROS%20DIMENSIONALES%20DEL%20BRAILLE%20V1.pdf, 4 Mar 2018

# Modeling and Simulation of Renewable Generation System: Tunisia Grid Connected PV System Case Study

Mansouri Nouha[1](✉), Bouchoucha Chokri[2], and Cherif Adnen[3]

[1] Analysis and Processing of Electric and Energetic Systems Unit, National School of Engineering Monastir, Tunis, Tunisia
nouha_eniml@yahoo.fr

[2] National School of Engineering Monastir, Monastir, Tunisia
cbouchoucha@steg.com.tn

[3] Tunisian Society of Electricity and Gas STEG, Tunis, Tunisia

**Abstract.** This paper seeks to evaluate and study Tunisia Grid-Connected system (PV/Wind Turbine), to improve the electricity production without interruption using renewable energy during daily as well as seasonally periods. In this vein, in order to manage and control the dynamic power stability, two modes have been treated, whether with or without integration of photovoltaics. To distinguish the variations between the two modes, the integrations of Renewable energy will be between 15% and 30%. To perform the correct system operations and to meet load requirements during the peak periods. Considering the impact of included modes and the Renewable energy source integrated into the grid, an accurate fault current characteristic analyzes have been treated. A novel protection scheme using PSSE simulation environment obtains the results and observations is then proposed. The obtained results proven that the loss of the generators have an undesirable effect on stability.

**Keywords:** Photovoltaic · Grid · Speed of generator · Frequency · Mechanical power of turbine

## 1 Introduction

Renewable energy system (like photovoltaic (PV), wind Turbine, etc.) has become a sustainable option for electric power generation in very sunny places [1, 2]. Photovoltaic integration is considered as an attracting solution to ensure the required energy during the peak times. The high penetration of Renewable energy Grid-connected prove its capability to reverse the power flow. Indeed, the integration of Renewable energy can introduce new technical challenges (such as: increase imbalance and voltage) to the distribution system even the peak power. For this reason, to solve the system fluctuations, an energy storage unit must be introduced [3, 4, 5]. To do this, Battery was selected as a potential solution. Most researches proven that the combination Solar energy with batteries to store the energy surplus, reduce the peak demand and minimize the photovoltaic production distributed issues on the grid-connected [6, 7]. Some of them have proposed a Grid-Connected (PV/Wind Turbine) system including a solar energy source and a wind Turbine that has long used Battery storage [8, 9]. Whereas, the work given by Ref. [8] discussed an autonomous hybrid generation system (PV/Wind/PEMFC) in which the subsystem sizes are optimized. An autonomous hybrid power system (PV/WT) is detailed and discussed in Ref. [10, 11]. The proposed system simultaneously links the grid. To maintain the energy demand when energy demand is insufficient, during peak periods a battery has been added as a backup power system [12, 13]. In Ref. [14], the authors presented a hybrid renewable energy (photovoltaic/diesel) system using battery as energy storage unit. To control the system operation and maintain the state of charge of the battery according to the weather conditions, an advanced energy management system has been developed using fuzzy logic control (FLC). The obtained results are proven the f reliability of the proposed Renewable system. An accurate comparative analysis between grid extensions and an autonomous power system (Photovoltaic/wind/Hydrogen) has been was carried out. The performed analyzes aimed to control the system option using [15]. As we are convinced that the proposed Renewable grid connected systems still unable to control and cooperate properly between sources, and to solve the issues power fluctuations during the peak periods. Our paper expands the ideas of previous works. Firstly, a solid method to analyze the effects of the PV integration on various grid levels is used. Then the stability of the power supply system has been evaluated and studied based on the discussed integration modes, which are with and without photovoltaic.

The remainder of this paper is organized as follows: Sect. 1 presents grid connected (PV/WT) system and equipment utilities, in Sect. 2, Finding and results is devoted to the analysis of the simulation results as well as the Photovoltaic integration study and the concluding remarks are discussed in Sect. 3. The work conclusion and the expected prospects are given by Sect. 4.

## 2 System Design: Tunisia Grid Connected PV System Case Study

The energy produced by a renewable energy (solar/WT) is of intermittent nature, due to its inability to vary the load constantly and to ensure the required energy demand. The combination of a Photovoltaic can be an efficient solution to provide energy demand at varying loads. To ensure the required energy demand, a grid connected PV system (Tunisia case study) is proposed. The adopted system design deals to minimize the power drawn from the solar system using battery (see Fig. 1a). The configuration of proposed system consists of the photovoltaic array, Inverter, AC loads and the associated controls (converter and over systems) and electrical utility.

In order to achieve the most accurate simulation, a dynamic PVs modeling have been included using PSSE. This later was chosen due to its efficiency that composed with 39 nodes, 46 lines, 9 generators, 45 loads, 13 transformers and a capacitor bank. Indeed, it can facilitate to evaluate the system stability according various constraints. An experimental database must be prepared, before using PSSE and are shown in (Fig. 1b. We have adopted this model for the dynamic modeling of photovoltaic generators In PSSE. This general model includes a WT4G (Converter/Generator model) and WT4E (Electrical Control Model). The collected files refereed to static and

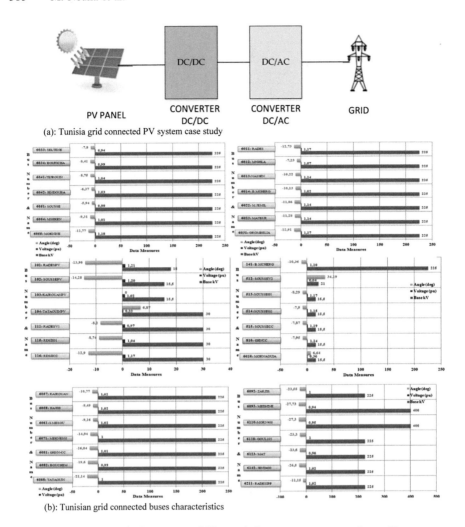

(a): Tunisia grid connected PV system case study

(b): Tunisian grid connected buses characteristics

**Fig. 1.** Typical structure of Photovoltaic generator connected to grid

dynamic files. To ensure a reference base point, an initial static simulation was running. Therefore, the voltage bus have been determinate based on Newton-Raphson iterative method. Newton-Raphson's iterative method was chosen to be the best solution for solving the power calculation flow (see Eq. 1). Indeed, this method is ranked well than Gauss-Seidel. Based on this method, the angle voltage values and magnitude correction matrix can be deducted. Tunisia Grid-Connected system (PV/Wind Turbine) can be based on the model connected to the network using a novel WT-Type 4 (Type 4 WTG).

$$\begin{pmatrix} \Delta P \\ \Delta Q \end{pmatrix} = \begin{pmatrix} \frac{\partial P}{\partial \theta} & \frac{\Delta P}{\Delta V} \\ \frac{\partial Q}{\partial \theta} & \frac{\partial Q}{\partial V} \end{pmatrix} \begin{bmatrix} \Delta \theta \\ \Delta V \end{bmatrix} \tag{1}$$

## 3  Findings and Results

Tunisian grid stability is evaluated with solar energy integration levels (between 15% and 30%). In this vein, the generated PV power were injected to the total power produced by the conventional group. Unfortunately, until now any power grid utility report has been registered for near future. For this reason, we have beginning to audit and study the Tunisia grid with these constraints. The treated grid is balanced between production and consumption, the integration of 15%–30% of solar PV can cause grid instability. To avoid this disruption and ensure a balance between production and consumption by increasing the loads, the adoption of a frequency study is needed (Eq. 2). To observe the generator stability effects without photovoltaic integration losses, a dynamic simulation system was performed (see Fig. 2).

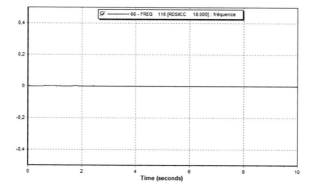

**Fig. 2.** frequency for Control System Dynamic Simulation

$$f(h(z)) = [1 + f(pu)] * 50 \qquad (2)$$

The speed variation is studied in disconnected generator case on the Tunisian center region: Sousse (see Fig. 3). Figure 3 shows the disconnection of the generator system generates internal systems fluctuations, which causes a shedding of the internal generator affects the stability of speed. Figure 4 shows that a 15% injection in the network can reduce the value of the frequency (box: the generator of the center (Sousse) is disconnected). Figures 5 and 6, the integration of 30% of the photovoltaic generation system to the grid we notice a speed variation with 30% PV is better (case: generator in the center (Sousse) is disconnected). Figures 7 and 8, shows the WT mechanical power using 15% and 30% of PV integration. Injecting 30% can provide more stability and advantage than 15%. Figure 9 shows that the PV generator was connected to bus 6012. The added bus was chosen due to its reliability. The PV system was sized to be 30% with a reference voltage of 1.012 p.u.

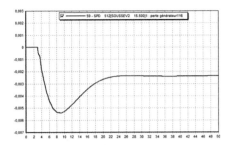

**Fig. 3.** Speed variation of intern generator without PV and in disconnected generator

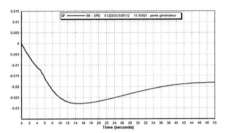

**Fig. 4.** Speed variation of intern generator with 15% PV integration

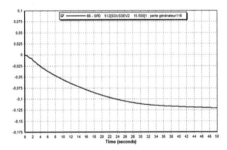

**Fig. 5.** Speed variation of intern generator with 30% PV integration

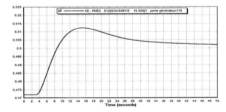

**Fig. 6.** Mechanical Power variation without PV

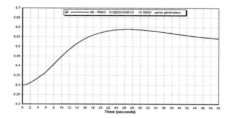

**Fig. 7.** Mechanical power variation PV in disconnection generator

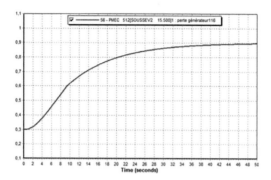

**Fig. 8.** Mechanical power variation turbine With 30% PV integration

## 4  Conclusion and Future Work

Grid-Connected system was seen as a better potential to ensure comfortable electrification services for customers and solve the load demand during the peak periods. In this context, renewable generation system as Photovoltaic is modeled and simulated through PSSE. To resolve power fluctuation during peak periods, a PV/WT Grid-Connected system has been treated according to several control scenarios. The Tunisian grid electricity dynamic behavior is studded. An examination of the potential influences of PV in the power grid is realized according to two modes. The obtained results prove that 30% of PV power injection in the electrical network had better maintain the stability and system. As future work, we tend to develop technical solutions available to integrate a large scale of PV-based microarrays to increase the penetration rate.

## References

1. Tang, X., Tsang, K., Chan, W.: Power conditioning system for grid-connected photovoltaic system. Sol. Energy **96**, 187–193 (2013)
2. Hejri, M., Mokhtari, H., Karimi, S., Azizian, M.: Stability and performance analysis of a single-stage grid-connected photovoltaic system using describing function theory. Int. Trans. Electr. Energy Syst. **26**(9), 1898–1916 (2016)

3. Jin, T., Xu, X.: Experimental investigation on a grid-connected photovoltaic air conditioning system. Appl. Mech. Mater. **672–674**, 54–60 (2014)
4. Rahnamaei, A., Salimi, M.: A novel grid connected photovoltaic system. Bull. Electr. Eng. Inform. **5**(2) (2016)
5. Augusto Oliveira Da Silva, S., Bruno Garcia Campanhol, L., Dário Bacon, V., Poltronieri Sampaio, L.: Single-phase grid-connected photovoltaic system with active power line conditioning. Eletrônica de Potência **20**(1), 8–18 (2015)
6. Castaneda, M., Cano, A., Jurado, F., Sanchez, H., Fernandez, L.M.: Sizing optimization, dynamic modeling and energy management strategies of a stand-alone PV/hydrogen/battery-based hybrid system. Int. J. Hydrog. Energy **38**(10), 3830e45 (2013)
7. Marzband, M., Ghazimirsaeid, S.S., Uppal, H., Fernando, T.: A realtime evaluation of energy management systems for smart hybrid home Microgrids. Electr. Power Syst. Res. (2017)
8. Ren, H., Wu, Q., Gao, W., Zhou, W.: Optimal operation of a grid connected hybrid PV/fuel cell/battery energy system for residential applications. Energy (2016)
9. Karami, N., Moubayed, N., Outbib, R.: Energy management for a PEMFC-PV hybrid system. Energy Convers. Manag. **82**, 154e68 (2014)
10. Brka, A., Al-Abdeli, Y.M., Kothapalli, G.: Predictive power management strategies for stand-alone hydrogen systems: operational impact. Int. J. Hydrog. Energy (2016)
11. Dursun, E., Kilic, O.: Comparative evaluation of different power management strategies of a stand-alone PV/Wind/PEMFC hybrid power system. Int. J. Electr. Power Energy Syst. **34**(1) (2012)
12. Paulitschke, M., Bocklisch, T., Bottiger, M.: Sizing algorithm for a PV-battery-H2-hybrid system employing particle swarm optimization. Energy Procedia **73**, 154e62 (2015)
13. Mohamed, A., Mohammed, O.: Real-time energy management scheme for hybrid renewable energy systems in smart grid applications. Electr. Power Syst. Res. **96**, 133e43 (2013)
14. Kalinci, Y., Hepbasli, A., Dincer, I.: Techno-economic analysis of a stand-alone hybrid renewable energy system with hydrogen production and storage options. Int. J. Hydrog. Energy **40**(24) (2015)
15. Roumila, Z., Rekioua, D., Rekioua, T.: Energy management based fuzzy logic controller of hybrid system wind/photovoltaic/diesel with storage battery. Int. J. Hydrog. Energy **42**(30) (2017)

# Privacy and Security Aware Cryptographic Algorithm for Secure Cloud Data Storage

B. Muthulakshmi$^{(\boxtimes)}$ and M. Venkatesulu

Kalasalingam University, Krishnankoil 626126, India
{selvamayil2010,venkatesulum2000}@gmail.com

**Abstract.** Cloud computing is an effective framework to enable on demand network access for sharing configurable computing resources such as applications, network, servers, services, and storages. It depends on two different technologies like distributed computing and virtualization. Cloud technology offers the users to store and retrieve their data easily in a remote manner. However, privacy and security are considered as the most challenging concerns in cloud data storage. Users are not ready to believe to store their data on any third party or companies without guaranteeing that will not access their data whenever they need. In this paper, a novel Privacy and Security aware Cryptographic Algorithm (PSCA) for data storage is proposed which is based on Invertible Non-linear Function (INF). In proposed technique, the encrypted data is only stored at the cloud storage without key and the encryption is done by the data authorities who also secure the decryption key. The end user has to send request to the data authority and after all verification process the encrypted data is given and with the obtained key and the knowledge of inverse INF the decryption is performed. Thus, both the internal and external attacks are avoided. The observed results are plotted and the proposed technique is compared with the existing AES and RSA techniques. From the results, it is observed that the proposed PSCA has better performance in terms of reduced time in all four parameters.

**Keywords:** Cloud computing · Privacy and security · Data authority

## 1 Introduction

Cloud computing is an emerging technology which becomes increasingly essential in storage and service provisioning of information in the Internet such as e-banking, e-billing, e-commerce, e-mail, e-transactions and so on. Online data interchange is the major requirement for all these services [1]. The transmitted data may comprise sensitive or private information, for example, credit/debit card details, business secretes, banking information etc. that needs more security as an expose of these secrete data to any unapproved users can produce extremely hazardous.

Cloud computing possess some of the important characteristics like location independent services, On-demand data services, access on wide network, frequent flexibility, estimated services, and pay as use. It also renders several benefits with increased security risks towards the outsourced data of the users. Cloud comprises an

© Springer Nature Switzerland AG 2019
C. Benavente-Peces et al. (Eds.): SEAHF 2019, SIST 150, pp. 323–337, 2019.
https://doi.org/10.1007/978-3-030-22964-1_37

external entity named as Cloud Service Providers (CSP) which is a separate unit. It verifies the accuracy and integrity of stored data files and the cloud users can obtain the proofs about stored data over cloud and can check whether it is altered or deleted. It is still facing wide range of threats including inside and outside attacks for data integrity. The data is stored at the remote location by the cloud users without any worries [2, 12]. Moreover, there is a high necessity for data privacy and security while exchanging the sensitive data of users throughout malicious or un-trusted networks. Thus, it is essential to formulate a novel and efficient security mechanism for converting confidential and sensitive data into a few unreadable formats while exchanging to create it more difficult for attackers for obtaining the original data. Cryptography is known as one of the effective techniques to accomplish this [3].

As users possess very less physical storage they require external storage to store their data. Conventional cryptographic primitives are employed for the purpose of data security which cannot be directly chosen [4–6]. Hence, how to effectively validate the accuracy of the outsourced data files of the cloud users without making any local copies of stored data files. In cloud computing, it becomes a major challenge for data privacy and security. Every cloud user has stored their data at the un-trusted third party that necessitates an efficient and effective security mechanism as data holders do not possess any physical access on their data. In a cloud environment, security and privacy of user's data are always considered as a significant issue [7]. Users can obtain several benefits by using cloud as an external storage such as low cost and easy data access however security issues should be concern while transferring or storing the confidential or sensitive data to cloud.

In our proposed mechanism, the data partitioning and encryption is performed by the data authority only who keeps the decryption key as well. The complete control on data is not given to the CSP that performs space allocation for obtained encrypted data. After verifying the originality of end user, the encrypted data is provided. By the way the internal attack is avoided. Each encrypted data cipher is stored at different locations of the same cloud storage or different cloud storages. So an intruder cannot able to obtain all the ciphers. In case if the intruder gathers all the encrypted ciphers of same data, he/she cannot perform decryption. Only with decryption key and the knowledge of inverse INF, the obtained encrypted data is decrypted. Thus, the external attack is also avoided.

This paper is organized as follows: In Sect. 2, layers of cloud computing is given and related work is reviewed in Sect. 2. In Sect. 3, the proposed system is explained and in Sect. 4 the research methodology is explained. Security analysis is demonstrated in Sect. 5 and the proposed security evaluation is analyzed in Sect. 6 in a detailed manner. In Sect. 7, the experimental results are given. Finally, this paper is concluded in Sect. 8.

## 2 Related Work

Dimitrios et al. (2012) have conducted a detailed analysis on cloud security [8]. They applied trusted third party to propose a novel security solution which is a combination of both the lightweight directory access protocol (LDAP) and the Single-Sign-On

(SSO) mechanisms. Especially, this cryptographic solution is based on public key infrastructure [6]. It is used to assure that the propose security solution comprising cryptography provides better confidentiality and integrity of stored data. Nevertheless, it never identifies the type of encryption algorithm involved. Besides, the proposed solution cannot identify the type of encryption algorithm to be applied.

Sumit and Joshi (2018) have presented a novel hybrid technique to enhance the security of cloud storage by combining AES and RSA algorithms with digital signature [9]. The authors clearly demonstrated the combination these three algorithms. Generally, AES and DES cryptographic algorithms are symmetric and asymmetric respectively in nature. But, in this paper, the private key generation is performed using these two AES and RSA algorithms. Then, digital signature is applied on the generated private key that provides data authentication in the cloud environment. The verification process is performed using RSA algorithm with 1024 bits of public key. Likewise the performance of the proposed algorithm is analyzed in terms of time consumption and compared with existing algorithms which shows that the proposed technique consumes very less time. But it is not applicable for large data.

Rahmani et al. [11] have presented a novel technique called Encryption as a Service (EaaS) for cloud computing which is based on the concept of XaaS for cryptography. This technique can prevent user-side encryption, inefficiency of CSP encryption and several security and privacy risks of cloud environment. Furthermore, the proposed technique does not demonstrate a comparative analysis of existing cryptographic techniques which can be incorporated.

Jiehong et al. [13] have evaluated the performance of AES, Blowfish, and GOST encryption algorithms. Comparison of these three algorithms has been performed at various sizes of data blocks and the comparison results are demonstrated to obtain performances of each algorithm. Among this, the blowfish algorithm provides the better performance regardless of the sizes of plain-text. It is observed that the key expansion process is the weakest segment of this Blowfish algorithm which could consume more time for key expansion than accomplish encryption/decryption if the size of the plaintext is small. Additionally, they specify that the key expansion of GOST algorithm consumes relatively same amount of time with respect to Blowfish, and it also costs relatively same time for both encryption and decryption to GOST algorithm. Finally, it is concluded that the AES algorithm has longest operation to perform decryption.

Wang et al. [14] have realized the arithmetic functions throughout the cipher texts of several users to perform addition and multiplication without the function learning of inputs/intermediate outputs. However, this technique requires for solving the discrete logarithmic issues that actively limits the input data length. Hence, it isn't appropriate to be implemented into the cases where there are numerous information providers and the given data size is huge.

Bhandari et al. [15] have specified that today the most difficult problems in cloud environment are for assuring the data privacy and security of the cloud users. The authors proposed a novel hybrid encryption RSA (HE-RSA) with Advanced Encryption Standard (AES) for efficiency, reliability, and consistency in cloud servers. They expected to utilize different cryptography ideas amid communication with its application in distributed computing and to upgrade the cipher-text security/encrypted data in cloud servers with reducing the utilization of time, cost and memory measure amid for both encryption and decryption. They noticed that the contrast between the execution time of the first RSA and Enhanced Algorithm utilizing Hybrid Encryption-RSA and AES is expanding definitely as the size of exponent is expanding.

In this paper, a novel privacy and security aware cryptographic algorithm (PSCA) is proposed to provide complete protection to the data stored in the cloud environment. It avoids both internal and external attacks and consumes less computational time.

# 3  Proposed System

## 3.1  Objective

The major objective of this work is to provide complete privacy and security to the data which are in need to store in the cloud storage. The cloud storage is attacked in two ways: internal attack and external attack. The proposed technique provides tight protection against these attacks. The general architecture of the proposed system is shown in Fig. 1 and the detailed architecture is demonstrated in Fig. 2. The participants of the proposed system are given in the following. They are,

**Data Authority**: The data authority creates/collects data in the form of text, images, audio and video. In other words, he/she is the owner of collected data that is to be shared with multiple cloud users who are in need, through the cloud platform. The gathered information of data authority should be secured from any illegitimate users of the cloud. It is done by checking the certificates of each data requesting end users. If the certificate of a user is original then it is denoted as authorized user which is allowed to access the data in the cloud storage.

**Cloud Storage**: Cloud storage is known as a kind of service provided by the service providers where one can store and retrieve data in a wireless manner. The stored data can also be controlled, backed up and preserved. Some of the examples of public cloud storage services are Amazon's S3, Amazon Elastic Compute Cloud (EC2), Sun Cloud, and Windows Azure services platform which permits the users to transfer their data to the cloud storage with no cost for creating and managing an infrastructure of private cloud. Rather the service providers may charge users based on the function of their requirements. It renders so many advantages like availability, confidentiality, and reliability at very low cost [11].

**Cloud Service Provider (CSP)**: Each essential data storage service is managed by the cloud service provider (CSP) which renders storage space for data storage along with few computational resources. The cloud users also gained some storage services from the CSP. In order to access/retrieve the data the corresponding user should perform together with cloud servers by cloud service providers. Both cloud data storage and servers are managed by the CSP over the cloud resources. The virtual infrastructure is permitted to perform host applications.

**End Users**: In cloud computing services, the end users are considered to be the most critical data users. The cloud environment has to meet some of their needs includes service provisioning has to be in an easy, reliable, and highly accessible manner. These end users are allowed to operate the stored data in the cloud depends on

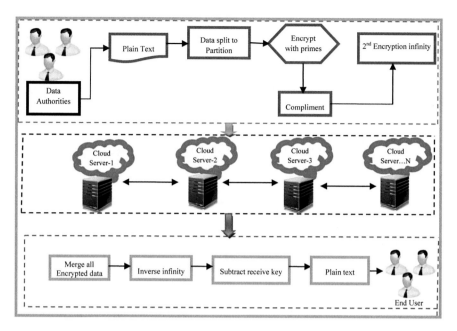

**Fig. 1.** Detailed architecture of the proposed technique

their permissions rendered by the data authorities and these permissions comprises of read/write e-stored revised data.

## 4   Research Methodology

In this research, we have proposed Privacy and security aware cryptographic algorithm (PSCA) which includes both encryption and decryption techniques. The proposed PSCA is mainly based on a novel technique called Invertible Non-linear Function (INF) that encrypts the data in a complicated manner and its inverse form is performed to decrypt the encrypted data. From the decrypted data the integrity of the proposed algorithm is obtained.

### 4.1   Mechanism

There are four entities involved in the proposed mechanism including data authority, cloud storage, cloud service provider (CSP), and end users. The proposed mechanism includes six steps which are demonstrated in Fig. 3. Initially, the collected data of data authority is large volume. Therefore, it should to be partitioned into multiple small partitions. Then, these partitions are encrypted using encryption algorithm of the PSCA. In our proposed technique, both data partitioning and encryptions are performed by the data authority to improve data privacy and security. Then the data authority sends the encrypted partitions to the cloud provider which commands the cloud storage to store theses partitions at multiple locations of same cloud or different cloud storages.

If an end user wants to access the cloud data, he/she request the data authority to provide authentication permission. After verifying the authorization of the end user the data authority provides the access certificate and key to the end user which has to be sent to the cloud provider. It verifies the originality of the obtained access certificate and if it is true or original, then it commands the cloud storage to provide the encrypted data. Using the obtained key, the end user decrypts the received encrypted data.

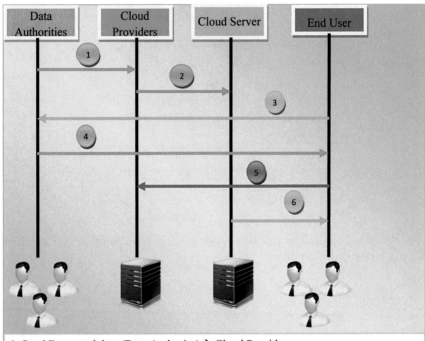

1. Send Encrypted data (Data Authority)→ Cloud Provider
2. Command to store in multiple locations (Cloud Provider) → Cloud storage
3. Request to access the data (End user)  → Data Authority
4. Send Access Certificate and key (Data Authority)→End User
5. Send the Access Certificate (End User)→Cloud Provider
6. Verify and Command to provide encryption data (Cloud Server)→End User

**Fig. 2.** Proposed mechanism

## 4.2   Data Partitioning

Data partitioning plays a vital role in cloud computing. It is defined as the process of splitting or partitioning a large data file into multiple small partitions. Therefore, these small fragments are easily encrypted and uploaded or stored in cloud storage at less time. It makes the entire operation easy and simple. It requires very less time to download the files and one can quickly access the data in cloud.

## 4.3 Encryption

Encryption is considered as an essential operation which is performed on outsourced data to cloud storage. An efficient and effective cryptographic system is required to store and maintain the outsourced data in cloud. The encryption is a part of cryptography. Data partitioning makes the encryption easier. In our proposed technique, the encryption is done in two steps. At first, prime numbers and its complements are taken to encrypt each data partition. Then invertible non-linear function is applied on each first encrypted data to perform second encryption. It produces the cipher of unreadable format. These ciphers are processed by the CSP to store at several locations of same cloud or several cloud storages. The general form of invertible non-linear function is given as (1),

$$y(x) = px + q \tag{1}$$

where $p$ and $q$ are integers and $x$ is the cipher texts produced by the first encryption. To perform second encryption, each cipher partition is multiplied by $p$ and added with $q$. The proposed encryption flow is shown in Fig. 3.

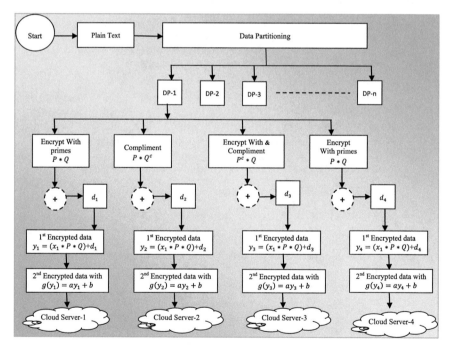

**Fig. 3.** Flow of encryption

**Algorithm for Encryption**

1.  *Random Data Partationing and distribute plaintext* $; X_i \rightarrow x_1, x_2, x_3 \ldots x_n$
2.  *Genearte prime numbers* $P_{mx}$ *and* $P_{mx} = 2X_i$
3.  *take coordination of* $P_{mx}(Q_i, R_i)$
4.  *take prime complement of* $P_{mx}\left(Q_i^{ci}, R_i^{ci}\right);$
5.  *compute* $K_i^{ci} = 2^{p+1} - K_i$
6.  *Generate random integer* $I = d_1 + d_2 + d_3 \ldots d_n$
7.  $I_n = x_n$ *we taken* $(N = 4);$
8.  *case* 1:
9.  *First encrypted data*
     $(i)\ y_1 = (x_1(Q_i * R_i) + d_1)$
     $(ii)\ y_2 = \left(x_1\left(Q_i * R_i^{ci}\right) + d_2\right)$
     $(iii)\ y_3 = \left(x_1\left(Q_i^{ci} * R_i\right) + d_3\right)$
     $(iv)\ y_4 = \left(x_1\left(Q_i^{ci} * R_i^{ci}\right) + d_4\right)$
10. *case* 2:
     $(i)\ g(y_1) = a(y_1) + b$
     $(ii)\ g(y_2) = a(y_2) + b$
     $(iii)\ g(y_3) = a(y_3) + b$
     $(iv)\ g(y_4) = a(y_4) + b$
     $(i)\ g(y_1) = a(y_1) + b$
     End

## 4.4   Decryption

Decryption is performed at the end user who gathers access certificate and key from the data authority and verified by the cloud service provider that command the storage to provide the encrypted data requested by the end user. The obtained encrypted data is then merged and decrypted using key with the knowledge of inverse INF. The inverse form of invertible non-linear function is given as (2),

$$y^{-1}(x) = \frac{x_i - q}{p} \tag{2}$$

where $p$ and $q$ are integers and $x$ is the data partition. The proposed flow of decryption is illustrated in Fig. 4.

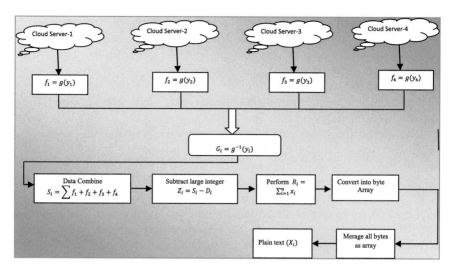

**Fig. 4.** Flow of decryption

**Algorithm for Decryption**

1. Step-1 Apply inverse from INF
2. $$G_i = g^{-1}(y_i) = \frac{y_i - b}{a}$$
3. Step-2 encrypted chipper text
4. $f_1 = g(y_1);$
5. $f_2 = g(y_2);$
6. $f_3 = g(y_3);$
7. $f_4 = g(y_4);$
8. Step-3 add all encrypted chipper text
9. $S_i = \sum f_1 + f_2 + f_3 + f_4;$
10. Step -4 subtract larger integer
11. $Z_i = S_i - I_i$
12. Delete zero pads
13. Step-5 Perform the above steps on all the four encrypted data parts (up to n x) and sum all of them.
14. Step 6: Convert into byte array
15. Step 7: Merge all Byte arrays
16. Step 8: Get original plaintext ( )

# 5 Security Analysis

While discussing about cloud security and privacy it necessary to consider a few concepts such as network security, hardware and control methodologies need to protect applications, frameworks, and information related to cloud computing. The essential characteristics of cloud computing is the idea of linkage with different elements that builds it hard and vital to secure these cloud environments. If a cloud platform is large and public, its privacy and security issues prompt to economic losses and terrible reputation which causes after the huge adoption of such novel solutions. The users generally store their vial information in the cloud environment. Therefore, the contravention of such information by an illegitimate user is unacceptable. Generally, the cloud is attacked in two ways such as inside and outside attacks.

The inside attacker is any of the administrator who can possess the possibility of hacking the data of cloud users. It is very hard to discover such type of attack. Therefore, while storing the information of users in the cloud storage they must be very cautious. It is necessary to develop techniques that prevent the data utilization though their information is operated by the third parties; he/she has not obtained the original data. Hence, the encryption of all the data must be performed before sent to the cloud storage [10]. The cloud security permits the above mentioned security requirements of data. The technological developments and their standardization create accessible some of the algorithms and protocols to respond to those problems.

# 6  Proposed Security Evaluation

This section presents the analysis of the proposed Privacy and security aware cryptographic algorithm (PSCA). The privacy and security are considered as the major issues in cloud storage. For cloud computing, there are several cryptographic techniques are available which provides privacy and security at its level best. In order to prove the efficiency of our proposed technique, we take two traditional and standard cryptographic techniques to compare the performances such as RSA and AES techniques.

As we know from security analysis, two types of security attacks such as inside and outside attacks are present which affects the privacy of the cloud environment. Therefore, performance evaluation is conducted in two scenarios: one is inside or internal attack and another one is outside or external attack.

At first, we consider inside attack. As the name itself defines that, this attack is performed by the network authorities or employees of the cloud. Usually self-interest on user's data will lead this type of attack. In existing AES and RSA techniques, the data authorities give complete control to the cloud service providers and the data is encrypted with decryption key where the data encryption is done by the CSP. Thus, they can easily perform the internal attacks. But in our proposed PSCA technique, the complete control on data is not given to the cloud service provider. In our mechanism, the data is partitioned and encrypted by the data authority and the encrypted data is only given to the CSP to allocate storage space in cloud storage and the data authority is only secure the decryption key. Without decryption key no one can perform decryption and use the stored data. Thus, the internal attack is completely avoided in our proposed technique.

Next, we discussed the mechanism to protect the user's data from outside or external attacks. First we consider the protection mechanisms of the existing AES and RSA techniques. As we mentioned above, both the techniques encrypt the user's data with corresponding decryption key and develop protection mechanism for cloud service providers only. From this, it is known that both the existing techniques provide complete protection control to CSP. In case if the CSP fails to provide protection mechanism or the attacker breaks the protection mechanism, the entire cloud data can be easily accessed by the attacker. In the proposed PSCA, if an end user wants to access the cloud data, he/she has to send request directly to the corresponding data authority. Then, the data authority verifies the requesting end user is authorized or legitimate. If the user is not a legitimate one, his/her request is rejected and if the user is authorized one, then the data authority provides access certificate and decryption key. After receiving the certificate and key, the end user has to send the access certificate to the CSP and it verifies the originality of the obtained certificate. If the certificate is fake then it observes its user id and intimates the data authority and they block such user. For original access certificate the CSP orders the cloud storage to provide the encrypted data. Then, the encrypted data is decrypted with the obtained decryption key. The major advantage of the proposed technique is that the end user cannot decrypt the encrypted data without having knowledge of inverse INF.

From the above analysis it is concluded that the proposed PSCA can be the efficient and effective technique for both internal attacks and external attacks.

## 7   Experimental Results

In this section, the implementation results of the proposed PSCA are given which is then compared with the existing AES and RSA techniques. The proposed technique is implemented in Java programming which is running under Windows 7 platform of 4 GB RAM and Intel(R) Core(TM)2 Duo CPU processor. In this experiment, there are four parameters such as Uploading time, Downloading time, Encryption time, and Decryption time of files are derived and each of them are plotted against the file size.

Figure 5 is plotted against file size (kb) and encryption time (ms) where the performance of the existing and the proposed techniques is compared. The encryption time is defined as the time needed to complete the encryption process of a given data

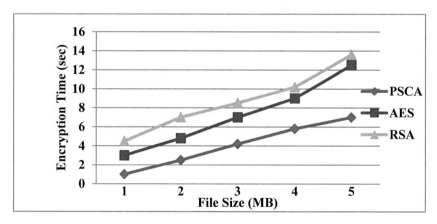

**Fig. 5.** File size versus encryption time

partition. It is otherwise termed as computational time. The complex functions of a technique needs more time to perform encryption. But the proposed technique performs simple arithmetic operations on data for encryption which is simple and requires very less time when compared to AES and RSA techniques. The obtained results of both existing and proposed techniques are plotted which shows the PSCA technique has very less time to perform encryption.

In Fig. 6 the file size (kb) is plotted against decryption time (ms) for AES, RSA and PSCA. It is the time required to decrypt the encrypted data and the inverse form of encryption is performed for decryption. As we stated above the proposed PSCA performs simple arithmetic operations which is easy and requires less time. From that it can be known that the inverse operation is also simple and consumes less time. The proposed technique performs well in decryption by reduced time compared to AES and RSA technique which is observed from the Fig. 6.

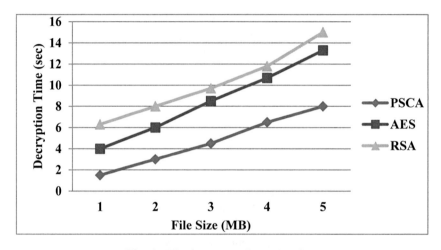

**Fig. 6.** File size versus decryption time

Figure 7 demonstrates the file uploading time which is defined as the time needed to convert the file into cipher text and upload to the cloud storage. The file uploading time is obtained by subtracting the end time of file uploading to the start time of file uploading. In this experiment, time needed to upload six files is taken and the graph is plotted for observed measures. Here the file size (kb) is plotted against the file uploading time (ms). The file uploading time depends on file size i.e. if the file size increased then the file uploading time is also increased. The plotted graph shows that the proposed technique has linear increase in uploading time while increasing the file size which is reduced compared to the existing AES and RSA techniques.

Finally, the file downloading time is measured and it is the time needed for end user verification and file downloading. Here the uploaded six files are downloaded and the time is observed which is then plotted. Figure 8. is plotted for the file size (kb) and the file downloading time (ms). As of in file uploading, the file downloading time is also

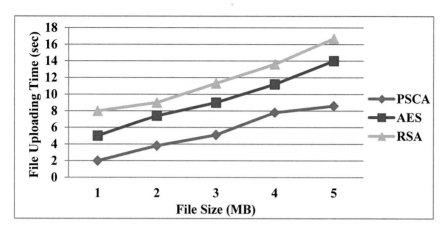

**Fig. 7.** File size versus file uploading time

increased for increasing file size. From Fig. 8, it is observed that the proposed technique has reduced downloading time compared to the other two techniques.

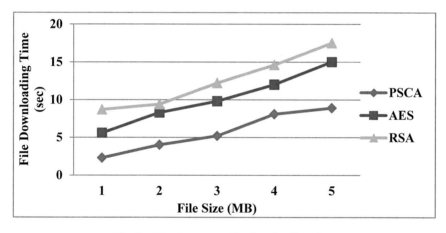

**Fig. 8.** File size versus file downloading time

The above comparisons proved that the proposed technique provides efficient and faster operations in all four parameters. The privacy and security levels of the proposed PSCA are clearly analyzed in security evaluation which states that the proposed security mechanism is better for both internal and external attacks than the existing ones.

# 8   Conclusion

In this paper, a novel privacy and security aware cryptographic algorithm is presented which aims to provide complete protection to data stored in the cloud environment. The cloud can offer simple and flexible storage services where privacy and security are the major concerns. The proposed encryption and decryption is based on invertible non-linear function and its inverse form respectively. The proposed technique is simulated and the obtained results are compared with the existing AES and RSA techniques. Detailed privacy and security analysis of the proposed and existing techniques is given in Sect. 6. There are four simulation parameters such as file uploading time, down-loading time, encryption time, and decryption time are taken and the measures are observed from simulation and they are then plotted against the file size. The obtained results are evident that the proposed PSCA provides effective privacy and security mechanism and it is efficient and faster in terms of less time in all four parameters than the existing techniques. In future, this work is extended for audio and video file encryption.

# References

1. Abed, F.S.: A proposed method of information hiding based on hybrid cryptography and seganography. Int. J. Appl. Innov. Eng. Manag. 2(4), 530–539 (2013)
2. Singh, A., Nandal, A., Malik, S.: Implementation of Ceaser cipher with rail fence for enhancing data security. Int. J. Adv. Res. Comput. Sci. Softw. Eng. 2(12), 78 –82 (2012)
3. Dey, S.: SD-AREE: an advanced modified Ceaser cipher method to exclude repetition from a message. Int. J. Inf. Netw. Secur. 1(2), 67–76 (2012)
4. Kester, Q.-A.: A Hybrid cryptosystem based on Vigenere cipher and columnar transposition cipher. Int. J. Adv. Technol. Eng. Res. 3(1), 141– 147 (2013)
5. Wang, C., Ren, K., Lou, W., Li, J.: Towards publicly auditable secure cloud data storage services. IEEE Netw. Mag. 24(4), 19–24 (2010)
6. Parmar, H., Nainan, N., Thaseen, S.: Generation of secure one-time password based on image authentication. J. Comput. Sci. Inf. Technol. 7, 195–206(2012)
7. Wang, H.: Proxy provable data possession in public clouds. IEEE Trans. Serv. Comput. 6(4), 551–559 (2013)
8. Zissis, D., Lekkas, D.: Addressing cloud computing security issues. Futur. Gener. Comput. Syst. 28(3), 583–592 (2012)
9. Sumit Chaudhary., Joshi, N.K.: Secured blended approach for cryptographic algorithm in cloud computing. Int. J. Pure Appl. Math. 118(20), 297–304 (2018)
10. A, L., Monikandan, S.: Data security and privacy in cloud storage using hybrid symmetric encryption algorithm. Int. J. Adv. Res. Comput. Commun. Eng. 2(8), 3064–3070 (2013)
11. Rahmani, H., Sundararajan, E., Ali, Z M., Zin, A.M.: Encryption as a service (EaaS) as a solution for cryptography in cloud. Procedia Technol. 11, 1202–1210 (2013)
12. Joshi, A., Joshi, B.: A randomized approach for cryptography. In: International Conference on Emerging Trends in Network and Computer Communications (ETNCC), pp. 293–296( 2011)
13. Wu, J., Detchenkov, I., Cao, Y.: A study on the power consumption of using cryptography algorithms in mobile devices. In: 2016 7th IEEE International Conference on Software Engineering and Service Science (ICSESS), (2016)

14. Wang, B., Li, M., Chow, S.S., Li, H.: A tale of two clouds: computing on data encrypted under multiple keys. In: 2014 IEEE Conference on Communications and Network Security (CNS), pp. 337–345. IEEE (2014)
15. Bhandari, A., Gupta, A., Das, D.: Secure algorithm for cloud computing and its applications. In: 2016 6th International Conference on Cloud System and Big Data Engineering (Confluence). IEEE (2016)

# Sequential Hybridization of GA and PSO to Solve the Problem of the Optimal Reactive Power Flow ORPF in the Algerian Western Network (102nodes)

I. Cherki[1($\boxtimes$)], A. Chaker[1], Z. Djidar[1], N. Khalfellah[1], and F. Benzergua[2]

[1] Laboratoire SCAMRE, Ecole National Polytechnique d'Oran, Oran, Algeria
cherki.imene@yahoo.fr
[2] Université des Sciences et de la Technologie d'Oran, Oran, Algeria

**Abstract.** In this study, we attempt to solve the problem of the optimal flow of the reactive power (ORPF) by the sequential hybridization of methaheuristics based on the combination of the two techniques on populations that are the genetic algorithm GA and the Particles Swarms Optimization PSO. The aim of this optimization is the minimization of the power losses while keeping the voltage, the generated power and the transformation ratio of the transformers within their limits.

**Keywords:** Reactive power flow · Metaheuristic methods · Metaheuristic hybridization · Genetic algorithm · Particles swarms · Electrical network

## 1 Introduction

The quick and great development of the electrical networks as well as the transit of the electrical energy for long distance led in searching of effective techniques and methods allowing the resolution of the problem of the optimal distribution of the powers and the control of the voltage of the electrical systems aiming at maintaining the voltage within the acceptable limits of operation.

They also make it possible to control and to minimize the transmission losses which consequently lead to a reduced cost. This is a difficult problem to solve specially for networks having big sizes like the Algerian western network with 102 nodes where many constraints of security must be taken into account such as assessment of the active and reactive powers, the transited powers and the voltage limits.

The aim of this paper is to contribute to the implementation of the methods of the resolution of the optimal distribution of the active and reactive powers that permit us to understand the problem in an interesting way.

In this paper, the methods of resolution of the problem of the optimal flow of the reactive power (ORPF) of the Algerian western networks (102 nodes) are based on the sequential hybridization of the two metaheuristical ones with population (Genetic

© Springer Nature Switzerland AG 2019
C. Benavente-Peces et al. (Eds.): SEAHF 2019, SIST 150, pp. 338–346, 2019.
https://doi.org/10.1007/978-3-030-22964-1_38

algorithm and the optimitization by particles swarms) in order to exploit the advantages of each one to obtain techniques with better performances.

A hybrid method can appear bad or interesting accordingly to the choice and the roles of its components. To define an effective hybrid method, it is necessary to know a well characterization of the advantages and the limits of each method [3].

## 2 The Metaheuristics

The metaheuristics are stochastic algorithms which progress towards an optimum by sampling of an objective function. They are generic methods being able to treat a broad range of different problems without requiring deep changes in the used algorithm. They appear of a great effectiveness to provide approximate solutions of good quality for a large number of classical problems of optimization and real applications of big sizes. For these reasons their investigations are in full development [2, 9].

### 2.1 The Genetic Algorithms

The genetic algorithms belong to the class of the evolutionary algorithms (Meta-heuristics with population). They consist in working simultneously with a set of solutions. That we make evolve gradually. The use of several solutions simultaneously makes it possible to naturally improve exploration of the space of the configurations their goal is to obtain an approached solution with a problem of optimization. They use three genetic operations, namely, the selection, the crossing and the change to transform an initial population of chromosomes in the objective to improve their quality. To carry out these phases of reproduction, it is necessary to code the individuals of the population [1, 11, 4].

### 2.2 Particle Swarm Optimization

Optimization by swarms of particles came out in the United States under the name of (PSO) Particle Swarm Optimization (PSO) due to the works of the two searchers Russell Eberhart (Electrical Engineer) and James Kennedy (Socio-psychologies) [2].

This algorithm belongs to the family of the evolutionary algorithms. It is inspired by the behaviour of the great regroupings of animals such as the clouds of birds, the fish's benches and the swarms of locusts. It is based in particular on a model developed the biologist Craig Reynolds at the end of 1980 making it possible to simulate the displacement of a group of birds. Another source of inspiration asserted by these Authors the socio-psychology [2].

## 3 Sequential Hybridization of Metaheuristics

The hybridization is a trend observed in many works completed on metaheuristics these ten last years. It makes it possible to benefit from the advantages cumulated from different metaheuristics. The origins of the hybrid algorithms of metaheuristics return to the work of Glover [2].

A hybrid method is research method that used at least two distinct search methods [3, 15].

It is possible to hybrid all the metaheuristics methods. In practice, the preoccupation with a performance or the constraints of computer resource is limiting the possibilities of hybridization. It is thus necessary to be careful in the choice of the methods used in order to obtain a good cooperation between the various components of the hybrid method [3, 15].

There exist several types of hybridization [5] like the hybridization in series (sequential), the hybridization in parallel and the hybridization in insertion (integrative).

The sequential hybridization is the most popular. It consists in applying several methods in such manner that the results of a given method are taken as initial solutions to the next method [3, 15].

In this study, we propose the technique of sequential hybridization which consists of the combination of both the genetic algorithm and the optimization by the swarm of particles.

## 4 Problematique

The problem of the optimal distribution of powers is relevant to the optimization process of the reactive power. It consists in minimizing a definite non linear objective function with non linear constraints. In our situation, the objective function represents the active losses in the electrical network.

$$P_L = \sum_i^n \sum_j^m -G_{ij} \cdot (V_i^2 + V_j^2 - 2.V_i.V_j.\cos\theta_{ij}) \tag{1}$$

The constraints of equality represent a balance between the production and the consumption:

$$\Delta P_J = \sum_{j=2}^n V_i V_j (G_{ij}\cos\theta_{ij} + B_{ij}\sin\theta_{ij}) - P_i^g + P_i^l = 0 \tag{2}$$

and

$$\Delta Q_J = \sum_{j=2}^n V_i V_j (G_{ij}\cos\theta_{ij} - B_{ij}\sin\theta_{ij}) - Q_i^g + Q_i^l - Q_i^{comp} = 0 \tag{3}$$

**Table 1.** Essential data for the Algerian western network

| | |
|---|---|
| Charges nodes | 92 |
| Nodes of generations | 10 |
| Lines | 119 |
| Transformers | 14 |

**Table 2.** Limits of the voltage of the nodes

| Voltage | Minimal value | Maximal value |
|---|---|---|
| 400 kV | 0.9 | 1.1 |
| 220 kV | 0.9 | 1.1 |
| 60 kV | 0.9 | 1.1 |

**Table 3.** Limits of the variables of control

| Variable | Minimal value | Maximal value |
|---|---|---|
| $a_i$ | 0.9 | 1.1 |
| $Q_1^g$ | −170 | 350 |
| $Q_6^g$ | −240 | 270 |
| $Q_{12}^g$ | −60 | 100 |
| $Q_{13}^g$ | −90 | 180 |
| $Q_{20}^g$ | −80 | 400 |
| $Q_{22}^g$ | −35 | 60 |
| $Q_{24}^g$ | −80 | 400 |
| $Q_{39}^g$ | −15 | 48 |
| $Q_{51}^g$ | −8 | 38 |
| $Q_{55}^g$ | −20 | 30 |

The constraints of inequality are the limits of the variables:

$$Q_{i,min}^g \leq Q_i^g \leq Q_{i,max}^g \quad i = 1\ldots n_g \tag{4}$$

$$Q_{i,min}^{comp} \leq Q_i^{comp} \leq Q_{i,max}^{comp} \quad i = 1\ldots n_{com} \tag{5}$$

$$a_{i,min} \leq a_i \leq a_{i,max} \quad i = 1\ldots n_T \tag{6}$$

$$V_{i,min} \leq V_i \leq V_{i,max} \quad i = 1\ldots n \tag{7}$$

where:
$P_i^g$            : active power generated at i-th node
$iQ_i^g$          : reactive power generated at i-th node
$P_i^l$             : active power consumption at i-th node

| | |
|---|---|
| $Q_i^l$ | : Reactive power consumption at i-th node |
| $Q_i^{comp}$ | : Reactive power of the compensator at node i |
| $\theta_{ij} = \theta_i - \theta_j$ | : Angles of the voltages at nodes i and j, |
| $G_{ij}$ | : Conductance between nodes i and j, |
| $B_{ij}$ | : Susceptance of the nodes $i$ and j |
| $a_i$ | : Transformation ratio of the i-th transformer |
| $n_g$ | : Number of generators |
| $n_T$ | : Number of transformers |
| $n_{comp}$ | : Number of t compensators |
| n | : Total number of nodes |

## 5  Illustration

We determined the effectiveness of the hybridization of the two metaheuristic methods with population in order to control the reactive power, the transformation ratio of the transformers and the voltage of the nodes as well as the values of these variables of control corresponding to the values of the minimum losses for the Algerian western network with 102 nodes, see Fig. 1.

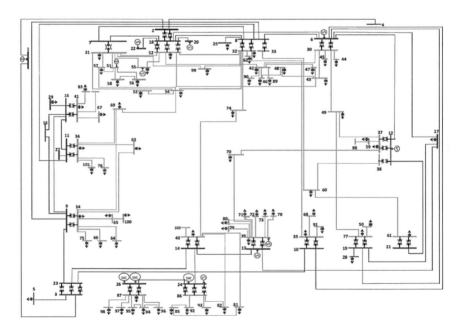

**Fig. 1.** Algerian western network 400/220/60 kW

To do this, we used sequential hybridization by carrying out the genetic algorithm and then the optimization by swarms of particles. This signifies that the solution given by the genetic algorithm is taken as the initial solution of the optimization by particles swarms.

The essential data for the Algerian western network and the limits of the variables of control are reported in the following tables, respectively [10] (Tables 1, 2 and 3).

To see the profile of the voltages of the nodes and the active losses, we carried out a study of burden-sharing by fast decoupled load flow method (FDLF). Then, we made the optimitization of the network by the GA and PSO algorithms. At the end, we introduced the sequential hybridization of the two algorithms.

This optimization consists of controlling the voltages of the nodes (See Fig. 2); the reactive power and the transformation ratio reported in Tables 4 and 5, respectively. The values of the minimal losses in the network and the duration of the execution of programs are given in Table 6.

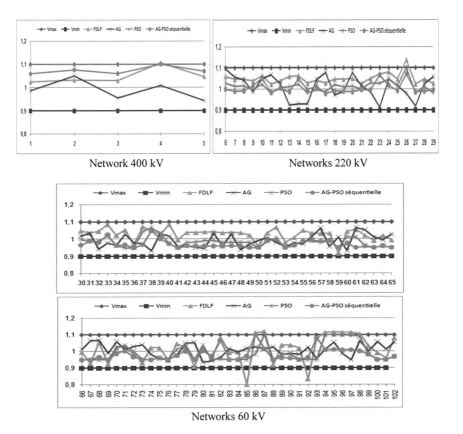

Network 400 kV

Networks 220 kV

Networks 60 kV

**Fig. 2.** The voltage of nodes in different cases before and after the optimization for GA, PSO and GA-PSO hybridization

**Table 4.** Generated powers Qg(MVAR)

| N° du Nœud | FDLF | GA | PSO | GA-PSO sequential hybridization |
|---|---|---|---|---|
| 1 | −236.56 | −118.01 | −118.78 | −170.53 |
| 6 | −76.28 | −203.16 | 590.04 | −200 |
| 12 | 78.47 | 72.16 | −44.91 | 75.61 |
| 13 | −46.1 | −84.42 | −58.4 | −32.93 |
| 20 | 7.72 | 109 | −29.71 | −25.8 |
| 22 | −24.07 | 11.18 | 66.56 | −3.29 |
| 24 | −45.18 | −5.58 | −77.74 | 37.91 |
| 39 | 116.85 | −8.31 | 139.29 | 38.86 |
| 51 | −20.44 | 4.41 | 87.2 | 20.43 |
| 55 | 6.97 | 25.75 | −2.09 | −0.48 |

**Table 5.** Transformation ratios

| N° du Nœud | FDLF | GA | PSO | GA-PSO sequential hybridization |
|---|---|---|---|---|
| 02→06 | 0.96 | 1 | 1.1 | 1.06 |
| 03→23 | 0.96 | 0.99 | 1.08 | 0.98 |
| 06→30 | 0.98 | 1 | 0.99 | 1 |
| 07→31 | 0.99 | 0.99 | 0.96 | 0.96 |
| 08 →32 | 0.98 | 1 | 0.99 | 0.98 |
| 08→33 | 0.95 | 0.99 | 0.99 | 0.94 |
| 09→34 | 0.98 | 0.99 | 0.99 | 1.01 |
| 10→35 | 0.98 | 0.99 | 1.09 | 1.05 |
| 11→36 | 0.99 | 0.99 | 1.01 | 1.01 |
| 12→37 | 0.96 | 1 | 0.95 | 1.02 |
| 12→38 | 0.99 | 0.99 | 0.94 | 0.94 |
| 13→39 | 1.07 | 0.99 | 1.01 | 0.97 |
| 14→40 | 0.95 | 1 | 1.02 | 1.05 |
| 15→41 | 0.98 | 0.99 | 1.05 | 1.03 |
| 18→52 | 0.98 | 1 | 1.01 | 0.99 |
| 19→77 | 1 | 0.99 | 1.01 | 1.01 |
| 21→61 | 0.97 | 1 | 0.98 | 1.06 |
| 24→86 | 0.97 | 0.99 | 1.06 | 0.91 |
| 26→87 | 0.99 | 1 | 0.98 | 1.02 |

**Table 6.** Active losses and times of execution

| | FDLF | GA | PSO | GA-PSO sequential hybridization |
|---|---|---|---|---|
| Active loses (MW) | 51.06 | 36.6 | 45.07 | 29.19 |
| Réduction (MW) | | 14.46 | 5.98 | 21.87 |
| Réduction (%) | | 28.31% | 11.72% | 42.82% |
| Exécution time | | 26.7 | 35.29 | 36.42 |

After the analysis of the results obtained from the two metaheuristic methods and their sequential hybridization, it appears clearly that the problem of the goings beyond lower and higher of the voltages of the nodes disappeared, (see Fig. 2). We notice that the active and reactive powers as well as the transformation ratio remain all within their limits.The comparative study of the solutions obtained by the basic metaheuristics GA and PSO and those obtained by their sequential hybridization shows that the last method gives better solutions in terms of the minimization of the losses as shown in Table 6. However, the sequential hybridization method seems to be less effective for the time of execution.

## 6 Conclusion

We have applied a combination of the two known methods (GA and PSO), namely the sequential hybridization method to the comprehension of the optimal flow of the powers for the case of the Algerian Western network (102 nodes).

It appears that this sequential method reproduces better solutions than the standard ones. All voltages were corrected with a better minimization on the losses of a percentage of 42.82%.

In another side, this study can be extended to the development of other types of hybridization like integrative hybridization.

## References

1. Goldberg, D.: Algorithme génétique -Exploration–optimisation et apprentissage automatique (1991)
2. Hachini, H.: Hybridation d'algorithmes métaheuristiques en optimisation globale et leurs applications, INSA de Rouen (2013)
3. Duvivier, D.: étude de l'hybridation des métaheuristique, application à un problème d'ordonnancement de type JOBSHOP, Université du littoral Cote D'OPALE (2000)
4. Yang, X.-S.: Engineering Optimization An Introduction with Metaheuristic Applications. Wiley, Hoboken, New Jersey, ISBN: 0470582464,0470640413, 378 p., April 2010
5. Lahdeb, M., Hellal, A., Arif, S.: Hybridations métaheuristiques en lots appliquées à l'écoulement optimal de la puissance réactive. Eur. J. Electr. Eng. 15(6), 587–612 (2012) https://doi.org/10.3166/ejee.15.587-612. ISSN 2103-3641
6. Chetih, S., Khiet, M., Chaker, A.: Optimal distribution of the reactive power and the voltage control in Algerian Network using the genetic algorithm method. IREE 4(4), July–August 2009. ISSN 1827-6660
7. Mohamed, L.: Théorie et application de méthodes d'hybridations métaheuristiques dans les réseaux électriques. Université Amar Telidji-Laghouat, février 2007
8. Sinsuphun, N., Leeton, U., Kwannetr, U., Uthitsunthorn, D., Kulworawanichpong, T.: Loss Minimization Using Optimal Power Flow Based on Swarm Intelligences. Trans. Electr. Eng. Electron. Commun. 9(1), 212–222 2011
9. Gerbex, S.: Métaheuristiques Appliquées au placement optimal de dispositifs facts dans un réseau électrique. Lausanne EPFL (2003)
10. Cherki, I.: Modélisation des FACTS et contrôle des puissances dans un réseau électrique par les méthodes métaheuristiques. ENP d'Oran Algérie (2012/2013)

11. Naima, K., Fadela, B., Imene, C., Abdelkader, C.: Use of genitic algorithm and particle swarm optimisation methods for the optimal control of the reactive power in Western Algerian power system. Energy Procedia **74**, 265–272 (2015)
12. Daoudi, R.: Optimisation des puissances réactives par la méthode Essaim Particules (PSO) dans un réseau d'énergie électrique. ENSET d'Oran Algérie (2011/2012)
13. Alliot, J.-M., Durand, N.: Algorithmes génétiques. 14 mars 2005
14. Bansal, R.C.: Otimization methods for electric power systems: an overview. Int. J. Emerg. Electr. Power Syst. **2**(1), 1–23 (2005)
15. Cotta, C., Talbi, E.G., Alba, E.: Parallel hybrid metaheuristics. In: Alba, E. (ed.) Parallel Metaheuristics, A New Class of Algorithms, pp. 347–370. Wiley, Hoboken (2005)
16. Abdelhakem-Koridak, L., Rahli,M.: Optimisation d'un dispatching environnement/économique de la production d'énergie électrique par un algorithme génétique. Département d'Electrotechnique, USTO, Oran, Algérie
17. Laufi, A., AllaouaL B.: Répartition optimale des puissances actives d'un réseau électrique utilisant l'algorithme de colonie de fourmis. Département d'électrotechnique, Université de Béchar Algérie
18. Mozafari, B., Amraee, T., Ranjbar, A.M., Mirjafari, M.: Particle Swarm optimization method for optimal reactive power procurement considering voltage stability. Scientia Iranica **14**(6), 534–545 (2007)

# Combined Use of Meta-heuristic and Deterministic Methods to Minimize the Production Cost in Unit Commitment Problem

Sahbi Marrouchi[(⊠)], Nesrine Amor, and Souad Chebbi

The Laboratory of Technologies of Information and Communication and Electrical Engineering (LaTICE), National Superior School of Engineers of Tunis (ENSIT), University of Tunis, Tunis, Tunisia
sahbimarrouchi@yahoo.fr, nisrine.amor@hotmail.fr, souad.chebbi@yahoo.com

**Abstract.** Solving the Unit Commitment problem (UCP) optimizes the combination of production units operations and determines the appropriate operational scheduling of production units to satisfy the expected consumption which varies from one day to one month. This paper represents a new strategy combining three optimization methods: Tabu search, Particle swarm optimization and Lagrangian relaxation methods in order to develop a proper unit commitment scheduling of the production units while reducing the production cost during a definite period. The proposed strategy has been implemented on a the IEEE 9 bus test system containing 3 production unit and the results were promising compared to strategies based on meta-heuristic and deterministic methods.

**Keywords:** Unit commitment · Optimization · Scheduling · TS-PSO-LR · PSO · Tabu search · Lagrangian relaxation

## 1 Introduction

The Continuity of service and the quality of electricity has become a strategic issue for the utility, the network manager and the equipment manufacturers and this is due to the need to increase economic competitiveness for businesses and the opening of the electricity market. In order to a better use of the electrical grid, it is necessary to solve the technical and economic problems. Operations scheduling production units or Unit Commitment (UC) improve operational planning of the electrical grid while ensuring continuity of service [1–3]. The main purpose of solving the Unit Commitment problem is to schedule production units to respond to the consumed power with the minimization of the total production cost. The optimal planning involves ensuring a better use of available generators subject to various constraints and guaranteeing the transfer of electrical energy from generating stations to the load. UC must satisfy the load demand, storage capability, minimum downtime, startup, and safety limits for each production unit. The production scheduling comprises determining startup and

© Springer Nature Switzerland AG 2019
C. Benavente-Peces et al. (Eds.): SEAHF 2019, SIST 150, pp. 347–357, 2019.
https://doi.org/10.1007/978-3-030-22964-1_39

each generation level for each unit in a given planning period [4–7]. Therefore, a study of literature [8] on methods which focus on unit commitment (UC) problem resolution shows that various optimization methods have examined this subject. Furthermore, Sasaki et al. demonstrated the possibility to use artificial neural network (ANN) to solve the UCP in which a large number of inequality constraints is processed. They have used the ANN to schedule generators and the dynamic programming to solve the load flow problem. The adopted strategy was compared to Lagrangian Relaxation (LR) and dynamic programming (DP) methods and the results offered a faster and cheaper solution compared to the LR and DP but it suffers from digital convergence because of the learning process. Certain works [9, 10] proposed a strategy based on tabu search method. They introduced new rules to generate an initial solution feasible to solve the Unit Commitment problem. This strategy consists on dividing the problem into two problems: the first combinatorial optimization problem is solved using tabu search algorithm and the second is a problem of nonlinear programming solved through the quadratic programming routine. The structure resolution through Tabu search method is similar to that used by simulated annealing [11] even though TS is provided with a simplified configuration, so it is easy to pass from one optimization to the other. Indeed, the main advantage of the adopted strategy is to extend the search space provided for the best optimal solutions which are stored in the tabu list. This method has provided a lower production cost solution, but it's slower compared to the Lagrangian relaxation. However, Logenthiran et al. [12] have proposed a new approach based on particle swarm optimization (PSO) algorithm for solving the unit commitment problem. They presented three versions of particle swarm: binary particle swarm optimization (BPSO), improved binary particle swarm optimization (IBPSO) and combined use of particle swarm optimization and Lagrangian relaxation programming (LR-PSO). The numerical results show that LR-PSO method has provided a lower production cost solution compared to LR, BPSO and IBPSO specially when the number of units increases. Whereas, when the number of units is small, BPSO is taken as the best method since it has the lowest production cost compared to other algorithms. Other works, [13, 14] presented a new approaches based on artificial intelligence to solve the UCP. The adopted approach combines two methods: tabu search and neural networks (ANN-TS) in order to get an optimal unit commitment scheduling allowing to have a minimal production cost in accordance with all constraints of the studied system. Artificial Neural networks provide a fast convergence to optimal solutions but it takes a lot of memory space because of the great number of constraints. According to our study, we thought to validate an approach to apprehend the whole unit commitment problem. To achieve this objective, our strategy for solving the Unit Commitment Problem is based on the combination of three stochastic optimization methods that are the Particle Swarm Optimization (PSO), the Tabu Search (TS) and Lagrangian Relaxation (LR) method in order to develop a proper unit commitment scheduling of the production units to minimize the production cost.

## 2   Problem Formulation

The objective of the Unit Commitment problem is to establish the best production unit plan that will be available to minimize the total operating cost of the generating units and to supply the forecasted load over a period of time $H$.

$$Min \left[ F_T(P_{ih}, U_{ih}) = \sum_{i=1}^{N_g} \sum_{h=1}^{H} [a_i P_{ih}^2 + b_i P_{ih} + c_i + ST_i(1 - U_{i(h-1)})] U_{ih} \right] \quad (1)$$

where;

$ST_i$:    The starting cost of the $i$th unit defined by:

$$ST_i = \begin{cases} HSC_i \text{ if } MDT_i \leq \tau_i^{OFF} \leq MDT_i + SC_i \\ CSC_i \text{ if } \tau_i^{OFF} \succ MDT_i + SC_i \end{cases} \quad (2)$$

$a_i, b_i$ and $c_i$:    Coefficients of the production cost,

$P_{ih}$:    Active power generated by the $i$th unit $h$th h, $i = 1, 2, 3, \ldots, N_g$

$U_{ih}$:    On/Off status of the $i$th production unit at the $h$th h, $U_{ih} = 0$ for the off state of one generating unit and $U_{ih} = 1$ for the operating status of one generating unit,

$HSC_i, CSC_i$:    Hot and Cold start-up cost of the $i$th unit,

$MDT_i, \tau_i^{OFF}, SC_i$:    Minimum down-time, continuously off-time and Cold start time,

$N_g, H$:    Number of generating units and time horizon for the UC

The minimization of the objective function is provided with the following constraints:

–   Power balance constraints

$$\sum_{i=1}^{N_g} P_{ih} U_{ih} = P_{dh} \quad (3)$$

–   Spinning reserve constraints

$$P_{dh} + P_{rh} - \sum_{i=1}^{N_g} U_{ih} P_{ih} \leq 0 \quad (4)$$

–   Generation limits

$$P_i^{min} . U_i \leq P_{ih} . U_i \leq P_i^{max} . U_i \quad (5)$$

- Minimum up-time constraint

$$U_{ih} = 1 \quad for \quad \sum_{t=h-up_i}^{h-1} U_{ih} \leq MUT_i \tag{6}$$

- Minimum down-time constraint

$$U_{ih} = 0 \quad for \quad \sum_{t=h-down_i}^{h-1} U_{ih} \leq MDT_i \tag{7}$$

With:

$P_{rh}$, $P_{dh}$, $P_{Lh}$:    System spinning reserve, amount of the consumed power and total active losses at the $h$th h,

$P_i^{min}$, $P_i^{max}$:    Minimum and maximum power produced by a generator,

$MUT_i$, $MDT_i$:    Continuously on and down-time of unit $i$.

# 3  Methodology of Resolution

Our strategy for solving the Unit Commitment Problem is based on the combination of three optimization methods that are the Particle Swarm Optimization (PSO), the Tabu Search (TS) and Lagrangian Relaxation (LR) method to find a good plan of the On/Off states of each unit over a period of time leading to obtain a good production cost. In this strategy, we thought to inquire about advantages of PSO method for solving complex and nonlinear problems, the flexibility of storage great memory of optimal solutions which is the major advantage of Tabu search and the convergence speed of the Lagrangian relaxation method. The proposed strategy will not only benefit from the advantages of each method but also reduces the disadvantages of each method.

## 3.1  Particle Swarm Optimization

Particle swarm optimization provides a population based search procedure in which individuals called particles change their positions with time. This method is able to generate high quality of solutions within shorter calculation time and stable convergence characteristic than other stochastic optimization methods. The PSO model consists of a swarm of particles moving in a definite dimensional real-valued space of possible problem solutions [14–16]. Every particle has a position $X_i = (x_i^1, x_i^2, \ldots, x_i^l)$ and a flight velocity $V_i = (v_i^1, v_i^2, \ldots, v_i^l)$. Indeed, each particle has its own best positions $P_{ibest} = (P_{ibest}^1, P_{ibest}^2, \ldots, P_{ibest}^l)$ and a global best position $G_{best} = (G_{best}^1, G_{best}^2, \ldots, G_{best}^l)$. Each time step is characterized by the update of the velocity and the particle is moved to a new position which is the sum of the previous position and the new velocity:

$$X_r^{k+1} = X_r^k + V_r^{k+1} \tag{8}$$

The update of the velocity from the previous velocity to the new velocity is obtained by:

$$V_r^{k+1} = w_k.V_r^{k+1} + c_1.rand.(P_{best}^k - X_r^k) + c_2.rand.(G_{best}^k - X_r^k) \tag{9}$$

where, $w_k$ is the inertia weight factor, $c_1$ et $c_2$ are acceleration constant, *rand* is a uniform random value between $[0, 1]$, $X_r^k$ and $V_r^k$ are respectively the position and the velocity of one particle $i$ at iteration $k$.

## 3.2   Tabu Search

Tabu search uses a local or neighborhood search procedure to iteratively move from a solution $X$ to a solution $X'$ in the neighborhood of $X$, until some stopping criterion has been satisfied. To explore regions of the search space that would be left unexplored by the local search procedure, TS modifies the neighborhood structure of each solution as the search progresses [9, 10].

The new neighborhood solutions $N^*(X)$ are determined through the use of memory structures. The search then progresses by iteratively moving from a solution $X$ to a solution $X'$ in $N^*(X)$. To determine the solutions admitted to $N^*(X)$, a tabu list (TL) memory is used, which is a short-term memory containing the solutions that have been visited in the recent past as less than the maximum number of iterations.

## 3.3   Lagrangian Relaxation

The Lagrangian relaxation solves the Unit commitment problem by relaxing or temporarily ignoring the constraints, power balance and spinning reserve requirements [15]. Therefore, to transform the complex nonlinear constrained problem into a linear unconstrained problem, we have considered the following Lagrangian function:

$$L(P_{ih}, U_i, \lambda_i) = \sum_{i=1}^{N_g} \sum_{h=1}^{H} [a_i P_{ih}^2 + b_i P_{ih} + c_i + ST_i(1 - U_{i(h-1)})]U_{ih} + \lambda_i.(P_d - \sum_{i=1}^{N_g} P_i U_{ih}) \tag{10}$$

where, $\lambda_i$ is the Lagrangian coefficient.

To establish our strategy, we have considered the partial derivatives of the Lagrange function (9) with respect to each of the controllable variables equal to zero.

$$\frac{\partial L}{\partial P_{ih}} = \frac{\partial[[a_i P_{ih}^2 + b_i P_{ih} + c_i + ST_i(1 - U_{i(h-1)})]U_{ih}]}{\partial P_{ih}} - \lambda_i\left(\frac{\partial P_{dh}}{\partial P_{ih}} - U_{ih}\right) = 0 \tag{11}$$

$$\frac{\partial L}{\partial \lambda_i} = P_{dh} - \sum_{i=1}^{N_g} P_i U_{ih} = 0 \tag{12}$$

Equations (11) and (12) represent the optimality conditions necessary to solve equation systems (1) and (3) without using inequality constraints (Eqs. (4) and (5)). Equation (11) can be written as follows:

$$\lambda_i = \frac{\frac{\partial\left[a_i P_{ih}^2 + b_i P_{ih} + c_i + ST_i(1-U_{i(h-1)})]U_{ih}\right]}{\partial P_{ih}}}{\frac{\partial P_{dh}}{\partial P_{ih}} - U_{ih}} ; i = 1,\ldots,N_G; h = 1,\ldots,H \qquad (13)$$

### 3.4   Proposed Strategy

The process of the Unit Commitment problem resolution by the combined use of Tabu search, Particle swarm optimization and Lagrangian Relaxation (TS-PSO-LR) methods is carried out according to the flowchart in Fig. 1.

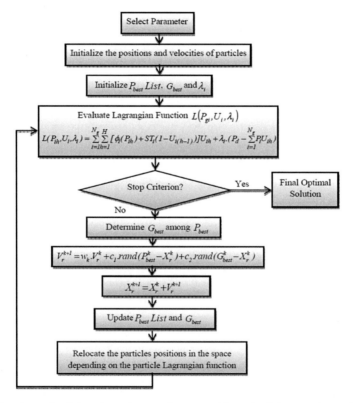

**Fig. 1.** Flowchart of solving the unit commitment problem via Tabu search, particle swarm optimization and Lagrangian relaxation

The proposed strategy not only helps to reach the optimal solution as quickly as possible using the speed of the Lagrangian relaxation but also to proceed through PSO method to search effective solutions corresponding to a minimum production cost and this is obtained through a specific determination of the new velocity and then the next best position corresponding to the best amount of generated power produced by each unit when it's in the ON state.

In the proposed method, it's notable that $P_{ibest}$ representing the best information of each particle and the history of each generated power $P_{gi}$ of each production unit is preserved in the list $P_{best}List$. Herein, in spite of the possibility of the PSO method with the solution better than $G_{best}^k$ around $P_{best}^k$, there is the possibility not to be searched enough. We pay attention to this possibility, and think use of history of $P_{ibest}^1$ in $P_{best}List$. Whenever the particles lose the searching ability when the velocity $V_r^k$ of one particle is very small, the TS-PSO-LR algorithm adapts the other $(P_{ibest}^2, \ldots, P_{ibest}^l)$ instead of $P_{ibest}^1$ to update equation of velocity. This action increases the searching ability and helps to find more optimal solutions enabling a minimal production cost while considering a best unit commitment scheduling.

# 4   Simulation and Results

In order to test the performance of the optimization proposed method; the strategy has been applied to an IEEE electrical network 9 buses, having 3 generators, over a period of 48 h. The characteristics of the different production units are given in Table 1.

In this paper, we have considered 48 successive periods in order to establish the temporal evolution of the power demand (Table 2).

Table 1.   Characteristics of production units

| U | $P_{gimax}$ (MW) | $P_{gimin}$ (MW) | a | b | c | M U T | M D T | $HSC_i$ ($) | $CSC_i$ ($) |
|---|---|---|---|---|---|---|---|---|---|
| 1 | 582 | 110 | 0.0756 | 30.36 | 582 | 8 | 8 | 4500 | 9000 |
| 2 | 330 | 74 | 0.00031 | 17.26 | 970 | 8 | 8 | 5000 | 10000 |
| 3 | 115 | 25 | 0.00211 | 16.5 | 680 | 5 | 5 | 560 | 1120 |

Table 2.   Amount of load required

| H | Load (MW) | H | Load (MW) | H | Load (MW) | H | Load (MW) |
|---|---|---|---|---|---|---|---|
| 1 | 353.2 | 13 | 993.2 | 25 | 833.2 | 37 | 682.2 |
| 2 | 378.5 | 14 | 913.2 | 26 | 813.2 | 38 | 715.2 |
| 3 | 463.2 | 15 | 853.2 | 27 | 763.2 | 39 | 773.2 |
| 4 | 573.2 | 16 | 725.2 | 28 | 713.2 | 40 | 843.2 |
| 5 | 628.2 | 17 | 613.2 | 29 | 626.2 | 41 | 883.2 |
| 6 | 693.2 | 18 | 580.2 | 30 | 547.2 | 42 | 911.2 |
| 7 | 713.2 | 19 | 673.2 | 31 | 503.2 | 43 | 945.2 |
| 8 | 753.2 | 20 | 730.2 | 32 | 473.2 | 44 | 960.2 |
| 9 | 843.2 | 21 | 835.2 | 33 | 433.2 | 45 | 1001.2 |
| 10 | 925.2 | 22 | 945.2 | 34 | 533.2 | 46 | 1003.2 |
| 11 | 963.2 | 23 | 1007.2 | 35 | 583.2 | 47 | 925.2 |
| 12 | 1013.2 | 24 | 893.2 | 36 | 627.2 | 48 | 823.2 |

Based on Table 3, the production cost found by the hybrid method based on the combination between Tabu search, Particle Swarm Optimization and Lagrangian Relaxation methods (TS-PSO-LR) method among 48 h is about 1.5800e + 06 $ lower compared to that obtained through Tabu search (PC = 2.2350e + 06 $) and through PSO (PC = 1.7813e + 06 $) or through Lagrangian Relaxation (PC = 1.6405e + 08$). This result shows the best performances of the adopted strategy in minimizing the production cost and proves that we can get promising results through hybridization.

Furthermore, concerning the resolution time, TS and PSO methods has presented the best time of convergence to an optimal solution compared to our strategy which requires 127.082 s to reach the global optimum. Besides, through Lagrangian Relaxation method, the unit commitment problem requires a lot of time to converge and this is explained by the complexity of the problem.

Therefore, based on Fig. 2, we can notice that the total amount of generated power by the production units is very similar to that consumed with a very limited amount of spinning reserve power compared to PSO, TS and LR methods where the generated powers are much higher than the amount requested.

This proves the effectiveness of tracking of the consumed power per each hour and shows the performance of the algorithms enabling to get a minimal production cost. Besides, this minimal production cost has been established thanks to a good On/Off statements scheduling set for each production units (Fig. 3). The organization is made through an estimation of the amount of load desired by the electric network, while taking into account of the allowable constraints.

**Table 3.** Production cost and time required to converge for each optimized method

|          | TS            | PSO           | LR            | TS-PSO-LR     |
|----------|---------------|---------------|---------------|---------------|
| PC ($)   | 2.2350e + 06  | 1.7813e + 06  | 1.6405e + 08  | 1.5800e + 06  |
| Time (s) | 1.536 s       | 2.432         | 4539 s        | 127.082 s     |

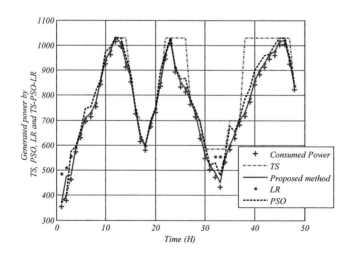

**Fig. 2.** Production cost through TS, PSO, LR and TS-PSO-LR methods

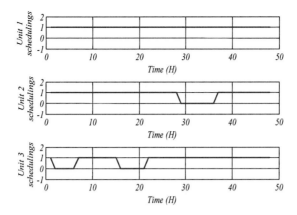

**Fig. 3.** Optimal binary combination of units operation

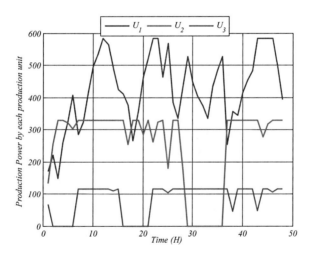

**Fig. 4.** Generated power by each production unit during 48 h

Furthermore, we confirm that our approach has allowed to select precisely the production units that should be available to respond to the load demand of the electrical network over a future period. In addition, the adopted approach was promising both in terms of convergence to get the best optimal solutions to minimize the cost production and for an efficient unit commitment scheduling for the different production units, Fig. 4.

Our strategy differs from other evolutionary computing techniques in providing an acceptable solution within a relatively short time and is likely to lead the search towards the most promising solution area.

# 5   Conclusion

The proposed strategy based on the use of the combined use of Tabu search, Particle swarm optimization and Lagrangian Relaxation (TS-PSO-LR) to solve the Unit Commitment problem has presented high performances in optimizing the production cost and capability of convergence to a global optimum as quick as possible compared to meta-heuristic (PSO, TS) and deterministic (LR) methods. In addition, our hybrid strategy has ensured a proper unit commitment scheduling leading to get a minimal production cost. Moreover, the right choice of the initial population suggests the possibility to obtain improvements in execution time. In addition, our strategy provides a fast enough time to converge to the optimal solution; which demonstrates the effectiveness of the adopted strategy compared to that obtained by Lagrangian Relaxation method.

# References

1. Merlin, A., Sandrin, P.: A new method for unit commitment at Electricity de France. IEEE Trans Power Appar. Syst. **102**, 1218–1225 (1983)
2. Lin, F.T., Kao, C.Y., Hsu, C.: Applying the genetic approach to simulated annealing in solving some NP-Hard problems. IEEE Trans Syst. **23**(6), 1752–1767 (1993)
3. Marrouchi, S., Chebbi, S.: New strategy based on fuzzy logic approach to solve the unit-commitment problem. In: Proceedings of the International Conference on Control, Engineering Information Technology (CEIT '14). Sousse, Tunisia (2014)
4. Garg, R., Sharma, A.K.: Economic generation and scheduling of power by genetic algorithm. J. Theor. Appl. Inf. Technol., JATIT (2005)
5. Aoki, K., Satoh, T., Itoh, M., Ichimori, T., Masegi, K.: Unit commitment in a large-scale power system including fuel constrained thermal and pumped-storage hydro. IEEE Trans. Power Syst. (PWRS) **2**(4) (1987)
6. Ouyang, Z., Shahidehpour, S.M.: An intelligent dynamic programming for unit commitment application. IEEE Trans Power Syst. **6**(3), 1203–1209 (1991)
7. Marrouchi, S., Ben Saber, S.: A comparative study of fuzzy logic, genetic algorithm, and gradient-genetic algorithm optimization methods for solving the unit commitment problem. Math. Probl. Eng. **2014**, Article ID 708275, 14 (2014)
8. Sasaki, H., Watanabe, M., Kubokawa, J., Yorino, N., Yokoyama, R.: A solution method of unit commitment by artificial neural network. IEEE Trans. Power Syst. **7**(3) (1992)
9. Mantawy, A.H., Shokri, S.Z., Selim, Z.: A unit commitment by Tabu search. Proc. Inst. Elect. Eng. Gen. Transm. Dist. **145**(1), 56–64 (1998)
10. Mantawy, A.H., Abdel Magid, L., Selim, S.Z.: A unit commitment by Tabu search. IEEE Proc. (1998)
11. Cheng, C.P., Liu, C.W., Liu, C.C.: Unit commitment by annealing-genetic algorithm. Electr. Power Energy Syst. **24**, 149–158 (2002)
12. Logenthiran, T., Srinivasan, D.: Particle swarm optimization for unit commitment problem. IEEE Trans. Power Syst. (2010)
13. Rajan, A.G.C.C.: Neural based tabu search method for solving unit commitment problem with cooling-banking constraints. Serb. J. Electr. Eng. **6**(1), 57–74 (2009) UDK: 004.832.2
14. Kumar, S.S., Palanisamy, V.: A new dynamic programming based hopfield neural network to unit commitment and economic dispatch. IEEE Trans. Power Syst. (2006)

15. Raglend, I.J., Raghuveer, C., Avinash, G.R., Padhy, N.P., Kothari, D.P.: Solution to profit based unit commitment problem using particle swarm optimization. Appl. Soft Comput. J. **10**(4), 1247–1256 (2010)
16. Marrouchi, S., Hessine, M.B., Chebbi, S.: Combined use of an improved PSO and GA to solve the unit commitment problem. In: The International Multi-Conference on Systems, Signals and Devices (SSD'2018), pp. 19–22. Hammamet-Tunisia (2018)

# Development and Experimental Validation of a Model to Simulate an Alkaline Electrolysis System for Production of Hydrogen Powered by Renewable Energy Sources

Mónica Sánchez[1](✉), Ernesto Amores[1], David Abad[1],
Carmen Clemente-Jul[2], and Lourdes Rodríguez[3]

[1] Centro Nacional del Hidrógeno (CNH2), Prolongación Fernando El Santo s/n,
13500 Puertollano, Ciudad Real, Spain
monica.sanchez@cnh2.es
[2] Dpto. Energía y Combustibles, ETSI Minas y Energía, Universidad Politécnica
de Madrid (UPM), Ríos Rosas 21, 28003 Madrid, Spain
[3] Universidad Europea de Madrid (UEM), Tajo s/n, Urbanización El Bosque,
28670 Villaviciosa de Odón, Madrid, Spain

**Abstract.** This work shows an ASPEN Plus model to simulate an alkaline electrolysis plant with the objective of evaluating the performance of a complete system under different operating conditions, such as temperature, pressure and current density. As this type of process modelling software is not able to simulate the behavior of the electrolysis cells, an electrochemical model has been previously developed. It is a semi-empirical model that allows determining the voltage cell, Faraday efficiency and gas purity as a function of the current, using both physical principles related to the electrolysis process and statistical methods. The different parameters defined in the equations of the model have been calculated by MATLAB, using a non-linear regression, on the basis of experimental data from different experiments obtained of an alkaline test bench. These equations have been used as a subroutine within an ASPEN Plus model including the rest of balance of plant (BoP) components to analyze a complete alkaline electrolysis plant to produce hydrogen. The obtained results allow determining which process variables are the most influential and hence, should be optimized in order to improve the performance and operation of an alkaline electrolyzer.

**Keywords:** Electrolysis · Simulation · Hydrogen

## 1 Introduction

The development of renewable energies has been spectacular in recent years. However, the increase of the share of renewable energy in electricity production affects adversely the energy system because of the unpredictability inherent in wind and solar power. The massive penetration of renewable energy sources (RES) will force energy providers to develop large amount of energy storage to improve power stability [1].

© Springer Nature Switzerland AG 2019
C. Benavente-Peces et al. (Eds.): SEAHF 2019, SIST 150, pp. 358–368, 2019.
https://doi.org/10.1007/978-3-030-22964-1_40

Hydrogen technologies can help to cope with these challenges and to contribute to the new energy system development. In this context, water electrolysis is considered as an important technology, as it represents the linkage between electrical and chemical energy, through the hydrogen production. Different water electrolysis technologies are present in the market. They can be divided into alkaline, proton exchange membrane (PEM) and solid oxide (SO) electrolysis cells, according to the electrolyte used [2].

Currently, alkaline water electrolysis (AWE) is the most mature technology and it is commercially available for large-scale hydrogen production, covering a broad range of hydrogen production capacities, from 1 to 760 $Nm^3/h$, with an acceptable manufacturing cost (1000–5000 $/kW) depending on the production capacity. Technically, alkaline water electrolysis is characterized by the use of an alkaline aqueous solution of 25–35 wt % KOH as electrolyte, operating under temperatures between 70–140 °C and pressures commonly close to 15–30 bar, although some systems can reach up to 200 bar. Therefore, the key of this technology is its maturity, availability, no platinum-group metals containing components materials and the comparatively low specific cost [2, 3].

Although AWE is already a well-established technology, it has some limitations such as the low current density and the problems derived from the influence of the dynamic operation on the gas purity, efficiency and durability [2]. For last years, R&D efforts have been carried out to improve the design and the performance of electrolysis system mainly focus on [4, 5]:

- Development of advanced electrocatalytic materials to reduce the electrode over-voltage, especially in the anode reaction.
- Minimization of the space between the electrodes in order to reduce the ohmic losses and allow working with higher current densities.
- Development of new advanced material to be used as separator. In this regard, the use of ion exchange inorganic membranes has become wide-spread due to the elimination of liquid electrolyte and the reduction of the crossover.
- Modelling of the cell voltage and hydrogen production as a function of the current applied to characterize the behavior of the electrolysis cell.

However, it is necessary to take into account that the design and configuration of auxiliary systems and balance of plant (BoP) components have a high influence on the performance of the electrolysis cells, as well as, cost, efficiency and lifetime. The BoP represents approximately 20–30% of the capital cost in an alkaline electrolyser and the stack efficiency can be reduced up to 15% if the design and configuration of the auxiliary systems are not optimized [3]. Identifying technical improvements related to auxiliary systems and developing models including balance of plant components, is necessary to optimize alkaline electrolysis systems powered by renewable energy sources and, also reduce parasitic loads during part load operation.

In this work, an ASPEN Plus model, including both an electrochemical cell model developed in a previous work [6] and balance of plant components, is proposed to simulate a complete alkaline electrolysis plant. This model will provide a tool to evaluate the different process variables in order to optimize the performance and the efficiency of the system and analyze its thermodynamic behavior to identify the losses sources and maximize the cost-effectiveness of an alkaline electrolysis plant.

## 2  Simulation of a 15 kW Alkaline Electrolyzer

### 2.1  Electrochemical Model for the Water Alkaline Electrolysis Stack

A mathematical model to evaluate the performance of a 15 kW alkaline electrolyzer have been developed in a previous work [6]. The model proposed includes polarization curve, Faraday efficiency and gas purity, considering temperature, pressure and current density. The equations used for the ASPEN Plus model are shown in Table 1.

The polarization curve describes the electrochemical behavior of an electrolyzer. In order to model the current-voltage curve accurately ($i$-$U$), it must be taken into account thermodynamics, kinetics and resistive effects. In this way, the voltage ($U_{cell}$) is defined as the sum of the reversible voltage ($U_{rev}$), which can be determined by the change in Gibbs energy, and the different overpotentials. One of the most widely used models to analyze the current-voltage relationship was proposed by Ulleberg [7], which describes the voltage of an electrolyzer according to the parameters "s" (V) and "t" ($m^2$/A) related to activation overpotentials and "r" ($\Omega m^2$) for the ohmic overpotentials. However, in this model the parameters only depend on the temperature (T). In order to obtain a wider model, the equation has been modified to include pressure (p). For this purpose, an additional parameter "d" ($\Omega m^2$) [6], which represents the variation in the ohmic overpotentials according to pressure has been incorporated in this model, Eq. (1).

On the other hand, the amount of gas produced by an electrochemical process can be related to the electrical charge consumed by the cell, which is described by Faraday's law. From this, it is possible to measure the effectiveness of the real process using the Faraday efficiency ($\eta_F$). This parameter is defined as the ratio between the volume of gas produced and the theoretical volume that should be produced during the same time. In this investigation, the Faraday efficiency has been modeled using Eq. (2), which includes a linear relation for temperature ("$f_1$", "$f_2$") [7, 8]. The pressure has not been included due to its slight influence.

Finally, a precise knowledge of the purity of gases is essential when the electrolyzer is dynamically operated using a renewable energy source. The contamination of the produced gases is of special importance when the electrolyzer is operated at low current densities. In particular, the diffusion of hydrogen to oxygen (HTO) is the most critical parameter. The HTO values can be calculated theoretically using diffusion coefficients and the solubility of gases. However, these parameters are not always easy to estimate. For this reason, the empirical Eq. (3) has been used considering the effect of pressure and temperature, based on previous studies [6, 8].

The model constants $r_1$, $r_2$, $d_1$, $d_2$, s, $t_1$, $t_2$, $t_3$, $f_{11}$, $f_{12}$, $f_{21}$, $f_{22}$, $C_1$–$C_9$ and $E_1$–$E_9$ (see Table 1) have been calculated by means of a non-linear regression using MATLAB, according to a previously established procedure [9]. To do this, the experimental data obtained in a test bench for the polarization curve, Faraday efficiency and gas impurities have been used [6].

The root-mean-square (RMS) error has been evaluated for the voltage with an average result of 5.67 mV per cell and for the Faraday efficiency and HTO with a value lower than 1% [6]. These results show an excellent correlation between experimental and modeled data.

**Table 1** Proposed equations for modeling a 15 kW alkaline electrolyzer stack [6]

| Model | Proposed Equations | Eq. |
|---|---|---|
| Polarization curve | $U_{cell} = U_{rev} + [(r_1 + d_1) + r_2 T + d_2 p]$ $i + s \cdot \log\left[\left(t_1 + \frac{t_2}{T} + \frac{t_3}{T^2}\right)i + 1\right]$ <br><br> Where: $r_1 = 4.45153 \cdot 10^{-5}$ $\Omega m^2$; $r_2 = 6.88874 \cdot 10^{-9}$ $\Omega m^2$ $°C^{-1}$; $d_1 = -3.12996 \cdot 10^{-6}$ $\Omega m^2$; $d_2 = 4.47137 \cdot 10^{-7}$ $\Omega m^2$ $bar^{-1}$; $s = 0.33824$ V; $t_1 = -0.01539$ $m^2 A^{-1}$; $t_2 = 2.00181$ $m^2$ $°CA^{-1}$; $t_3 = 15.24178$ $m^2$ $°C^2 A^{-1}$ | (1) |
| Faraday efficiency | $\eta_F = \left(\frac{i^2}{f_{11} + f_{12} \cdot T + i^2}\right) \cdot (f_{21} + f_{22} \cdot T)$ <br><br> Where: $f_{11} = 478645.74$ $A^2 m^{-4}$; $f_{12} = -2953.15$ $A^2 m^{-4}$ $°C^{-1}$; $f_{21} = 1.03960$; $f_{22} = -0.00104°$ $C^{-1}$ | (2) |
| Gas impurities | $HTO = \left[C_1 + C_2 T + C_3 T^2 + (C_4 + C_5 T + C_6 T^2) \cdot \exp\left(\frac{C_7 + C_8 T + C_9 T^2}{i}\right)\right]$ $+ \left[E_1 + E_2 p + E_3 p^2 + (E_4 + E_5 p + E_6 p^2) \cdot \exp\left(\frac{E_7 + E_8 p + E_9 p^2}{i}\right)\right]$ <br><br> Where: $C_1 = 0.09901$; $C_2 = -0.00207°$ $C^{-1}$; $C_3 = 1.31064 \cdot 10^{-5}$ $°C^{-2}$; $C_4 = -0.08483$; $C_5 = 0.00179°$ $C^{-1}$; $C_6 = -1.13390 \cdot 10^{-5}$ $°C^{-2}$; $C_7 = 1481.4$ $Am^{-2}$; $C_8 = -23.6$ A $m^{-2}$ $°C^{-1}$; $C_9 = -0.257$ $Am^{-2}$ $°C^{-2}$; $E_1 = 3.71417$; $E_2 = -0.93063$ $bar^{-1}$; $E_3 = 0.05817$ $bar^{-2}$; $E_4 = -3.72068$; $E_5 = 0.93219$ $bar^{-1}$; $E_6 = -0.05826$ $bar^{-2}$; $E_7 = -18.38$ $Am^{-2}$; $E_8 = 5.87$ $Am^{-2}$ $bar^{-1}$; $E_9 = -0.4642$ $Am^{-2}$ $bar^{-2}$ | (3) |

## 2.2   Simulation of the Balance of Plant

The simulation diagram of the alkaline electrolyzer plant is shown in Fig. 1.

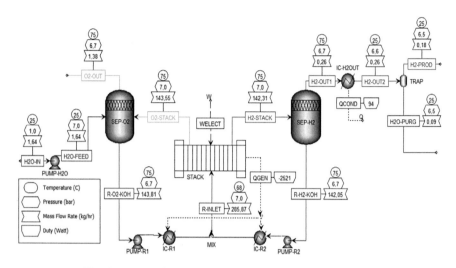

**Fig. 1.** Process flow diagram of an alkaline electrolysis plant

Simulations have been implemented in ASPEN Plus with the objective of evaluating the performance of a complete system including the balance of plant under different operating conditions, such as temperature, pressure and current density.

The electrochemical reaction takes place on the electrochemical cells (STACK) and according to Fig. 1, it is simulated trough a personalized block that integrate the equations for modelling the cell stack introduced in Sect. 2.1. These equations (see Table 1) are integrated in ASPEN Plus as a subroutine, using a tool called ASPEN Custom Modeler (ACM) which is able to create its own operation unit. The electrochemical model is implemented in ACM through the definition of the different process variables and the selection of the thermodynamic and physical property package. Once it has been compiled, simulations can be performed at different operating conditions.

This sub-routine calculates the cell stack voltage, the hydrogen and oxygen production and the quantity of hydrogen in the oxygen (HTO) due to the crossover for an electric power input (WELECT), depending on the number of cells and the active area of the electrode, maintaining the stack temperature and pressure system at the set-point. The results are returned to ASPEN Plus to continue with the procedure. The STACK unit also calculates the excess energy supplied (difference between cell voltage and the thermoneutral voltage). That results in the production of waste heat (QGEN) and an increase of the temperature of the electrolyte and the gases produced in the stack. This waste heat must be removed to ensure a constant operating temperature.

Hydrogen (H2-STACK) and oxygen (O2-STACK) flow produced in cell stack are led with the electrolyte (KOH, 35% wt) to the liquid-gas separation vessel (SEP-H2 and SEP-O2, respectively), where the electrolyte is separated from the gas and returned back by recirculation pumps (PUMP-R1 for anode circuit and PUMP-R2 for cathode circuit). Both KOH recycles (R-H2-KOH and R-O2-KOH) pass through a heat exchanger (IC-R1and IC-R2) to cool down the electrolyte before entering to the stack and thus control the cells temperature. For calculating the temperature of the electrolyte inlet the model assumes that the heat to dissipate in the heat exchangers must be at least the heat generated (QGEN) in the cell stack due to the irreversibilities.

The hydrogen produced and separated in the biphasic separation vessel (H2-OUT1) is dried through a condenser (IC-H2OUT) and a water trap (TRAP) to cool the hydrogen stream up to 25 °C (H2-OUT2) and eliminate the maximum amount of water (H2O-PURG). The simulation takes into account the part of the hydrogen produced diffused at the cell stack to the oxygen flow (HTO) due to crossover phenomena.

Finally, deionized water at the desired conductivity of maximum 5 µS/cm (H2O-IN) is fed from a water tank into the oxygen separator (SEP-O2) by a pump (PUMP-H2O to provide water to electrolysis process (H2O-FEED).

# 3   Experimental Set Up

An alkaline water electrolysis stack has been characterized at different operating conditions for carrying out this study. The cell stack is composed of 12 bipolar electrolysis cells of 1000 $cm^2$ surface area connected electrically in series. Each electrolysis cell contains a pair of electrodes (anode and cathode) separated by a porous diaphragm and assembled to obtain a zero-gap configuration. This means that the distance between the different elements of the cell stack is extremely low in order to achieve a more efficient electrolysis process. The electrolyte circulating inside of the cells is a concentrated KOH solution in water (35% wt). It can operate up to pressure of 10 bar and temperatures between 50–80° C.

## 3.1   15 kW Alkaline Electrolysis Test Bench

A test bench developed by *Centro Nacional del Hidrógeno* (CNH2) in the framework of DESPHEGA project [10] has been used for carrying out the experimental tests. This system is a flexible tool designed to test alkaline water electrolysis (AWE) stacks in a wide range of conditions. Figure 2 shows a picture of the Alkaline Electrolysis Laboratory in CNH2 where the 15 kW alkaline electrolysis test bench is located.

The general facility, see Fig. 2, is composed by the following components:

1. A balance of plant (BoP) that includes all the equipment and instrumentation to ensure the electrolysis stack operation as well as the control system (National Instruments, COMPACT RIO) to monitor and control the main process variables and guarantee the safety of the plant.
2. A chiller (CTA Refrigeration Industrial, IPE M4/SMART), to maintain an optimum operating temperature in the cell stack.
3. A gas cabinet which provides nitrogen to inertia the system.
4. A deionized water system (SIEMENS, ULTRACLEAR RO 10) including a 30 L tank, which provides water type ASTM II.

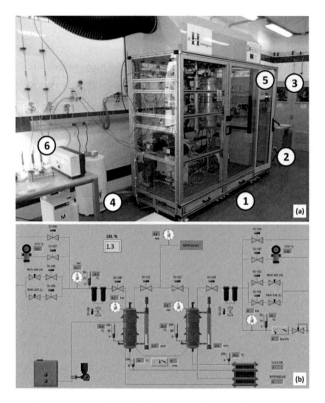

**Fig. 2.** Experimental set-up: (a) 15 kW alkaline water electrolysis test-bench developed by CNH2; (b) simplified process flow diagram of the balance of the plant (BoP)

5. Two programmable DC switching power supplies (AMETEK, AMREL SPS60-250) to provide energy to the cells, with the possibility of being connected in series or parallel depending on the number of cells of the stack.
6. A micro-GC to analyze the composition of the gases produced (AGILENT, 490 MICRO GC).

## 3.2 Test Protocol

Polarization curves have been used to compare the performance of the electrolysis stack under different operation temperatures and pressures. The experimental methodology to perform the polarization curve has been developed to ensure the repeatability and reproducibility of the obtained data. The procedure followed is based on the well-established FCTESTNET protocol for fuel cell [11], but adjusted and modified to be used with electrolysis cells/stacks. In each of these tests the polarization curve, the purity of the gases produced and the Faraday efficiency have been studied.

## 4   Results

### 4.1   Influence of Temperature and Pressure in the Electrolysis System

Figures 3a, b show the polarization curve, power required and heat generated by the AWE electrolysis stack at different temperatures and pressure, respectively.

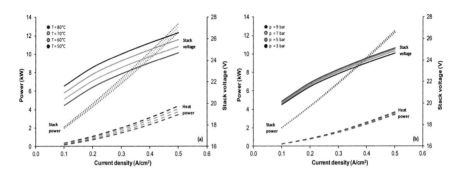

**Fig. 3.** Polarization curve, power required and heat generated in the alkaline electrolysis system at different a) temperature b) pressure

According to the polarization curve model, when the temperature increases from 50 °C to 80 °C, the voltage progressively reduces. As a consequence, the power required in the electrolysis and the heat generated decrease when the temperature is higher. On the other hand, the pressure has much less influence on the potential and power of the electrolyzer. In fact, it is observed that in the modeled range, an increase in pressure from 5 bar to 9 bar has practically no perceptible influence on the stack voltage.

Figures 4a, b shows the hydrogen production and the content of hydrogen in oxygen (HTO) at different temperatures and pressures, respectively.

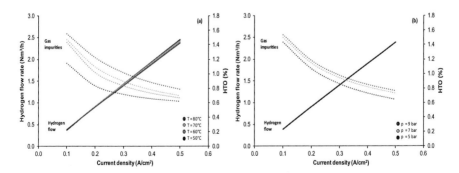

**Fig. 4.** Hydrogen flow rate and gas impurities in the alkaline electrolysis system at different: (a) temperatures, (b) pressures

With reference to the hydrogen production, the temperature shows a greater influence, since the hydrogen flow rate increases when the temperature is reduced from 80 °C to 50 °C, especially at low current densities (Fig. 4a). About the pressure (Fig. 4a), it has not been included in the model because it shows very little effect on the Faraday efficiency within the range of pressures considered during this study.

Regarding to the content of hydrogen in oxygen, as can be seen in and, it increases strongly with decreasing the current density for all temperatures (Fig. 4a) and pressures (Fig. 4b). This result must be considered when the electrolyzer operates at partial loads, using renewable energy sources [6]. Therefore, for safety reasons, the minimum applied current density in the tests was 0.1 A/cm$^2$. Also, the results show that high pressures and temperatures have a great influence in the generated impurities, because the diffusion phenomena and the gas migrations produce increase.

## 4.2   Energy Efficiency Analysis

In this study, an energy analysis has been conducted to investigate the performance of an alkaline electrolyzer plant. The effect of parameters such as operating temperature and current density on the stack efficiency has been studied. The pressure has not been included due to its slight influence. The efficiency stack can be defined as voltage efficiency that represents the difference between the real cell voltage and the thermoneutral voltage, Faraday efficiency that is the ratio between the volume of gas produced and the theoretical volume that should be produced during the same time and the stack efficiency, Eq. (4), that is the relation between la energy contained in the produced hydrogen and the energy consumed by the process.

$$n_{Stack} = \frac{Q_{H2,prod} \cdot LHV}{W_{elec}} \tag{4}$$

where, $Q_{H2}$ is the hydrogen outlet flow rate of $H_2$, LHV is the hydrogen lower heating value and $W_{elec}$ is the rate of electric energy input.

Figure 5a shows the efficiencies obtained with the model at different temperatures. As can be seen, the voltage efficiency increases with temperature, since the potential required for electrolysis decreases. On the other hand, an increase in temperature leads to a lower Faraday efficiency, mainly due to more parasitic currents losses [6]. Also, as the current density decreases, the efficiency is lower because the effect of parasitic currents is greater. Therefore, Faraday efficiency and voltage efficiency show opposite trend with respect to temperature. However, when the overall efficiency is analyzed, it is observed that at low current densities the efficiency is slightly higher for low temperatures, but when the current density increases, the highest overall efficiency is obtained with high temperatures. Figure 5b shows the overall efficiency and energy consumption for 400 mA/cm$^2$ and 7 bar at different temperatures. As can be seen, when the temperature increases, the overall efficiency improves and the energy consumption is reduced.

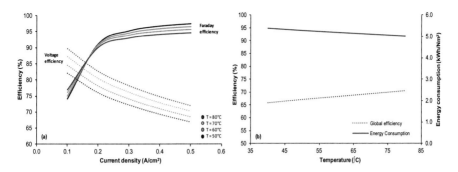

**Fig. 5.** Energy efficiency analysis at different temperatures: (a) voltage and Faraday efficiency, (b) energy consumption and overall efficiency

## 5  Conclusions

Alkaline water electrolysis is the most mature technology and it is commercially available for large-scale hydrogen production. As a consequence, in recent years, a significant effort has been made to model the operation of AWE, especially when they are powered by RES. In this paper, an ASPEN Plus model of an AWE plant has been proposed with the objective of evaluating the performance of a complete system, including the balance of plant, under different operating conditions, such as temperature, pressure and current density. The electrochemical model of the electrolysis cells developed and experimentally validated in a previous work has been integrated in Aspen Plus as a subroutine for calculating the voltage cell, hydrogen and oxygen production and gas purity as a function of the current.

According to the results of the model, when the temperature increases, the voltage and the hydrogen production progressively reduce, while the pressure has much less influence on these variables. On the other hand, the content of hydrogen in oxygen (HTO) increases strongly with decreasing the current density for all temperatures and pressures. Also, when the temperature increases, the overall efficiency improves and the energy consumption is reduced.

In conclusion, Aspen Plus provide a useful design tool that can be applied to maximize the cost-effectiveness of an alkaline electrolysis system powered by renewable energy sources for hydrogen production. The obtained results allow determining which process variables are the most influential and hence, should be optimized in order to improve the performance and operation.

**Acknowledgements.** This work has been developed in the facilities of the *Centro Nacional del Hidrógeno* (CNH2), whose financial supporters are *Ministerio de Economía, Industria y Competitividad* (MINECO, Spain), *Junta de Comunidades de Castilla-La Mancha* (JCCM, Spain) and *European Regional Development Fund* (ERDF).

# References

1. Kotowicz, J., Jurczyk, M., Wecel, D., Ogulewicz, W.: Analysis of hydrogen production in alkaline electrolyzers. J. Power Technol. **96**(3), 149–156 (2016)
2. Lehner, M., Tichler, R., Steinmüller, H., Koppe, M.: Power-to-Gas: Technology and Business Model. Springer, Cham (2014)
3. Bertuccioli, L., Chan, A., Hart, D., Lehner, F., Madden, B., Standen, E.: Development of water electrolysis in the European Union. FCH JU funded studies. Cambridge (UK): element energy, Lausanne (CH): E4tech Sarl (2014)
4. Ursua, A., Gandía, L.M., Sanchis, P.: Hydrogen production from water electrolysis: current status and future trends. Proc. IEEE **100**(2), 410–426 (2012)
5. Michal, J., Daniel, W., Zeng, K., Zhang, D.: Recent progress in alkaline water electrolysis for hydrogen production and applications. Prog. Energy Combust. Sci. **36**(3), 307–326 (2010)
6. Sánchez, M., Amores, E., Rodríguez, L., Clemente-Jul, C.: Semi-empirical model and experimental validation for the performance evaluation of a 15-kW alkaline water electrolyzer. Int. J. Hydrog. Energy **43**(45), 20332–20345 (2018)
7. Ulleberg, Ø.: Modeling of advanced alkaline electrolyzers: a system simulation approach. Int. J. Hydrog. Energy **28**(1), 21–33 (2003)
8. Hug, W., Bussmann, H., Brinner, A.: Intermittent operation and operation modeling of an alkaline electrolyzer. Int. J. Hydrog. Energy **18**(12), 973–977 (1993)
9. Amores, E., Rodríguez, J., Carreras, C.: Influence of operation parameters in the modeling of alkaline water electrolyzers for hydrogen production. Int. J. Hydrog. Energy **39**(25), 13063–13078 (2014)
10. Sánchez, M.: Desarrollo de sistemas de producción de hidrógeno energético por generación alcalina: proyecto DESPHEGA. Rev. Energética XXI **136**, 26–27 (2013)
11. The fuel cell testing and standardization network. EU FP5-EESD funded project ENG2-CT-2002-20657, FCTESTNET test procedures (draft v1.4) (2006)

# Building a Digital Business Technology Platform in the Industry 4.0 Era

Maurizio Giacobbe[1](✉), Maria Gabriella Xibilia[2], and Antonio Puliafito[2]

[1] University of Messina, Piazza S. Pugliatti 1, 98122 Messina, Italy
mgiacobbe@unime.it

[2] Department of Engineering, University of Messina, C.da Di Dio 1, 98166 Messina, Italy
mxibilia@unime.it, apuliafito@unime.it

**Abstract.** The current trend of automation and data exchange in manufacturing technologies is commonly referred to as the fourth industrial revolution, and known as Industry 4.0. Its "pillars" are Cyber-Physical Systems (CPS), Cognitive Computing (CC), Internet of Things (IoT) and Cloud. In such a scenario, to know both technologies and processes supporting digital business is mandatory. In this paper we present our "vision", i.e., an approach integrating Business Intelligence and Machine Learning and addressed to CIOs and IT leaders in order to build a complete digital business technology platform with generation of *linked data knowledge*.

**Keywords:** Big data · Data analytics · Digital business · IoT · Linked data knowledge · Smart manufacturing

## 1 Introduction and Motivation

Nowadays the growth in technology innovation for business has solid foundation in the use of predictive analytics, user behavior analytics, and advanced data analytics methods that are able to extract value from data. All these processes and their applications are generally summarized with the term *Big Data*, e.g., about government and science communities [1], smart grids [2], open data applications [3], etc. The fourth industrial revolution, known as **Industry 4.0**, represents the current trend of automation and data exchange in manufacturing technologies. The main "pillars" of Industry 4.0 are *Cyber-physical systems (CPS)* [4], *Cognitive Computing (CC)* [5], *Internet of Things (IoT)* and *Cloud*, thus pursuing new levels of efficiency in delivering *IoT-as-a-Service (IoTaaS)* [6].

The result is the growth of new business perspectives (Industrial Internet of Things, Industrial Analytics, Cloud Manufacturing) by pushing private, public, and hybrid Cloud providers to integrate their systems with IoT devices (managing sensors and actuators) in order to provide smart services. Moreover, traditional *Operational Technologies* (Advanced Automation, Advanced Human Machine Interface, Additive Manufacturing) can be adopted together with the

© Springer Nature Switzerland AG 2019
C. Benavente-Peces et al. (Eds.): SEAHF 2019, SIST 150, pp. 369–375, 2019.
https://doi.org/10.1007/978-3-030-22964-1_41

above-mentioned new perspectives by businesses to mitigate the initial economic impact due to new investments.

In this paper we provide an overview on *Big Data Analytics* and *Business Intelligence* topics thus presenting our "vision", i.e., how to build a *digital business platform* with generation of *linked data knowledge*. Moreover, we present a case study concerning a first application of the proposed strategy.

**Data analytics** is generally the process (or the set of processes) of examining data sets in order to obtain conclusions about the information they contain focused on the topics of interest (e.g., focused on the business). Because we watch to its use to better address both *decision support* and *enterprise executive* systems, we prefer to specify it as *Data analysis*, that is the process (or the set of process) of inspecting, cleansing, transforming, and modeling data finalized to "transform" *customer data* in *information*. **Business Intelligence (BI)** usually refers to a set of business processes that collect and analyze strategic information, the technology to execute these processes, and the results obtained from their execution. If Data analysis allows business to have information from customer data, the BI supports it in to "transform" information in *knowledge* to support decision-making. **Machine learning (ML)**, instead, differs from BI because it is better focused on industrial topics. ML can be considered as a branch of *artificial intelligence* to give CPS the ability to "learn" from data, and progressively improve performance about specific tasks and without being explicitly programmed.

## 2 Proposed Strategy

In order to provide a systematic vision of the concepts presented so far, in this section we introduce the proposed digital business platform through its functional architecture. To this end, the Algorithm 1 summarizes the main activities that should be solved by a company/business in order to develop new marketable services, typically Business-to-Customer (B2C) and Business-to-Government (B2G), and to provide low-cost services addressed to other companies/businesses, typically Business-to-Business (B2B) and Business-to-Business-to-Customer (B2B2C). The approach is Cloud-based, i.e., the "normal" one in order to offer scalable and cost-effective infrastructure, to combine unstructured data from social media with available customers structured data, to enable fast predictive analysis through the real-time data collecting, and not least to enable open innovation for medium and small businesses.

Figure 1 shows the *data value chain* and represents in a graphical form the steps reported in the Algorithm 1. Compared with a traditional business workflow, the proposed approach includes the generation of **linked data knowledge** that, in turn, can be accessible in an *open innovation* ecosystems to other businesses or governmental entities.

**Algorithm 1** Big Data-oriented Activities for Digital Business

---
1: **Inputs:** Data from IoT, Mobile, Social Networks, Business, Linked Open Data;
2: **Data collection:** data are captured from the input data sources and transmitted to the Cloud through configurable gateways.
3: **Data Value Chain:**
4:    **Data storage and management:** data are stored and organized at Cloud enabling the management of *terabytes* of stored data. Operations result in *clean datasets* available for Data Analysis;
5:    **Data analysis (DA):** data are inspected, clean, transformed, and modeled in order to discover information useful for *decision-making* through DA software.
6:    **DA Output:** Business data.
7:    **Machine Learning (ML):** statistical techniques are used to give CPS the ability to "learn" from data, and progressively improve performance about specific tasks without being explicitly programmed.
8:    **ML Output:** improvement in supply chain management, thus resulting in a "smart" manufacturing in the Industry 4.0 scenario.
9:   **Business Intelligence:** exploring data analysis for *business* information. Extraction of *insights*, also combining internal data with external data, and identification of new areas of opportunities or inefficiencies/costs to be eliminated (insight services). Generation of linked data knowledge [6] [7].
10:    **BI Output:** New marketable services, Low-cost services for other businesses, linked data knowledge.

---

# 3 Case Study: From IoT to Cognitive Computing with OpenStack

In this section we present the implemented digital business platform for an entity that uses IoT data from the monitoring and the control of power consumptions and environmental parameters (temperature, brightness, presence) in geo-distributed buildings in the city of Messina [8,9]. Connectivity is provided by *Bluetooth piconets* inside offices and through the *Ethernet* inside buildings in a hierarchical star configuration of the network. Data are exchanged with sensors and actuators both locally and through the Internet by gateway nodes and therefore with the Cloud. This case study follows the indications provided by the Algorithm 1 and depicted in Fig. 1. The digital business IoT platform is optimized for time-series data, streaming data and to handle real-time ingestion and real-time streaming analytics. Main architectural components are:

- the OpenStack Cloud;
- the IoTronic resource management service;
- the Grafana analytics platform;
- the InfluxDB time-series database;
- the Cognitive machine learning service on top of Openstack.

**OpenStack**[1] is a Cloud operating system that controls large pools of compute, storage, and networking resources, thus enabling the passage from *Infrastructure-*

---
[1] https://www.openstack.org/, last accessed November 9, 2018.

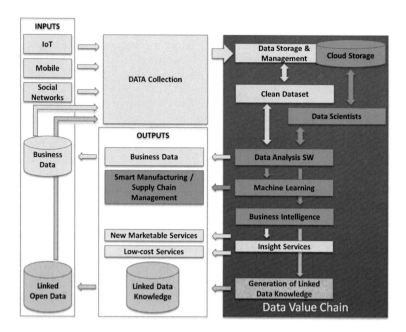

**Fig. 1.** Data value chain in a digital business technology platform including the generation of linked data knowledge.

*as-a-Service (IaaS)* to *IoT-as-a-Service (IoTaaS)*. **IoTronic** is an IoT resource management service for OpenStack Clouds. It allows to manage IoT resources as part of an OpenStack datacenter. Basic scenario for the installation of the Iotronic Service is provided at the official GitHub repository[2]. Infrastructure management and interaction services are exposed as RESTful APIs. **Grafana**[3,4] is an open source, feature rich metrics dashboard and graph editor for Graphite, Elasticsearch, OpenTSDB, Prometheus and InfluxDB. **InfluxDB**[5] is a custom high-performance data store and analytics written in *Go* specifically for time series data. The repository [10] contains code for benchmarking InfluxDB against other databases (also including NoSQL category) and time series solutions, thus motivating our choice. The open source core allows InfluxDB to be integrated at Cloud side based on the OpenStack to exchange time series data with the controlled smart environments, e.g., smart manufacturing. Continuous Queries (CQ) and Retention Policies (RP) enable the automatic process of down-sampling data and expiring old data. **Cognitive**[6] is a multi-tenant interactive machine learning service on top of Openstack. It has python bindings for api access.

---

[2] https://github.com/openstack/iotronic, last accessed November 9, 2018.

[3] https://grafana.com/, last accessed November 10, 2018.

[4] https://github.com/grafana/grafana, last accessed November 10, 2018.

[5] http://docs.grafana.org/features/datasources/influxdb/, last accessed November 10, 2018.

[6] https://wiki.openstack.org/wiki/Cognitive, last accessed November 9, 2018.

Figure 2 summarizes the implemented digital business platform.

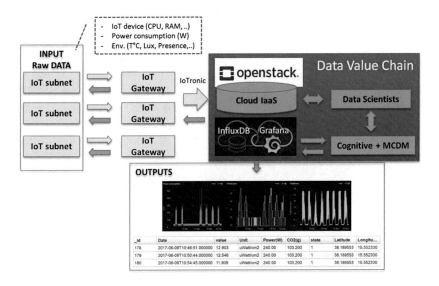

**Fig. 2.** General scheme of the implemented digital business technology platform.

Figure 3 shows the result of one-week "learning" period by implementing a self-controlled software engine that, based on a *multi-criteria decision making* algorithm, allows to dynamically update the best brightness level taking into account presence and IT equipments use. The proposed digital business environment is completely open source and allows IT leaders interested in to adopt a no-vendor lock-in strategy to address business in an open innovation context. Technological components are full integrable in order to assure interoperability among businesses operating in a Cloud OpenStack environment.

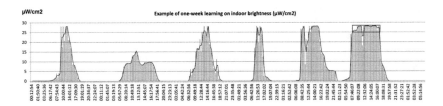

**Fig. 3.** One-week "learning" period with dynamic update of the best brightness level taking into account presence and IT equipments use.

# 4    Conclusion

Industry 4.0 is the "new" era of industrial and business applications. As a consequence, companies may need to adapt their systems to be competitive on the market. This work wants to be a support for *Chief Information Officers (CIOs)* and *IT leaders* to better understand "how to" build a digital business technology for the above-mentioned purpose. The proposed approach overcomes traditional business platforms that, although they are able to provide insight services, are not able to enrich their wealth of knowledge from *linked open data*. Data analysis and Business Intelligence processes are used to generate "new" knowledge resulting in new marketable services and low-cost services. Machine Learning addresses predictive analysis for smart manufacturing and optimized supply chain management.

# References

1. Aron, J.L., Niemann, B.: Sharing best practices for the implementation of Big Data applications in government and science communities. In: 2014 IEEE International Conference on Big Data (Big Data), Washington, DC, pp. 8–10 (2014). https://doi.org/10.1109/BigData.2014.7004469
2. Han, X., Wang, X., Fan, H.: Requirements analysis and application research of big data in power network dispatching and planning. In: 2017 IEEE 3rd Information Technology and Mechatronics Engineering Conference (ITOEC), Chongqing, pp. 663–668 (2017). https://doi.org/10.1109/ITOEC.2017.8122436
3. Ciancarini, P., Poggi, F., Russo, D.: Big Data quality: a roadmap for open data. In: 2016 IEEE Second International Conference on Big Data Computing Service and Applications (BigDataService), Oxford, pp. 210–215 (2016). https://doi.org/10.1109/BigDataService.2016.37
4. Cyber-Physical Systems (Document Number: nsf18538): National Science Foundation. https://www.nsf.gov/pubs/2018/nsf18538/nsf18538.htm (February 13, 2018)
5. Valipour, M., Wang, Y.: Sentence comprehension and semantic syntheses by cognitive machine learning. In: 2018 IEEE 17th International Conference on Cognitive Informatics & Cognitive Computing (ICCI*CC), Berkeley, CA, pp. 38–45 (2018). https://doi.org/10.1109/ICCI-CC.2018.8482024
6. Giacobbe, M., Di Pietro, R., Longo Minnolo, A., Puliafito, A.: Evaluating information quality in delivering IoT-as-a-service. In: 2018 IEEE International Conference on Smart Computing (SMARTCOMP), Taormina, pp. 405–410 (2018). https://doi.org/10.1109/SMARTCOMP.2018.00037
7. Espinosa, R., Garriga, L., Zubcoff, J.J., Mazn, J.: Linked open data mining for democratization of big data. In: 2014 IEEE International Conference on Big Data (Big Data), Washington, DC, pp. 17–19 (2014). https://doi.org/10.1109/BigData.2014.7004479
8. Giacobbe, M., Pellegrino, G., Scarpa, M., Puliafito, A.: The ESSB system: a novel solution to improve comfort and sustainability in smart office environments. In: 2017 IEEE 14th International Conference on Networking, Sensing and Control (ICNSC), Calabria, pp. 311–316 (2017). https://doi.org/10.1109/ICNSC.2017.8000110

9. Giacobbe, M., Pellegrino, G., Scarpa, M., Puliafito, A.: An approach to implement the smart office idea: the #SmartMe Energy system. J. Ambient Intell. Humaniz. Comput. (2018). https://doi.org/10.1007/s12652-018-0809-0
10. Code for comparison write ups of InfluxDB and other solutions. https://github.com/influxdata/influxdb-comparisons (last committed 2018)

# Single-Phase Grid Connected Photovoltaic System Using Fuzzy Logic Controller

Abdelaziz Sahbani[1,2(✉)]

[1] Medina College of Technology, Al Madinah al Munawwarah, Kingdom of Saudi Arabia
abdellazizsahbani@yahoo.fr,
a.sahbani@mct.edu.sa
[2] Laboratory of Automatic (LARA), National Engineering School of Tunis, Tunis El Manar University, BP37, 1002 Tunis, Tunisia

**Abstract.** This paper is about the study and simulation of a single-phase grid connected photovoltaic system. Maximum power point tracking with DC-DC boost converter and a single phase full-bridge DC-AC converter are the main parts of this system. To obtain the maximum power from the photovoltaic system, fuzzy logic controller has been applied. After the DC boosting stage the DC link voltage is connected with a single phase full-bridge inverter in order to be converted in AC voltage source, which is coupled with the single phase grid. The switching control of this inverter is done by an additional fuzzy logic controller (FLC). A Sinusoidal Pulse width modulation (SPWM) is used to switch the transistors used in the full-bridge inverter. A phase locked loop (PLL) technique use the voltage grid as reference to generate an estimated phase angle in order to have synchronization between the grid and the inverter. MATLAB Simulink has been used to get the simulations results.

**Keywords:** Photovoltaic (PV) · Maximum power point tracking (MPPT) · Fuzzy logic controller (FLC) · DC-DC boost converter · Grid connected inverter · Phase locked loop (PLL)

## 1 Introduction

Energy come from the photovoltaic (PV) effect is one of the most important and prerequisite resource because solar energy is free, extensively available. Recently, photovoltaic system can produce electric power deprived of environmental impact and has low maintenance costs.

The output electric power from the PV module depends on the temperature and the solar insolation [1]. To have the better efficiency of the PV module a lot of research propose to use maximum power point tracking (MPPT) algorithm.

There are different MPPT techniques used and described in literature [2–10]. Authors show that Perturb and Observe (P&O) and incremental conductance (IC) techniques, are the most widely used and it can be implemented very simply. But,

© Springer Nature Switzerland AG 2019
C. Benavente-Peces et al. (Eds.): SEAHF 2019, SIST 150, pp. 376–385, 2019.
https://doi.org/10.1007/978-3-030-22964-1_42

there are other MPPT technique which are more fast and precise such as fuzzy logic or neural network techniques. Some work [6–10] show that MPPT using fuzzy logic controllers present good performance when we have a change in the atmospheric conditions and it show a better results than the other methods.

In the literature, for the PV-grid connection there different structures have been proposed [11, 12], the most topologies are realized around a cascade of DC-DC boost converter used to increase the PV voltage and DC-AC converter (inverter) used to converter DC voltage to AC voltage with a sine wave output current. A phase locked loop (PLL) is added in aim to get a synchronization of inverter output voltage with the voltage grid [17, 18].

In this paper, a system description is presented in first section in which a present a modeling of PV module, maximum power point tracking method (MPPT), 2.3.

DC-DC Boost converter and the inverter modeling, in second section we present the simulation results using Matlab Simulink.

## 2   System Description

In this work, we present a single-phase grid connected PV system. Figure 1 shows that the system is composed of a PV array, a DC-DC boost converter to increase the output voltage of the PV array, a maximum power point tracking (MPPT) is used to switch the DC-DC boost converter and a single phase inverter.

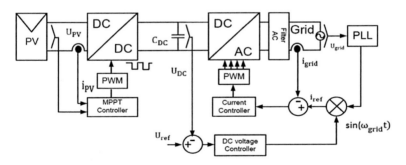

**Fig. 1.**  Bloc diagram of the single-phase grid connected PV system

### 2.1   Modeling of Photovoltaic Panel

Figure 2 show the equivalent circuit of the PV cell. We noticed that the equivalent circuit is composed of a, series resistance Rs, a shunt resistance Rsh, diode and D and a photo-current source Iph.

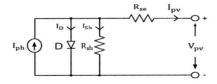

**Fig. 2.**  Equivalent circuit of the PV cell

The PV cells are assembled in larger units called PV panels which are grouped in a parallel-series to give the PV arrays.

To give the model of the PV array, the PV cell model is multiplied by the number in series and in parallel cells.

The equation for the current and voltage of the PV array is written as follows [1, 2]

$$I_{PV} = N_P I_{ph} - N_P I_0 \left( e^{\frac{q(\frac{V_{PV}}{N_s} + \frac{I_{PV} R_S}{N_p})}{KTA}} - 1 \right) - \frac{\left( \frac{V_{PV}}{N_s} + \frac{I_{PV} R_s}{N_p} \right)}{R_{sh}} \qquad (1)$$

We noticed from this equation that PV efficiency is sensitive to small change in RS but insensitive to variation in RSh for that raison the most simplified model [2] of PV module is described by the Eq. (2)

$$I_{PV} = N_P I_{ph} - N_P I_0 \left( e^{\frac{q(\frac{V_{PV}}{N_s} + \frac{I_{PV} R_S}{N_p})}{KTA}} - 1 \right) \qquad (2)$$

where,

Vpv :   output voltage cell (V).
$I_0$ :   diode saturation current(A).
$I_{Pv}$ :   current cell (A).
$I_{Ph}$ :   photo current (A).
q :   charge of electron
K :   Boltzmann constant
T :   cell temperature (K).
$R_{Sh}$ :   cell shunt resistance ($\Omega$).
$R_s$ :   cell series resistance ($\Omega$).
Ns :   number in series cells.
Np :   number in parallel cells.
A :   ideal factor which is depend on PV cell technology

## 2.2   Maximum Power Point Tracking Method (MPPT)

From model described above (Eqs. 1 and 2) we noticed that the maximum power and the output voltage are depend on irradiation and cell temperature so to produce a maximal power and maintains the output voltage at a constant value despite of the irradiance and the temperature a maximum power point tracking method (MPPT) must be used.

To maximize the output power an increasing or decreasing duty ratio of the DC-DC converter is need according to the output power of the PV cell versus the voltage curve.

There are different MPPT techniques used and described in literature [2–5]. Authors show that Perturb and Observe (P&O) and incremental conductance (IC) techniques, are the most widely used and it can be implemented very simply. But, there are other MPPT technique which are more fast and precise such as fuzzy logic or neural network techniques. Some work [6–8] show that MPPT using fuzzy logic controllers present good performance when we have a change in the atmospheric conditions and it show a better results than the other methods.

The proposed fuzzy MPPT controller has one output and two inputs.

The output of the FLC is duty cycle, which is used to control the boost DC-DC converter operation. The inputs of the fuzzy controller are Error signal determined from the PV cell's Power versus Voltage (P-V) curve (Fig. 3) and variation of the error signal.

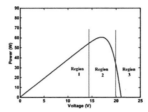

**Fig. 3.** (P-V) curve

where

$$e(k) = \frac{P_{PV}(k) - P_{PV}(k-1)}{V_{PV}(k) - V_{PV}(k-1)} \tag{3}$$

$$\Delta e(k) = e(k) - e(k-1) \tag{4}$$

The output signal is the control increment $\Delta U(k)$ and it's defined as follows:

$$U(k) = \Delta U(k) + U(k-1) \tag{5}$$

The proposed fuzzy MPPT Controller use a trapezoidal and triangular membership functions, by Negative BIG (NB), Negative Middle (NM), Zero (Z), Positive Middle (PM) and Positive BIG (PB), were used for both the two inputs and either for the output signal.

Table 1 show the rules assigned to output for the different values of $e(k)$ and $\Delta e(k)$

**Table 1.** Table captions should be placed above the tables.

| $e(k)$ | $\Delta e(k)$ | | | | |
|---|---|---|---|---|---|
| | NB | NM | Z | PM | PB |
| NB | Z | Z | NB | NB | NB |
| NM | Z | Z | NM | PM | PM |
| Z | NM | Z | Z | Z | PM |
| PM | PM | PM | PM | Z | Z |
| PB | PB | PB | PB | Z | Z |

## 2.3  DC-DC Boost Converter Modeling

The DC-DC boost converter is included between the PV panel and the AC-DC converter, the converter is standard unidirectional boost topology as presented by Fig. 4. If the switch is in ON state, the diode is on inverse polarization and the current increases. When the switch is in OFF state, the diode is directly polarized. Consequently, the inductor's current decreases and discharges its energy to the load.

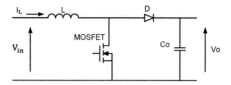

**Fig. 4.** DC-DC boost converter

Therefore, the studied DC–DC boost converter has two topologies corresponding to its switch states. The first topology corresponds to the ON state of the switch and the second topology corresponds to the off state of the switch.

The dynamic behavior the corresponding converter is described by the following equations:

$$\begin{cases} \frac{di_L}{dt} = \frac{1}{L}\left(v_{in} - r_1 i_L - (1-d)v_0\right) \\ \frac{dv_0}{dt} = \frac{1}{C_0}\left((1-d)i_L - \frac{i_0}{C_0}\right) \end{cases} \tag{6}$$

where $i_L$ is the inductor current, $v_0$ and $v_{in}$ are respectively the output and the input voltage, $C_0$ is the filter capacitance and $d$ denote the state of the Switch $Sw$, so $d = 1$ if the switch is ON and $d = 0$ if the switch is OFF.

The choice of the state vector $x = \begin{bmatrix} i_L \\ v_0 \end{bmatrix}$ allows the formulation of the following nonlinear state space representation:

$$\begin{cases} \dot{x} = \begin{bmatrix} \dot{i}_L \\ \dot{v}_0 \end{bmatrix} = A \begin{bmatrix} i_L \\ v_0 \end{bmatrix} + BU \\ v_0 = Cx \end{cases} \tag{7}$$

where

$$A = \begin{bmatrix} -\frac{r}{L} & -\frac{1-d}{L} \\ \frac{1-d}{C_0} & 0 \end{bmatrix}, \ B = \begin{bmatrix} \frac{1}{L} & 0 \\ 0 & \frac{1}{C_0} \end{bmatrix},$$

$$U = \begin{bmatrix} v_{in} \\ i_0 \end{bmatrix}, \ C = \begin{bmatrix} 0 & 1 \end{bmatrix}$$

## 2.4  DC-AC Converter (Inverter) Modeling

DC-AC converter (inverter) operating is used to connect the PV generator with the grid, the inverter works as a source of controlled current when it is connected to the grid and it gives an output considering reference current. The quantity of active power exchange between PV module and grid is depends upon the regulation of DC link voltages.

In literature, there are several topologies and control methods [11–16]. Figure 5 presents the structure of the single-phase DC-AC converter. A Sinusoidal Pulse width modulation (SPWM) is used to switch the transistors used in this. When PV inverter and grid having same phase angel, amplitude & frequency then we can say both are synchronized. To get this synchronization, There are several approaches to extract from a current or voltage signal the phase information. Zero-crossing method, αβ and dq Filtering Algorithm, and phase locked loop (PLL).

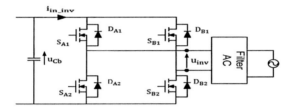

**Fig. 5.** H-bridge structure inverter

PLL methods are used a lot in literature especially when we need to have synchronization between grid and inverter and this ca be done by measuring the AC voltage from the grid to generate an estimated phase angle and an estimated frequency to control the output signal of the inverter. Figure 6 shows the basic structure of PLL technique, in this structure we can find a phase detection (PD) bloc used in order to have the phase difference between the generated output signal and the input one, a voltage controlled oscillator (VCO) to generates a signal which have the same output frequency and loop pass filter (LF) to reduce the ripple noise in the estimated frequency.

More details concerning the idea and application of this synchronization technique can be found in literature [17, 18]

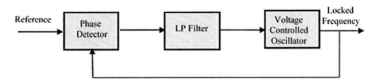

**Fig. 6.** PLL technique

The output current of the proposed inverter is not sinusoidal in nature, so, an LCL filter is used to change it to a perfectly sinusoidal wave and to reducing the harmonics and noise, these filters are commonly used in grid connections applications. The LCL filter can reduce the current harmonics and the distortions in injected voltage perfectly if inductance values is low [19].

## 3   Simulation Results

The system presented above has been tested by simulation using MATLAB/SIMULINK. Figure 7 shows that the DC bus voltage is stable when the irradiation changes from 1000 to 600 W/m$^2$ (Fig. 8) at 0.4 s and when we have temperature changes from 25 to 40 °c (Fig. 9).

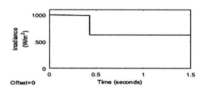

**Fig. 7.** DC bus voltage (voltage inverter input)

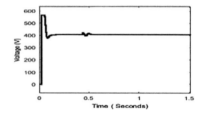

**Fig. 8.** Step change in irradiance

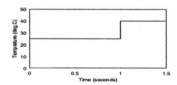

**Fig. 9.** Step change in temperature

Figure 10 represents the output inverter voltage before filter. From Figs. 11 and 12 we noticed that the grid voltage and the output voltage after filtering are sinusoidal and in phase. Figures 13 and 14 represent respectively the grid current and the inverter current we observed that when we connected the inverter to the grid current decrease and the load is fed at the same time by inverter and grid. Figures 15 and 16 represent the same results as Figs. 13 and 14 but the inverter is disconnected from the grid.

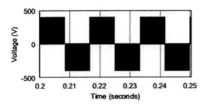

**Fig. 10.** Output inverter voltage before filter

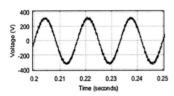

**Fig. 11.** Output inverter voltage before filter

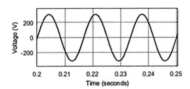

**Fig. 12.** Grid voltage

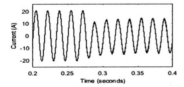

**Fig. 13.** Grid current decrease

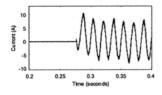

**Fig. 14.** Inverter current (connection state)

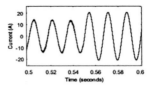

**Fig. 15.** Grid current increase

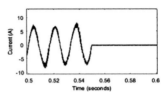

**Fig. 16.** Inverter current (disconnection)

# 4  Conclusion

In this paper, we presented a single-phase grid connected photovoltaic inverter. This system consists of a non-isolated DC-DC boost and a single-phase full-bridge DC-AC converter. The MPPT technique has been realised based on FLC technique and an additional FLC was used to control the full-bridge inverter. The synchronization between the grid and the inverter was done by using PLL technique witch use the voltage grid as reference to generate an estimated phase angle. The system is simulated by MATLAB-Simulink. The simulations results prove that the single phase solar power inverter is able to supply to the grid at different solar cell temperature and solar irradiance and the output inverter current is always in phase with the grid current.

# References

1. Krishan, R., Sood, Y.R., Uday Kumar, B.: The simulation and design for analysis of photovoltaic system based on MATLAB. In: International Conference on Energy Efficient Technologies for Sustainability (ICEETS), pp. 647–651, 10–12 (2013)
2. Hussein, K.H., Muta, I., Hoshino, T., Osakada, M.: Maximum photovoltaic power tracking: an algorithm for rapidly changing atmospheric conditions. IEEE Proc. Gener., Transm. Distrib. **142**(1), 953–959 (2005)
3. Esram, T., Chapman, P.L.: Comparison of Photovoltaic Array Maximum Power Point Tracking Techniques. IEEE Trans. Energy Convers. **22**(2), 439–449 (2007)
4. Femia, N., Petrone, G., Spagnuolo, G., Vitelli, M.: Optimization of perturb and observe maximum power point tracking method. IEEE Transaclions Power Electron. **20**(4), 963–973 (2005)
5. Kjaer, S.: Evaluation of the "hill climbing" and the "incremental conductance" maximum power point trackers for photovoltaic power systems. IEEE Trans. Energy Convers. **27**(4), 922–929 (2012)

6. Khan, S.A., Hossain, M.I.: Design and implementation of microcontroller based fuzzy logic control for maximum power point tracking of a photovoltaic system. In: International Conference on Electrical and Computer Engineering (ICECE 2010), pp. 322–325 (2010)
7. Nabipour, M., Razaz, M., Seifossadat, S., Mortazavi, S.: A new MPPT scheme based on a novel fuzzy approach. Renew. Sustain. Energy Rev. **74**, 1147–1169 (2017)
8. Bendib, B., Krim, F., Belmili, H., Almi, M.F., Boulouma, S.: Advanced Fuzzy MPPT Controller for a stand-alone PV system. Energy Procedia **50**, 383–392 (2014)
9. Robles Algarín, C., Taborda Giraldo, J., Rodríguez Álvarez, O.: Fuzzy logic based MPPT controller for a PV system. Energies **10**, 2036 (2017)
10. Reddy, D., Ramasamy, S.: A fuzzy logic MPPT controller based three phase grid-tied solar PV system with improved CPI voltage. In: International Conference on Innovations in Power and Advanced Computing Technologies (i-PACT) (2017)
11. Kjaer, S., Pedersen, J., Blaabjerg, F.: A review of single-phase grid-connectedinverters for photovoltaic modules. IEEE Trans. Ind. Appl. **41**(5), 1292–1306 (2005)
12. Blaabjerg, F., Teodorescu, R., Liserre, M., Timbus, A.: Overview of control and grid synchronization for distributed power generation systems. IEEE Trans. Ind. Electron. **53**(5), 1398–1409 (2006)
13. Dasgupta, S., Sahoo, S.K.. Panda S.K.:, A new control strategy for single phase series connected PV module inverter for grid voltage compensation, In: Power Electronics and Drive Systems, International Conference on. IEEE, PEDS 2009, pp. 1317–1322 (2009)
14. Roy, S., Malkhandi, A., Ghose, T.: Modeling of 5 kW single phase grid tied solar photovoltaic system. In: International Conference on Computer Electrical and Communication Engineering (ICCECE) (2016)
15. Arafa, O.M., et al.: Realization of single-phase single-stage grid-connected PV system. J. Electr. Syst. Inf. Technol. **4**(1), 1–9 (2017)
16. Mchaouar, Y.: Nonlinear control of single-phase grid-connected photovoltaic systems via singular perturbation. In: Proceeding of the 5th IEEE International Conference on Smart Energy Grid Engineering, pp. 210–216 (2017)
17. Timbus, A., Liserre, M., Teodorescu, R., Blaabjerg, F.: Synchronization methods for three phase distributed power generation systems—an overview and evaluation. In: Proceedings of 36th IEEE Power Electronics Specialists Conference (PESC), pp. 2474–2481 (2005)
18. Behera, R.R., Thakur, A.: An overview of various grid synchronization techniques for single-phase grid integration of renewable distributed power generation systems. In: International Conference on Electrical Electronics and Optimization Techniques (ICEEOT)-2016, pp. 2876–2880 (2016)
19. Reznik, A., Simoes, M.G., Al-Durra, A., Muyeen, S.M.: LCL filter design and performance analysis for grid-Interconnected systems. IEEE Trans. Ind. Appl. **50**(2), 1225–1232 (2014)

# Synthesis of Sit-to-Stand Movement Using SimMechanics

Samina Rafique$^{(\boxtimes)}$ ⓘ, M. Najam-l-Islam ⓘ, and A. Mahmood ⓘ

Bahria University, 46000 Islamabad, Pakistan
samina.rafique@bui.edu.pk

**Abstract.** Biomechanical movements have been an area of interest to the researchers from various disciplines. Using mathematical models of human musculoskeletal structure motion analysis is done in order to understand and improve body movement mechanism. For people having motion disorders due to disease or aging, sit-to-stand (STS) is the minimum of the motion that may keep them from being fully handicapped. The human factor involved in studying STS is therefore very high. This paper analyses human STS movement using four-segment nonlinear model realized in SIMULINK's SimMechanics environment. The role of kinematic variables like joint positions and velocities in human motion is specially studied. Since the measurements are noise contaminated and neurofeedback to Central Nervous System (CNS) is subject to time delay, estimation of true states is challenging. To estimate position and velocity profiles and achieve a smooth human-like STS motion a Linear Quadratic Regulator (LQR) based compensator is employed. The full order state feedback controller has proved to be robust and thus the STS motion achieved is close to natural motion. All segments contribute to the task in physiologically relevant movement coordination. This scheme bears the potential to achieve the controlled STS motion for diagnosis, rehabilitation and humanoid robotics applications.

**Keywords:** Biomechanics · Motion control · Neuro-feedback · State estimation

## 1 Introduction

STS movement takes place in coordination between physical components and central nervous system (CNS) generates movement commands in accordance with feedback from muscles and vestibular sensors. The components and roles of different feedbacks to CNS in STS movement is still not known completely [1]. Research in relative fields include human/ animal locomotion and gait analysis, designing and control of prosthetics and implants (like crutches), diagnosis and rehabilitation processes [2, 3] and equipment's design. Human motion analysis and improvement techniques are now being extensively used in sports coaching and ergonomics [4, 5]. In [6] Maxim Raison

---

Member IEEE.

© Springer Nature Switzerland AG 2019
C. Benavente-Peces et al. (Eds.): SEAHF 2019, SIST 150, pp. 386–392, 2019.
https://doi.org/10.1007/978-3-030-22964-1_43

et el, have utilized 5-repetition-STS test technique to construct inverse dynamic model. Postural stability in elderly people is studied in [7]. Inter-joint dynamic interaction during quiet standing is also studied by [8]. In [9] the human STS stable movement has been modeled by four link biomechanical model for various phases of STS. In [10] fuzzy modeling and fuzzy controller are employed to study mechanics of human musculoskeletal system during STS movement. In [11] an LMI based control is used by mixed H_2/H_∞ controller. The PID controller has been used in [12] and STS motion control using reduced measurement is studied in [13]. The paper has been organized as follows: In Sect. 2, a brief description of 4-segment non-linear model employed is described. Specifications used to linearize the model and construction of linear quadratic controller (LQR) and observer design then follows. Simulations and results are discussed next and conclusions have been derived in the end.

## 2   Methodology

The model used for human STS motion analysis is a nonlinear 2D four-segment model with 6 degree of freedom (DOF) as shown in Fig. 1. The model is defined in sagittal plane. Body above hip comprises of head-arm-trunk (HAT) and modeled as single segment. The model used has been extensively studied [9–11, 13] for STS motion, various controllers and regions of stability.

### 2.1   Mathematical Model

The non-linear model for STS motion [9] is

$$D(\theta)\ddot{\theta} + H(\theta, \dot{\theta})\dot{\theta} + G(\theta) = \vec{\tau} \tag{1}$$

Where $D(\theta)$ are the inertial component of joint moments, $H(\theta, \dot{\theta})$ are the Coriolis components and $G(\theta)$ are gravitational components. $\theta$ is joint angle in body frame and $\vec{\tau}$ are the joint torques. Define state vector

$$x = \begin{bmatrix} \Delta\theta_1 & \Delta\theta_2 & \Delta\theta_3 & \Delta\dot{\theta}_1 & \Delta\dot{\theta}_2 & \Delta\dot{\theta}_3 \end{bmatrix}^T = [x_1 \ x_2 \ x_3 \ x_4 \ x_5 \ x_6]^T$$

where $\Delta\theta_i$ is error between current and desired position. Also define

$$u = \tau = [\tau_1 - \tau_2, \tau_2 - \tau_3, \tau_3]$$

Non-linear state space equation is

$$\dot{x}_1 = x_4, \dot{x}_1 = x_5, \dot{x}_3 = x_6$$

$$\begin{bmatrix} \dot{x}_4 \\ \dot{x}_5 \\ \dot{x}_6 \end{bmatrix} = [D]^{-1}\left[ -H \begin{bmatrix} x_4 \\ x_5 \\ x_6 \end{bmatrix} - G + \vec{\tau} \right] \tag{2}$$

Linearizing in upright standing position

$$x_0 = [\pi/2, 0, 0, 0, 0, 0]^T \ and \ u_0 = [0, 0, 0]^T$$

Linearized state space model is

$$\dot{x} = Ax + Bu\big|_{x_0, u_0}$$

## 2.2   SimMechanics Model

Using SimMechanics environment the STS model of (2) has been implemented (see Fig. 2) in terms of physical quantities involved [14]. The model is defined in body frame. Each joint is equipped with sensors ($J_i$) to measure angular positions ($\theta_i$) and angular velocities ($\dot{\theta}_i$). Each joint is individually driven by an actuator ($T_i$) which is controlled by control input ($u_i$).

## 2.3   Measurement Noise and Delay

The measurements of joint positions and velocities using sensors are always contaminated with noise. Also the neurofeedback to CNS is subjected to time delay; 10, 15 and 30 m sec delays are typical for hip, knee and ankle joint positions respectively. Both noise and time delays have been modeled in this paper.

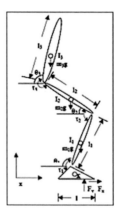

**Fig. 1.** 2D four segment STS mode

## 2.4   Linearized Model

As shown in Fig. 2, three joint positions and velocities comprise the state vector x. In the presence of measurement noise and transport delay, measurements are not suitable for state feedback control. Hence states are reconstructed using an observer, hence a linear compensator is realized (Fig. 3).

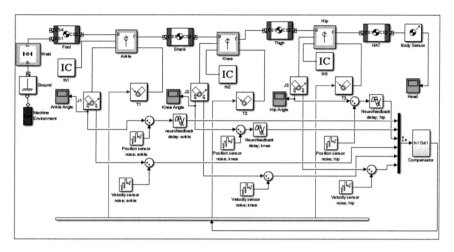

**Fig. 2.** STS motion synthesis developed in SimMechanics

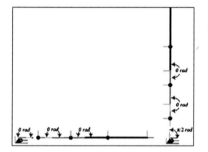

**Fig. 3.** Model defined in body frame

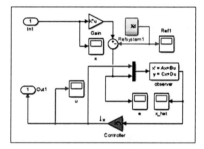

**Fig. 4.** LQR based compensator

## 2.5 LQR Robust Controller

In order to control STS movement an LQR controller is employed to gain more flexibility of tuning the gains individually [15] using elements of state weighting matrix Q and control weighting matrix R. Controller thus obtained is robust and optimum. The control law to be designed is

$$u = -G\hat{x}$$

Where $\hat{x}$ is the state estimated by observer and $G$ is the LQR controller gain. Control input $u$ comprises of passive torques used to run three actuators at the joints.

## 2.6 Observer

An observer has been incorporated in the system (Fig. 4) for estimation of actual states that are buried in measurement noise and cannot be used for feedback control for observer shown in Fig. 4. The dynamic equation for observer is

$$\dot{\hat{x}} = A\hat{x} + [B\ K]\begin{bmatrix} u \\ x_e \end{bmatrix}$$

The linearized model of plant [A, B, C, D] is used in designing the observer gain K and compensator gain G. Both the controllers are LQR. Observer model is $[A_o, [B_o\ K_o], C_o, D_o]$, where $A_o = A$, $B_o = B$, $C_o = I_{6x6}$ and $D_o = O_{6x9}$. Observer output is state estimate. K is designed by tuning matrices $R_o$ and $Q_o$ having compatible dimensions.

### 2.7 Reference Trajectories

Joint angle profiles are generated from a technique borrowed from literature [16]. Figure 5 depicts required angle profiles at three joints in STS movement. $X_{di}$ are the desired angle profiles of ankle, knee and hip. It is interesting to note that no velocity references have been provided for tracking (Figs. 6 and 7).

## 3  Simulations and Results

The 4-segment STS model is defined in lying position. From initial condition of the model, the sitting position at the start of STS movement, to the final standing, all phases shown in Fig. 9. Since the states converge to equilibrium condition smoothly it is established that the estimates of states are quite close to the actual states and compensator is controlling the system as satisfactorily as it would have done using actual states. The motion is physiologically relevant and closely resembles that of actual human STS motion. By fine tuning LQR gains for compensator and observer gains cheap control technique has been adapted which results in higher torques (see Fig. 8).

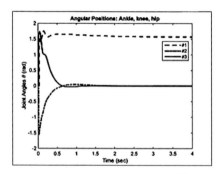

**Fig. 5.** Reference trajectories

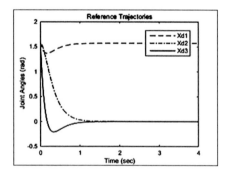

**Fig. 6.** Angular positions of three joints during STS

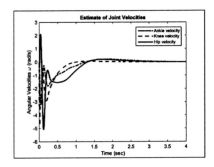

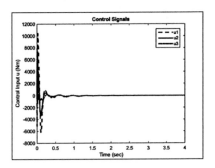

**Fig. 7.** Joint velocity estimates

**Fig. 8.** Torque inputs to actuators at three joint

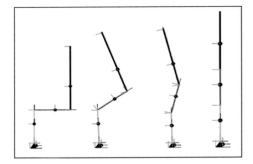

**Fig. 9.** Physiologically relevant STS motion achieved by LQR controller

## 4 Conclusion

STS motion using 2D four segment non-linear model developed in SimMechanics with full order observer using LQR design has beensynthesized. All the six measurements were subject to sensor noise as well as neurofeedback time delay. The system tracks state trajectories to carryout natural and physiologically relevant STS motion. The motion obtained is smooth and close to human STS. This shows the contribution of two kinematic variables i.e., joint positions and velocities in carrying out STS task in collaboration with CNS modeled by LQR controller. In the next stage the system is intended to be modified to a system with fewer sensors for position orientation using robust controller.

# References

1. Matthis, J.S., Fajen, B.R.: Humans exploit the biomechanics of bipedal gait during visually guided walking over complex terrain (2013)
2. Schenkman, M., Berger, R.A., Riley, P.O., Mann, R.W., Hodge, W.A.: Whole-body movements during rising to standing from sitting. Phys. Ther. **70**(10), 638–648 (1990)
3. Kerr, A., Deakin, A.H., Clarke, J.V., Dillon, J.M., Rowe, P., Picard, F.: Biomechanical analysis of the sit-to-stand movement following knee replacement: a cross-sectional observational study. In: XXIV Congress of the International Society of Biomechanics (ISB 2013) (2013)
4. Knudson, D.: Fundamentals of Biomechanics. Springer Science & Business Media (2007)
5. Veeraraghavan, A., Chowdhury, A.R., Chellappa, R.: Role of shape and kinematics in human movement analysis. In: Proceedings of the 2004 IEEE Computer Society Conference on Computer Vision and Pattern Recognition, 2004. CVPR 2004. (Vol. 1, pp. I–I). IEEE (2004)
6. Raison, M., Laitenberger, M., Sarcher, A., Detrembleur, C., Samin, J.C., Fisette, P.: Methodology for the assessment of joint efforts during sit to stand movement. In: Injury and Skeletal Biomechanics. InTech (2012)
7. Pai, Y.C., Wening, J.D., Runtz, E.F., Iqbal, K., Pavol, M.J.: Role of feedforward control of movement stability in reducing slip-related balance loss and falls among older adults. J. Neurophysiol. **90**(2), 755–762 (2003)
8. Sasagawa, S., Shinya, M., Nakazawa, K.: Interjoint dynamic interaction during constrained human quiet standing examined by induced acceleration analysis. J. Neurophysiol. **111**(2), 313–322 (2013)
9. Mughal, A.M., Iqbal, K.: Fuzzy optimal control of sit-to-stand movement in a biomechanical model. J. Intell. Fuzzy Syst. **25**(1), 247–258 (2013)
10. Mughal, A.M., Iqbal, K.: Fuzzy modeling and optimal control of biomechanical sts movement. In: The 10th International Conference on Variable Structures, Istanbul, Turkey (2008)
11. Mughal, A.M., Perviaz, S., Iqbal, K.: LMI based physiological cost optimization for biomechanical STS transfer. In: 2011 IEEE International Conference on Systems, Man, and Cybernetics (SMC), pp. 1508–1513. IEEE (2011)
12. Roy, A., Iqbal, K.: Synthesis of stabilizing PID controllers for biomechanical models. In: Proceedings of 2005 IFAC World Congress, Praha (2005)
13. Rafique, S., Mahmood, A., Najam-ul-Islam, M.: Robust control of physiologically relevant sit-to-stand motion using reduced order measurements. In: Proceedings of the Future Technologies Conference, pp. 783–796. Springer, Cham (2018)
14. Iqbal, K., Pai, Y.C.: Predicted region of stability for balance recovery: motion at the knee joint can improve termination of forward movement. J. Biomech. **33**(12), 1619–1627 (2000)
15. Friedland, B.: Control system design: an introduction to state-space methods. Courier Corporation (2012)
16. Mughal, A.M., Iqbal, K.: Synthesis of angular profiles for bipedal sit-to-stand movement. In: 40th Southeastern Symposium on System Theory, 2008. SSST 2008. pp. 293–297. IEEE (2008)

# Comparative Study of MPPT Algorithms of an Autonomous Photovoltaic Generator

Troudi Fathi[✉], Houda Jouini, and Abdelkader Mami

Laboratory LAPER, University Tunis El Manar, 2092 Tunis, Tunisia
troudi.fathi@gmail.com

**Abstract.** This paper presents a comparative study of several MPPT techniques for extracting the maximum power of a photovoltaic generator. Three MPPT control algorithms have been developed including a network-based MPPT algorithm at the ANN neurons, the FVCO fraction method of Vco and the FCC method that fraction of Icc The simulations were made on the MATLAB-SIMULINK environment and the results of simulations are promising in order to choose the most appropriate technique in the envisaged sunshine conditions.

**Keywords:** P&O · ANN · MPPT · INC · FVCO · FCC

## 1 Introduction

This context, this paper focuses on a comparative study of the different extracting the maximum MPPT point methods of an autonomous photovoltaic generator [1, 2, 5]. Thus, the paper consists of three parts, the first presents the model of the study system, the second part mentions the MPPT methods studied as well as their theoretical studies, the third part is devoted to the simulations results of the photovoltaic generator characteristics of the as well as the various MPPT methods mentioned, the simulations results of the MPPT methods are compared in the last part.

## 2 Study System Model and Proposed MPPT Methods

### 2.1 Study System

The synoptic diagram describing the system is presented by the following figure (Fig. 1).

Where the Photovoltaic generator characteristic and the boost converter characteristics are given respectively by the following equivalent diagram and the equivalent diagram of a photovoltaic cell [2, 3] (Figs. 2 and 3).

© Springer Nature Switzerland AG 2019
C. Benavente-Peces et al. (Eds.): SEAHF 2019, SIST 150, pp. 393–400, 2019.
https://doi.org/10.1007/978-3-030-22964-1_44

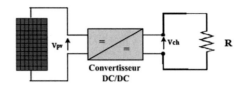

**Fig. 1.** Synoptic diagram of the study system

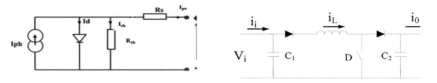

**Fig. 2.** Equivalent circuit diagram of a    **Fig. 3.** Boost converter
photovoltaic cell

## 2.2    Proposed MPPT Methods

Three methods are proposed and studied to ensure the MPPT control such as Artificial Neural Network algorithm for MPPT control, MPPT control by FVCO method controlled by ANN method and MPPT command by the application of the FCC controlled by ANN method.

## 3    Simulation and Results

We implement under the MATLAB-SIMULINK environment the study model with the MPPT method techniques, while presenting the response of the model under standard lighting conditions in order to elaborate e comparative study of its.

### 3.1    Photovoltaic Generator Simulation in the MATLAB-SIMULINK Environment

Based on the mathematical model of the developed solar cell, we present the schematic SIMULINK blocks of shown in the figure be (Fig. 4) [4].

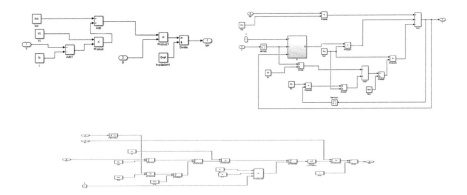

**Fig. 4.** Simulink block of solar module

The characteristics of the chosen photovoltaic generator are presented in the following table (Table 1).

**Table 1.** Photovoltaic generator characteristics

| Variable | Value | Unit |
|---|---|---|
| Standard reference insolation **G** | 1000 | W/m² |
| Temperature standard temperature **T** | 25 | °C |
| Open circuit voltage **Voc** | 22.5 | V |
| Short circuit current **Icc** | 6 | A |
| Optimum tension **Vmpp** (standard condition) | 18.54 | V |
| Optimum courant **Impp** | 5.63 | A |
| Peak power requested **Pmpp** (standard condition) | 104.38 | W |

### 3.2 Simulation of the MPPT Method Based on ANN

The values of the voltage **Vmpp** of the photovoltaic generator is at the temperature T = **25** °C are presented in the following table (Table 2).

**Table 2.** Vmpp table for different insolation

| G (W/m²) | Vmpp |
|---|---|
| **600** | 17.82 |
| **800** | 18.42 |
| **1000** | 18.54 |
| **1200** | 19.25 |
| **1400** | 17.78 |

We chose a neural network of an input layer, a hidden layer and an output layer, Input layer with two inputs, such as Temperature $x1 = T$ and insolation $x2 = G$, $X = [x1; x2]$, Hidden Layer Consists of ten neurons and the output layer to an output, $y = \mathbf{Vmpp}$ [1, 2] (Fig. 5).

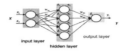

**Fig. 5.** Synoptic of neural network [2]

The duty cycle is generated by the control algorithm of this method. The voltage delivered by the photovoltaic generator under the standard conditions $G = 1000$ W/m$^2$, $T = 25$ °C and by the application of the method based on the artificial neural network algorithm of the MPPT is represented in following figure.

Neural network algorithm under the standard conditions $G = 1000$ W/m$^2$ and $T = 25$ °C is presented in the Fig. 6. Note that under the standard operating conditions (Fig. 7).

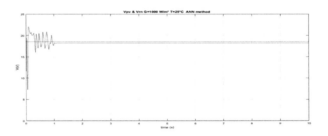

**Fig. 6.** **Vpv** and **Vrn** voltage of ANN algorithm

**(a)**                                   **(b)**

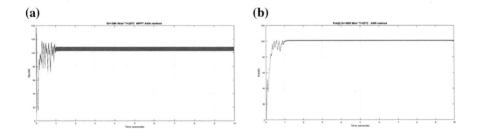

**Fig. 7. a** $P_{pv}$ curve of photovoltaic panel commanded by ANN algorithm, **b** $P_{ch}$ curve of photovoltaic panel commanded by ANN algorithm

G = 1000 W/m$^2$   and   T = 25   °C   the   maximum   power   is   extracted **Ppv = 104 W** after one second, with a fluctuation of **3.5 W**.

### 3.3   Simulation of FVCO Method Controlled by ANN Algorithm

We chose a neural network of an input layer, a hidden layer and an output layer, Input layer with two inputs, x1 = **T** Temperature and **x2 = G** insolation, X = [x1; x2], Hidden Layer Consists of **five neurons** and the output layer to an output, **y = k = Vmpp/Vco** (Table 3).

The voltage **Vpv** and the current **Ipv** delivered by the photovoltaic panel under the standard conditions is presented by the following figure (Figs. 8 and 9).

**Table 3.** Table of values of the ratio k of the tension Vm pp/Voc of photovoltaic generator

| G (W/m$^2$) | Voc | Vmpp | k |
|---|---|---|---|
| **600** | 21.5 | 17.82 | 0.8288 |
| **800** | 21.8 | 18.42 | 0.8450 |
| **1000** | 22.5 | 18.54 | 0.8240 |
| **1200** | 22.7 | 19.25 | 0.8480 |
| **1400** | 22.8 | 19.25 | 0.7798 |

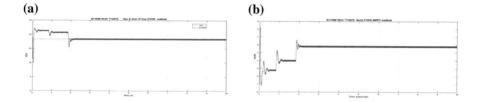

**Fig. 8.** a: V$_{pv}$ and K*V$_{co}$ curve of FVCO method MPPT, b: I$_{pv}$ curve of FVCO method MPPT

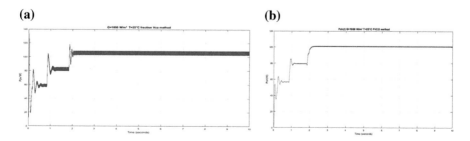

**Fig. 9.** a: P$_{pv}$ curve of FVCO method MPPT, b: P$_{ch}$ curve of FVCO method MPPT

### 3.4  Simulation of FCC Method Controlled by ANN Algorithm

We chose a neural network of an input layer, a hidden layer and an output layer, Input layer with two inputs, x1 = **T** Temperature and **x2 = G** insolation, X = [x1; x2], Hidden Layer Consists of **five neurons** and the output layer to an output, **y = k = Impp/Icc** (Table 4).

**Table 4.**  Table of values of the ratio k of the tension **Impp/Icc** of photovoltaic generator

| G (W/m$^2$) | Icc | Impp | K |
|---|---|---|---|
| 600 | 5.1 | 3.24 | 0.6353 |
| 800 | 5.6 | 4.6 | 0.8214 |
| 1000 | 6 | 5.15 | 0.8583 |
| 1200 | 6.5 | 6.09 | 0.9369 |
| 1400 | 7.2 | 6.85 | 0.9514 |

Let's present the pace of the current **Ipv** generated by the PV model in standard condition (Figs. 10 and 11).

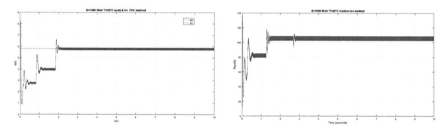

**Fig. 10.**  Tension **Vpv** and **Vrn** of ANN algorithm

**Fig. 11.**  $I_{pv}$ and $I_m$ = k*Icc curve of FCC MPPT method G = 1000 W/m² T = 25 °C

The power delivered by the study system controlled by the FCC method is presented by the following figure (Fig. 12).

**(a)**                **(b)**

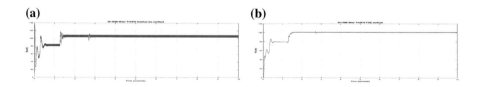

**Fig. 12.**  a: $P_{pv}$ curve of FCC method MPPT, b: $P_{ch}$ curve of FCC method MPPT

Note that under the standard operating conditions **G = 1000 W/m$^2$** and **T = 25 °C** the maximum power extracted is **Ppv = 104 W** after more than one second, with a fluctuation of 3.5 W.

For low insolation levels of 600 W/m$^2$, medium insolation G = 1000 W/m$^2$ and high insolation G = 1400 W/m$^2$ and at a constant temperature T = 25 °C the power generated by the photovoltaic panel is given by the following figures. The maximum power delivered by the low insolation photovoltaic generator **G = 600 W/m$^2$** and **T = 25 °C** is **Ppv = 60 W** applying the MPPT technique. The study system converges towards the maximum power whatever the method applied except the fastest method is INC and the slowest method is FCC. These results are represented by the Fig. 13. According to the simulation the stabilization time of the two FVCO and FCC methods is greater, and the other three methods will not be strongly influenced (Figs. 14 and 15).

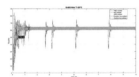

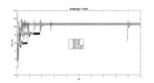

**Fig. 13.** Ppv curve G = 600 W/m$^2$ T = 25 °C for different MPP algorithm

**Fig. 14.** Ppv curve G = 1000 W/m$^2$ T = 25 °C for different MPP algorithm

**Fig. 15.** Ppv curve G = 1400 W/m$^2$ T = 25 °C for different MPP algorithm

The following table presents a summary of comparison of the cited algorithms (Table 5).

**Table 5.** Summary table of comparisons of the cited MPPT algorithms

| MPPT algorithm | P&O | INC | FVCO | FCC |
|---|---|---|---|---|
| Convergence speed | Fast | Fast | Medium | Slow |
| Precision | 97% | 98% | 96% | 96 |
| Sensor type | Voltage, current | Voltage, current | Voltage | Current |
| Identification of PV parameters | Not necessary | Not necessary | Necessary | Necessary |

# 4 Conclusion

This paper presented the model of the photovoltaic generator with the different tracking algorithms of the MPP in different lighting conditions to compare the behavior of the system operation (Panel & chopper) in the case of low sun, medium sun and strong sunshine. The results prove that the INC algorithm is more efficient than the P & O method, the method based on the ANN algorithm performs better than the FVCO method which is in turn more efficient than the FCC method, in order to choose the algorithm to apply according to the speed and precision chosen. The perspectives of this work and implement practice these algorithms and compare their performance under practical conditions.

## References

1. Bouselham, L., Hajji, M., et al.: A new MPPT-based ANN for photovoltaic system under partial shading conditions. In: 2016 8th International Conference on Sustainability in Energy and Buildings, SEB-16, 11–13 September 2016, Turin, Italy (2016)
2. Essefi, R.M., Souissi, M., Abdallah, H.H.: Maximum power point tracking control using neural networks for stand-alone photovoltaic systems. Int. J. Mod. Nonlinear Theory Appl. Sci. Res. (2014)
3. Abbes, H., Abid, H., Loukil, K., et al.: Etude comparative de cinq algorithmes de commande MPPT pour un système photovoltaïque. Revue des Energies Renouvelables **17**(3), 435–445 (2014)
4. Zainudin, H.N., Mekhilef, S.: Comparison study of maximum power point tracker techniques for PV systems. In: Proceedings of the 14th International Middle East Power Systems Conference (MEPCON'10), Cairo University, Egypt, 19–21 December 2010, Paper ID 278 (2010)
5. Yang, J.-L., Su, D.-T., Shiao, Y.-S.: Research on MPPT and single-stage grid-connected for photovoltaic. WSEAS Trans. Syst. **7**(10) (2008)

# A Comparative Analysis of DPC and SMC-DPC-SVM Control Approaches in Three-Phase Electrical Power Systems

Maha Zoghlami[1,2(✉)] and Faouzi Bacha[1,2]

[1] Laboratory of Computer Science for Industrial Systems (LISI), University of Carthage, Tunisia, 676 INSAT Urban Center North BP, 1080 Cedex, Tunis, Tunisia
zoghlami.maha@yahoo.fr

[2] Department of Electrical Engineering, High National School of Engineering of Tunis (ENSIT), University of Tunis, Tunisia, 5 Av Taha Hussein Montfleury, 1008 Tunis, Tunisia

**Abstract.** The purpose of this article is to compare two different control structures which are direct power control (DPC) and sliding mode control based on direct power control with space vector Modulation (DPC-SMC-SVM) for two-levels conversion applications. Finally, we present a study of the robustness of the (SMC-DPC-SVM) in three-phase power supply systems.

The first approach DPC approach has developed to control the active and reactive power from switching table by selecting the optimum value of commutation state.

The second strategy (SMC-DPC-SVM) has been studied widely used applications due to its insensitivity to parameter variations, and robustness against external disturbances. It is shown that DPC-SMC-SVM has several advantaged; good dynamic response, constant chopping frequency.

According to that reason, we will focus on the first-order sliding mode control with a suitable parameter slip surface, which is presented in this paper to control the converter with infinite load. The active filter has the role of compensating for the main types of current disturbances in the power supply systems, it is also recognized that they generate undesirable components, caused by the switching frequency of the converter. Moreover, the variation of the active line filter aims to verify the robustness and efficiency of our order. The main objective of the active filter variation of line, is to check the robustness and the efficiency of our order. However, the performance of this control is verified by the simulation results with the software MATLAB/SIMULINK.

**Keywords:** Direct power control (DPC) · Sliding mode control (SMC) · Space vector modulation (SVM)

## 1 Introduction

Several problems to the diode rectifiers have been observed in recent years, which like the low input power factor, and the presence of harmonics in the input currents. However, the Pulse Width Modulation (PWM) converters are adopted in applications

© Springer Nature Switzerland AG 2019
C. Benavente-Peces et al. (Eds.): SEAHF 2019, SIST 150, pp. 401–413, 2019.
https://doi.org/10.1007/978-3-030-22964-1_45

that require less distortion in the current waveforms, [1]. Thus, the unity power factor operation can be easily performed by regulating the currents in phase with the power-source voltages [2]. In fact, the development of control methods for PWM rectifiers was possible, thanks to advances in power semi-conductor devices and digital signal processors. These methods allows fast operation and cost reduction. In addition, they offer possibilities for implementation of sophisticated control algorithms. The electrical system is composed of the use of voltage source DC/AC converters connected to the power grid [3], and power conditioning and transmission equipment (active power filter, Voltage Source Converter transmission (VSC), etc…) [4]. The authors propose in [5] the Direct Power Control (DPC) control strategies. The control vector of the inverter is computed using instantaneous active and reactive power values. Key features of this control strategy are a fast response and a good tracking capability. General classification of the DPC strategy is based on the constant switching frequency versus a variable switching frequency implementation [6]. This DPC operate with a constant switching frequency. In [7], Space Vector Modulation (SVM) achieves the DPC approaches use the constant switching frequency. Using SVM, a switching sequence is generated from a control vector, which may be calculated using various control methods. The DPC approaches based on the predictive control, but without a modulator [7]. The switching sequence is generated by minimizing a cost function of the desired performance of the control system. It requires complicated online calculation, and that requires a more complex digital hardware.

In [7] propose a comparative study between a DPC_calssique and DPC_developed and the advantage to develop this approach. In fact, general classification of the DPC_developed strategy is based on the constant switching frequency versus of variable switching frequency implementation.

In fact, general classification of the DPC_developped strategy is based on the constant switching frequency versus of variable switching frequency implementation [8]. Afterwards, for a variable structure system like a grid connected inverter we can we can found the Sliding Mode Control (SMC) approach. This strategie in a continuous-time (CT) system is robust with respect to matching external disturbance and parameter changes [9]. However, the variable and even theoretically infinite, switching frequency was the main disadvantage of the SMC. For that, the first work with a constant switching frequency SMC for power rectifier application are presented in the paper [10]. The SMC based DPC approaches are described in [7]. The SMC block directly calculates the desired control vector, which eliminates the actual errors of active and reactive powers. Besides, the inverter must also switch the state with a high frequency, and therefore the switching losses may be substantial. Thus, a control strategy with a switching time reduction can significantly improve the efficiency of a power inverter. We proposed than SMC_ DPC_SVM allows a lower chattering frequency. In contemporary applications, we went from the Direct Power Control Developed (DPC_developed) to sliding mode control based on direct power control with space vector modulation (SMC-DPC-SVM). This combinaison was developped for many raison. In the first hand, DPC is proposed in [2], is recognized as the most promising strategy. It is known that SMC is robust with respect to matching external disturbance and parameter changes [11]. The proposed strategy directly is regulates the instantaneous active and reactive power injected into grid, and it can reasonably reject system uncertainties, [12]. Using a prediction method, the influence of the system uncertainties are eliminated by introducing the correction of the control vector [13].

## 2   Mathematical Model of Three-Phase Rectifier

The Grid Side Converter (GSC) provides a constant DC bus voltage and suppresses harmonic distortion of grid currents. It also has the ability to recover energy and has a wide field of view in the DC power supply and reactive power compensation. The equivalent circuit of the GSC connected to the grid is given in Fig. 1, [14].

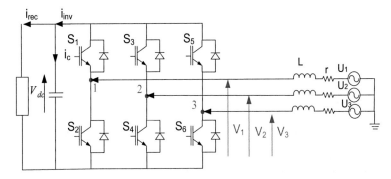

**Fig. 1.**   Topology of a three-phase PWM rectifier voltage converter

The most convenient structure of inverter topologies is that of three switching arms. In this section, we are particularly interested in the general structure of the inverter. Diagram of the converter on the grid side

$$\begin{bmatrix} U_1 \\ U_2 \\ U_3 \end{bmatrix} = \begin{bmatrix} U_{\text{eff}} \cdot \sqrt{2} \cdot \sin(\omega t) \\ U_{\text{eff}} \cdot \sqrt{2} \cdot \sin\left(\omega t - \frac{2\pi}{3}\right) \\ U_{\text{eff}} \cdot \sqrt{2} \cdot \sin\left(\omega t + \frac{2\pi}{3}\right) \end{bmatrix} \tag{1}$$

$$\begin{cases} U_1, U_2, U_3 : \text{ Voltage of the grid} \\ U_{\text{eff}} : \text{Root Mean Square} \\ V_1, V_2, V_3 : \text{ Voltage of the converter} \\ S_1, S_3, S_5 : \text{ Switching signals of the switches} \\ (r,L) : \text{ RL Filter} \end{cases}$$

The voltage of the inverter as a function of the switching state is given by the following matrix:

$$\begin{bmatrix} V_1 \\ V_2 \\ V_3 \end{bmatrix} = \frac{V_{dc}}{3} \cdot \begin{bmatrix} 2 & -1 & -1 \\ -1 & 2 & -1 \\ -1 & -1 & 2 \end{bmatrix} \begin{bmatrix} S_1 \\ S_3 \\ S_5 \end{bmatrix} \tag{2}$$

## 3   The Grid-Connected Filter Modeling

Filtering is a form of signal processing, obtained by sending the signal through a set of electronic circuits, which modify its frequency spectrum and phase, also thus its time form, [15]. It can be either to eliminate or unwanted parasitic frequencies, and to isolate it in the frequency band or bands of interest.

In a converter connected to the electrical grid, the presence of mains coupling inductances is essential to ensure the control of the currents injected into this grid. These inductors act as a low pass filter that limit the current ripple at the switching frequency and reduce the propagation of harmonics in the power grid. The model of the filter (RL) connected between the inverter and the grid is showing in Fig. 2.

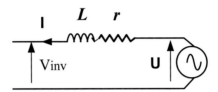

**Fig. 2.**   Model of the RL filter connected to the electrical grid

The voltage on each phase is written under Eq. (3):

$$L.\frac{dI}{dt} + r.I = -V_{inv} + U \tag{3}$$

## 4   The Direct Power Control Strategy

The overall DPC structure is based on instantaneous active and reactive power control loops. The switching states of the converter are selected by a switching table based on the calculation instantaneous active and reactive powers, error signals, respectively, provided by fixed band hysteresis comparators and the position of the source voltage vector. Depending on the value of this position, the axis $(\alpha\beta)$ is divided into twelve sectors where each sector must be associated with a logical state of the converter (Fig. 3).

Using the simplified model of the GSC grid converter, we can determine the active and reactive powers injected in the grid that can be presented in the following equations [16]

$$\begin{cases} V_\alpha = r\,I_\alpha + L\frac{dI_\alpha}{dt} + U_\alpha \\ V_\beta = r\,I_\beta + L\frac{dI_\beta}{dt} + U_\beta \\ C\frac{dV_{dc}}{dt} = I_{inv} - I_{dc} = (d_1I_1 + d_2I_2 + d_3I_3) - I_{dc} \end{cases} \tag{4}$$

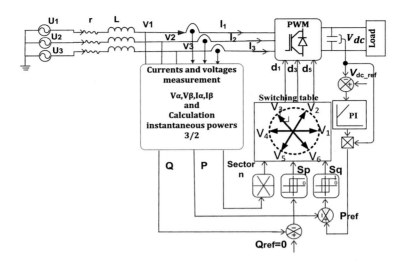

**Fig. 3.** DPC_developed block diagram of the converter connected to the grid

The instantaneous active and reactive powers injected on the grid can be written as:

$$
\begin{cases}
S_g = P_g + jQ_g \\
P_g = \frac{3}{2}\left(U_\alpha I_\alpha + U_\beta I_\beta\right) \\
Q_g = \frac{3}{2}\left(U_\beta I_\alpha - U_\alpha I_\beta\right)
\end{cases}
\tag{5}
$$

$$
\begin{cases}
\dfrac{dP_g}{dt} = \dfrac{3}{2}\left(U_\alpha \dfrac{dI_\alpha}{dt} + I_\alpha \dfrac{dU_\alpha}{dt} + U_\beta \dfrac{dI_\beta}{dt} + I_\beta \dfrac{dI_\beta}{dt}\right) \\
\dfrac{dQ_g}{dt} = \dfrac{3}{2}\left(U_\beta \dfrac{dI_\alpha}{dt} + I_\alpha \dfrac{dU_\beta}{dt} + U_\alpha \dfrac{dI_\beta}{dt} + I_\beta \dfrac{dU_\alpha}{dt}\right)
\end{cases}
\tag{6}
$$

### 4.1   The Line Voltage Vector Position

The phase of the power-source voltage vector is converted to the sector signal $\theta$. However, the stationary coordinates are divided into twelve sectors, as shown in Table 1, and the angle can be deduced from Eq. (7) [5].

$$
\delta = \arctan\frac{V_\beta}{V_\alpha}
\tag{7}
$$

### 4.2   Switching Table

The selection of the vector is determined by the following table according to the variation in the active and reactive powers with the position of voltage vector.

In the conventional switching table, the extensive use of vectors $V_0$ and $V_7$ weakens the control of the reactive power. Zero vectors can increase the active power.

The main purpose of the developed switching table is to reduce the switching frequency. Compared to the conventional table, this developed switching table improves its ability to regulate the reactive power. However, the zone exists where the active power is out of control (Table 1).

**Table 1.** Switching table of DPC_developed

| $S_p$ | $S_q$ | $d_1, d_3, d_5$ | | | | | | | | | | | |
|-------|-------|------------|------------|------------|------------|------------|------------|------------|------------|------------|---------------|---------------|---------------|
|       |       | $\theta_1$ | $\theta_2$ | $\theta_3$ | $\theta_4$ | $\theta_5$ | $\theta_6$ | $\theta_7$ | $\theta_8$ | $\theta_9$ | $\theta_{10}$ | $\theta_{11}$ | $\theta_{12}$ |
| 1     | 0     | $v_6$      | $v_1$      | $v_1$      | $v_2$      | $v_2$      | $v_3$      | $v_3$      | $v_4$      | $v_4$      | $v_5$         | $v_5$         | $v_6$         |
| 1     | 1     | $v_2$      | $v_3$      | $v_3$      | $v_4$      | $v_4$      | $v_5$      | $v_5$      | $v_6$      | $v_6$      | $v_1$         | $v_1$         | $v_2$         |
| 0     | 0     | $v_1$      | $v_1$      | $v_2$      | $v_2$      | $v_3$      | $v_3$      | $v_4$      | $v_4$      | $v_5$      | $v_5$         | $v_6$         | $v_6$         |
| 0     | 1     | $v_2$      | $v_2$      | $v_3$      | $v_3$      | $v_4$      | $v_4$      | $v_5$      | $v_5$      | $v_6$      | $v_6$         | $v_1$         | $v_1$         |

$$v_1 = (100), v_2 = (110), v_3 = (010), v_4 = (011), v_5 = (001), v_6 = (101)$$

## 5   The Sliding Mode Control Based on Direct Power Control with Space Vector Modulation

The Fig. 4 shows Sliding Mode Control based on Direct Power Control with Space Vector Modulation structure of the voltage converter (SMC-DPC-SVM). SMC consists of forcing the path of the system to follow the reference quantities. The SMC_DPC supplies the reference voltages ($V_\alpha$, $V_\beta$) to the SVM modulator to generate the states of the converter switches, [17]. The operation of the SM is effected by the choice of the sliding surface, so that all the trajectories of the system obey a behavior of tracking, regulation and stability. This choice is obtained by the linear combination of the error variables, which are defined as the difference between the state variables and their references. In the case of DPC, the sliding surfaces can be designed using the error variables of the active and reactive powers. Using the simplified model of the GSC grid converter, we can determine the active and reactive powers injected into the grid.

### 5.1   The Sliding Surface

The objectives of the converter control is to follow or slide along the predefined active and reactive power paths. The sliding surface is defined as follows:

$$S = [S_P \quad S_Q]^T \tag{8}$$

In order to maintain the improved transient response and to minimize the steady-state error, the switching surfaces can be in integral form, [17]. They can also be designed using back-stepping and non-linear damping techniques:

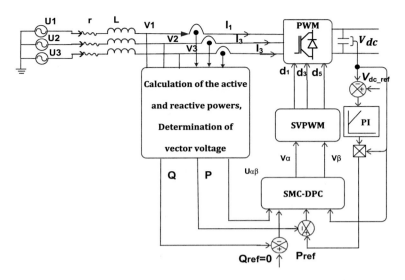

**Fig. 4.** Diagram of SMC_DPC_SVM of the converter connected to the grid

$$\begin{cases} S_P = e_P(t) + \lambda_P \int\limits_0^t e_P(\tau)\, d\tau - e_P(0) \\ S_Q = e_Q(t) + \lambda_Q \int\limits_0^t e_P(\tau)\, d\tau - e_Q(0) \end{cases} \tag{9}$$

$$\text{where :} \begin{cases} e_P\text{: error of the active instantaneous powers} \\ e_P\text{: error reactive instantaneous powers} \\ K_p, K_q\text{: the positive control gains} \end{cases}$$

$$\begin{cases} e_P = P_{g\_ref} - P_g \\ e_Q = Q_{g\_ref} - Q_g \end{cases} \tag{10}$$

Varieties $S_P = 0$ and $S_Q = 0$ represent accurate tracking of the active and reactive powers of the converter. When the system states reach the collector and slide along the surface, we have:

$$S_P = S_Q = \frac{dS_P}{dt} = \frac{dS_Q}{dt} = 0 \tag{11}$$

For:

$$\begin{cases} \dot{e}_P = -\lambda_P . e_P(t) \\ \dot{e}_Q = -\lambda_Q . e_Q(t) \end{cases} \tag{12}$$

## 5.2    Order Law by Sliding Mode

$$\begin{cases} \dot{S}_P = \dot{e}_P + \lambda_P\, e_P = \dot{P}_g + \lambda_P\, e_P \\ \dot{S}_Q = \dot{e}_Q + \lambda_Q\, e_Q = \dot{Q}_g + \lambda_Q\, e_Q \end{cases} \tag{13}$$

$$\begin{cases} \dfrac{dS_P}{dt} = \underbrace{\dfrac{3}{2L}\left(U_\alpha^2 + U_\beta^2\right) + \dfrac{r_g}{L_g}P_g + \omega\, Q_g + \lambda_p\left(P_{g\text{-ref}} - P_g\right)}_{E_P} \\[4pt] \qquad\qquad -\dfrac{3}{2\,L_g}\left(U_\alpha V_\alpha + U_\beta V_\beta\right) \\[4pt] \dfrac{dS_Q}{dt} = \underbrace{\dfrac{r_g}{L_g}Q_g - \omega\, P_g + \lambda_Q\left(Q_{g\text{-ref}} - Q_g\right)}_{E_Q} + \tfrac{3}{2L_g}\left[-\left(U_\beta V_\alpha + U_\alpha V_\beta\right)\right] \end{cases} \tag{14}$$

$$\begin{bmatrix} \dfrac{dS_P}{dt} \\ \dfrac{dS_Q}{dt} \end{bmatrix} = \begin{bmatrix} E_P \\ E_Q \end{bmatrix} + Z\begin{bmatrix} V_\alpha \\ V_\beta \end{bmatrix} \tag{15}$$

The substitution Eq. 14 in 15 leads to:

$$\dfrac{dS}{dt} = E + ZV; \ \text{Avec}: \begin{cases} E = \begin{bmatrix} E_P \\ E_Q \end{bmatrix} \\ V_g = \begin{bmatrix} V_{g\alpha} \\ V_{g\beta} \end{bmatrix} \\ Z = -\dfrac{3}{2L_g}\begin{bmatrix} U_\alpha & U_\beta \\ U_\beta & -U_\alpha \end{bmatrix} \end{cases} \tag{16}$$

The next quadratic Lyapunov function given by Eq. 9 is considered [22].

$$W = \frac{1}{2}S^T S \tag{17}$$

The time quadratic Lyapunov function is then:

$$\frac{dW}{dt} = S^T\frac{dS}{dt} = S^T\left(A + BV_g\right) \tag{18}$$

The variation of the time of this function must be strictly negative with $S \neq 0$. This is the condition of the path attraction to the sliding surface [18]. Therefore, the control law of the controller must be correctly chosen for this condition to be verified.

The proposed SMC control law is described by the following expression:

$$V = -Z^{-1}\left\{ \begin{bmatrix} E_P \\ E_Q \end{bmatrix} + \begin{bmatrix} K_P & 0 \\ 0 & K_Q \end{bmatrix}\begin{bmatrix} \text{sign}(S_P) \\ \text{sign}(S_Q) \end{bmatrix} \right\} \tag{19}$$

Then:

$$\begin{cases} Z^{-1} = \frac{3}{2L_g}\begin{pmatrix} U_\alpha & U_\beta \\ U_\beta & -U_\alpha \end{pmatrix} \\ V_{\alpha\beta\_eq} = -Z^{-1}.E_{PQ} \\ V_{\alpha\beta\_n} = -Z^{-1}.K_{PQ}.sign(S_{PQ}) \end{cases} \tag{20}$$

We can proof the stability if $\frac{dW}{dt} < 0$, $(S_P.sgn(S_P) > 0)$ and $(S_Q.sgn(S_Q) > 0)$ [18]:

$$\frac{dW}{dt} = S^T\frac{dS}{dt} = -S^T\begin{bmatrix} K_P & 0 \\ 0 & K_Q \end{bmatrix}\begin{bmatrix} sgn(S_P) \\ sgn(S_Q) \end{bmatrix} \tag{21}$$

## 6   Robustness of the Control Approaches

This part presents the tests of the impact of parametric variations on the behavior of the system with the strategies developed based on the algorithms of the commands. Figure 5 illustrate the scenario of the variation of line inductance. During this test, the value of the line inductance variation varies from the initial value.

## 7   Simulations Results

Figure 6 shows the system responses to parametric variation for line inductance. We tested our GSC system connected to the network based on two commands DPC_developed and DPC_SMC-SVM for different variation of the line inductance.

For an error equal to, we notice that the variables of the system are robust and slightly varied over time with the command DPC_SMC-SVM by contribution to the

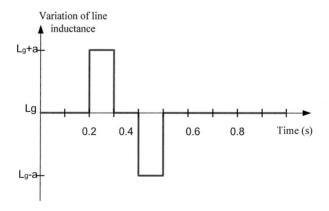

**Fig. 5.**  Scenario of the variation of line inductance

command DPC_developed that was not robust. Consequently, the robustness of the
DPC_SMC-SVM sliding mode control to the variations of the line inductance is verified and it is found that this type of control is insensitive to the parametric variation.

Figure 7 presents a comparative study between the commands developed
DPC_developed and DPC_SMC_SVM for the variation of the line inductance.

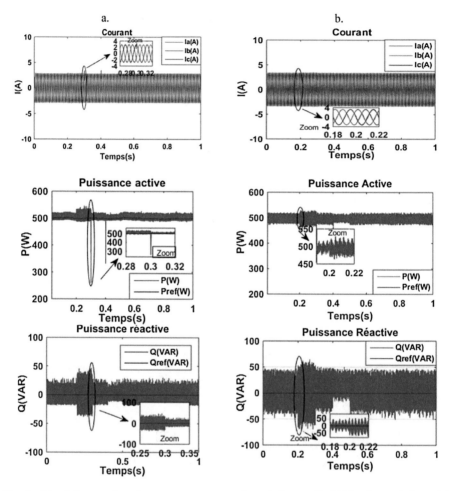

**Fig. 6.** Simulation result and response to parametric variation. a. Grid induction variation with
DPC_developed, b. grid induction variation with SMC-DPC-SVM

Good performance is obtained for the operating regime, which is much higher in
terms of harmonic distortion rates (THD 2.61%). There is also a precise adjustment of
the instant active and reactive power.

The advanced control SMC-DPC-SVM is able to ensure a perfect trajectory
tracking steady state and transient references with a very fast dynamics. With these

experimental simulations (Fig. 6), we were able to show the performance of this developed strategy.

a.                                                        b.

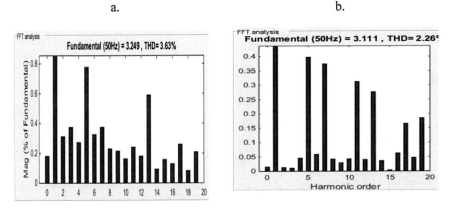

**Fig. 7.** Comparative study with FTT between DPC_developed and DPC_SMC_SVM for the variation of the line inductance. a. FFT of DPC_developed, b. FFT of SMC-DPC-SVM

Although the SMC-DPC-SVM command for this hybrid system is a robust and efficient controller, there are still grid disruption issues. As a result, we will exploit the different methods of compensation.

## 8   Conclusion

This paper presents an approach to Direct Power Control (DPC) strategy for a three-phase grid-connected converter. This control system is simulated and compared with other similar approach in the discrete-time sliding mode control combining direct power control strategy and the space vector modulation.

The active and reactive power are directly controlled by converter switching states using the value of the delivered power error calculated from previous samples of three phase voltages and currents, which output is an optimal control vector defined in "dq" reference frame minimizing the ripple is obtained using predicted values of three-phase currents, in order to minimizes the instantaneous active and reactive power displacement from their reference values.

The control vector is computed from the samples of voltages and currents and then converted to a switching sequence using space vector modulation.

An appropriate switching sequence is generated for each optimal vector using direct and indirect method. In the indirect method, the switching vector is the nearest to the optimal control vector, and the direct method uses space vector modulation.

The period of modulated signal is equal to the sample period. A correction of the control vector is defined with aim to eliminate the influence of the system uncertainties using predicted values of the active and reactive powers. Thus, the proposed strategy is robust even there is variations in parameters system. Moreover, hysteresis controllers, look-up tables, and pulse-width modulators are not necessary.

In the perspective work, we will present an experimental robustness study of the grid induction variation with SMC-DPC-SVM. After that, we will injected a reactive power on GSC connected to the grid. All the experimental will be implemented with the DSpace1004 card.

# References

1. Burlaka, V.V., Gulakov, S.V., Podnebennaya, S.K.: A three-phase high-frequency AC/DC converter with power-factor correction. Elektrotekhnika **4**, 43–47 (2017)
2. Kim, J.-H., Jou, S.-T., Choi, D.-K., Lee, K.-B.: Direct Power Control of Three-Phase Boost Rectifiers by using a Sliding-Mode Scheme, JPE 13-6-9 ISSN(Print): 1598-2092. ISSN (Online): 2093-471
3. Izanlo, A., Asghar Gholamian, S., Verij Kazemi, M.: Using of four-switch three-phase converter in the structure DPC of DFIG under unbalanced grid voltage condition. Electr. Eng. **100**(3), 1925–1938 (2018)
4. Bouzid, A.M., Guerrero, J.M., Cheriti, A., Bouhamida, M., Sicard, P., Benghanem, M.A.: Survey on control of electric power distributed generation systems for microgrid applications. J. Renew. Sustain. Energy Rev **44**, 751–766 (2015) (Elsiever)
5. Ohnishi, T.: Three phase PWM converter/inverter by means of instantaneous active and reactive power control. In: Proceedings of the International Industrial Electronics and Control Instrumentation, pp. 819–824, Oct/Nov 1991
6. Hu, J., Zhu, Z.Q.: Investigation on switching patterns of direct power control strategies for grid-connected DC–AC converters based on power variation rates. IEEE Trans. Power Electron. **26**(12), 3582–3598 (2011)
7. Pande, V.N., Mate, U.M., Kurode, S.: Discrete sliding mode control strategy for direct real and reactive power regulation of wind driven DFIG. Electric Power Syst. Res. (2013)
8. Zoghlami, M., Bacha, F.: Implementation of different strategies of direct power control. In: Proceedings of the 6th International Renewable Energy Congress (IREC 2015), Sousse, Tunisia, 24–26 Mar 2015
9. Šabanović, A., Fridman, L., Spurgeon, S. (eds.): Variable Structure Systems: From Principles to Implementation. IET Press (2004)
10. Pinto, S.F., Silva, J.F.: Constant-frequency sliding-mode and PI linear controllers for power rectifiers: a comparison. IEEE Trans. Ind. Electron. **46**(1), 39–51 (1999)
11. Huseinbegović, S., Peruniþiü-Draženoviü, B.: Discrete-time sliding mode direct power control for three-phase grid connected multilevel inverter. In: 4th International Conference on Power Engineering, Energy and Electrical Drives, Istanbul, Turkey, 13–17 May 2013
12. Barkat, S., Tlemçani, A., Nouri, H.: Direct power control of the PWM rectifier using sliding mode control. Int. J. Power Energy Convers. **2**(4) (2011)
13. Elnady, A., Al-Shabi, M.: Operation of direct power control scheme in grid-connected mode using improved sliding mode observer and controller. Int. J. Emerg. Electric Power Syst. 20180041 (2018)
14. Zaimeddine, R., Undeland, T.: Direct power control strategies of a grid-connected three-level voltage source converter VSI-NPC. In: Proceedings of 14th European Conference on Power Electronics and Applications, EPE, pp. 1–6 (2011)
15. Wong, M.-C., Dai, N.-Y., Lam, C.-S.: Parallel Power Electronics Filters in Three-Phase Four-Wire Systems. Electronics & Electrical Engineering

16. Zoghlami, M., Bacha, F.: Implementation of different strategies of direct power control. In: 6th International Renewable Energy Congress (IREC'2015). IEEE 978-1-4799-7947-9/15 (IEEE Explore)
17. Zoghlami, M., Kadri, A., Bacha, F.: Analysis and application of the sliding mode control approach in the variable-wind speed conversion system for the utility of grid connection. Energies **11**(4), 720 (2018). ISSN: 1996-1073, Impact Factor: 2.676
18. Zeng, B., Zou, J.X., Li, K., Xin, X.S.: A novel sliding mode control based low voltage ride through strategy for wind turbine. In Applied Mechanics and Materials, vol. 548–549, pp. 890–894. Trans Tech Publications, Zürich, Switzerland (2014)

# Design of Intelligent Controllers, for Sun Tracking System Using Optical Sensors Networks

Abdelaziz Sahbani[1]([✉]), Ali Hamouda Ali Saeed[2],
and Abdullah ALraddadi[1]

[1] Medina College of Technology, Al Madinah al Munawwarah, Saudi Arabia
abdellazizsahbani@yahoo.fr, araddadi@mct.edu.sa
[2] Interserve Learning & Employment International, Jizan, Saudi Arabia
Hamoudaali959@yahoo.com

**Abstract.** Clean and renewable energy is the goal of most companies, governments, the United Nations and organizations that they are plan to reduce pollution from industry, cars and other machinery or by mechanical engines that contribute. In increasing the amount of carbon dioxide resulting from (oil and gas). This steady increase leads to pollution of the environment and the atmosphere where changes occur in the atmosphere and the composition of the atmosphere of the Earth's, atmosphere result in a steady increase in the temperature of the earth, on the other hand, Oil and gas in the near future so the need for green energy and recyclable is one of the most important challenges in this paper, we proposed a synthetic system to track the sun depends on the features of optical sensors, where a three-dimensional model was designed in which we installed resistors light on the perimeter The shape of the ball and the coordinates of the points and make one of these optical sensors reference point for the rest of the points on the ball surface has been sustained

**Keywords:** Clean and renewable energy · Solar energy · Optical sensors · Fuzzy logic controller · Sun tracking

## 1 Introduction

Renewable energy is the key to our future. Energy is a form of matter where it can be transformed from one form to another in many ways. Many papers discussed the conversion mechanism, especially the conversion of energy into electrical energy. Mechanical motors are the largest source of energy, from thermoelectric to kinetic and then electromechanical. Nuclear energy has also been used to produce electrical energy. However, pollution affects our plant and environment and causes environmental damage caused by genetic change. Temperature increase per year due to increased carbon dioxide in the atmosphere. The needs of green energy or alternative energy, sometimes called renewable energy, have gained much attention at present time and

© Springer Nature Switzerland AG 2019
C. Benavente-Peces et al. (Eds.): SEAHF 2019, SIST 150, pp. 414–420, 2019.
https://doi.org/10.1007/978-3-030-22964-1_46

have become the focus of future technology. Cars, roads, airplanes, and other machines use solar energy, wind energy or other sources. Green energy, wind energy, biomass energy, geothermal temperature, sea temperature variations, sea waves, morning and evening waves, etc. [1, 2]. The use of solar energy began in Saudi Arabia in 1960. King Abdulaziz City for Science and Technology (KACST) began in 1977 systematic research on solar energy and green energy. The KACST Energy Research Institute and the US National Renewable Energy Laboratory produce the Saudi Solar Radiation Atlas in 1994 as a joint research and development project [3]. The solar village project site is located 50 km northwest of Riyadh and provides between 1 and 1.5 MWh of electricity.

## 2   Solid Angle Modules

The intensity of the sun's radiation, defined as the intensity of the radiation falling vertically on the surface area and the intensity of the light falling with the angle of fall, where the highest value when the radiation will be vertically point to the area of interest., a plane circle (2D) turns into a sphere (3D); the length of an arc of a circle (2D) turns into the surface area of a sphere (3D). The solid angle $\Omega$ as in Fig. 1, used to calculate sphere's segment area (a), which is proportional to the square of the sphere's radius R:

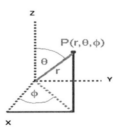

**Fig. 1.**  Spherical coordinate $(r, \theta, \varphi)$.

$$a = \Omega \cdot R^2 \quad \text{or} \quad \Omega = a/R^2$$

For small solid angles, the segment area of a sphere can be approximated to be considered as a flat surface [1].

So solid angles likewise can be expressed as differentials:

$$d\Omega = da/R^2$$

by converting the tow dimension to three dimension spherical coordinate $(r, \theta, \varphi)$.

$$d\Omega = \frac{da}{R^2} = \frac{(R\sin\theta\,d\phi)(R\,d\theta)}{R^2} = \sin\theta\,d\phi\,d\theta \qquad (1)$$

or

$$\int_{\phi=0}^{2\pi} (R\,sin\theta\,d\phi)((R\,d\theta) = R^2\,sin\theta\,d\theta \int_{\phi=0}^{2\pi} d\phi = 2\pi\,R^2\sin\theta\,d\theta \qquad (2)$$

The resulting differential segment area $da$ is similar to a belt having the width $R\,d\theta$ winded around the sphere as in Fig. 2.

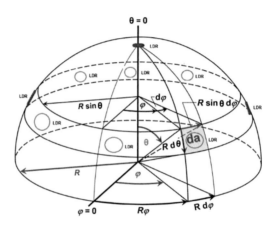

**Fig. 2.** LDR on spherical coordinate.

## 3   Proposed System

Light sensor used to detecting light intensity, or position of specific objects, measuring color, or detecting spectrum of specific chemical compound, this paper used light dependent resistance (LDR) as intensity measurement, hear used as optical network sensor distributed around the surface of spherical shape as the shape in the Fig. (2) one LDR fixed as reference point any point can be represented as function of spherical coordinate $(r, \theta, \phi)$. Additional LDR's fixed at specific points each one separated by 30° from the reference, and 45° from each other. The LDR or LDRs which perpendicular to the Sun radiation receive maximum intensity the its values of resistance its less than the other which read less intensity, each resistance measured as spherical coordinate and the position of the sun can be tracked with respect to the reference on and its relation with the LDR with maximum intensity. If the position of the reference $LDR_0$, the distance between any as angle change because at the surface the R is constant. The value of $\theta$ angle is $(0 \le \theta \le 180)$, and the values of $\phi$ angle is $(0 \le \phi \le 360)$.

## 3.1  Acquisition of Crisp Inputs

The crisp inputs (sun lite intensity via LDR as spherical coordinate) are two signals generated by ramp signal followed via transfer function for temp and transfer function with gain form sensors were read.

## 3.2  Fuzzification

Fuzzification Interface all LDR value measured and converted to the fuzzy logic value (from 10 ohms at high light intensity up to 10 mega ohms) at completely darkness. Computed values are high, medium, low. The position determined by the LDR spherical coordinate, the guide sensor that which indicated by minimum resistance value because LDR resistance it inversely proportional to the light intensity.

## 3.3  Rule Evaluation

The Rule evaluation in MATLAB used to represent the relation between input the Sun radiation intensity witch converted by geometrical design to spherical coordinate, and the output in this case it is the two motors, one rotating in the vertical and the second rotated on the horizontal directions [3]. Rule evaluation is performed one rule at a time, using the membership grade of each condition obtained. The output of the fuzzy logic controller depends on the membership grades of the rules. Depending on the current input values. The inferencing methods may be classified according to the nature of the action part of each rule. The Mamdani method used, the result is collection of term or fuzzy set related via min-product method or min-correlation.

## 3.4  Fuzzy Logic Controller

Fuzzy logic membership function designed form Fig. (2) and Tables 1 and 2, the input is the generated from input guide LDR, and its position to the reference LDR. The sensor placed around the sphere as in Fig. 2 one on the top it is the reference sensor normally this sensor is perpendicular to the PV cells, the LDR which received maximum light intensity it become the guide sensors the rules represented as

**Table 1.**  Table of the fuzzy rules input and output membership.

| NO | INPUT sensor position | Output 1 Motor 1 | Output 2 Motor 2 |
|----|----------------------|------------------|------------------|
| 1  | U   | H  | NC   |
| 2  | UR  | H  | PSR  |
| 3  | R   | H  | PSSR |
| 4  | RD  | H  | NSR  |
| 5  | D   | H  | NC   |
| 6  | UL  | H  | NSR  |
| 7  | L   | H  | NSSR |
| 8  | LD  | H  | PSR  |
| 9  | CE  | NC | NC   |

**Table 2.** Table of the fuzzy second rules input and output membership.

| NO | INPUT sensor position | Output 1 Motor 1 | Output 2 Motor 2 |
|---|---|---|---|
| 1 | U | L | NC |
| 2 | UR | L | PSR |
| 3 | R | L | PSSR |
| 4 | RD | L | NSR |
| 5 | D | L | NC |
| 6 | UL | L | NSR |
| 7 | L | L | NSSR |
| 8 | LD | L | PSR |
| 9 | CE | NC | NC |

If …… then ……. And ………..

For output number one which control the horizontal rotation by 30 degree each steps, if the position of the sensor as same axis with the reference sensor they are No change (NC), in case of the guide sensor is in the position left to the axes by angle 30 degree then it represented as positive step right (PSR) in this case the reference sensor tracking the position of maximum intensity.

## 4  Results and Analysis

The simulation was performed in MATLAB/ SIMULINK environment and fuzzy logic tools box as in Fig. 3. In which fuzzy controller with one input signal represented as random in put in place of LDR and two output, the fuzzy controller with two output motor one and motor two used to control PV system Fig. 4.

*Mamdani Fuzzy controller*

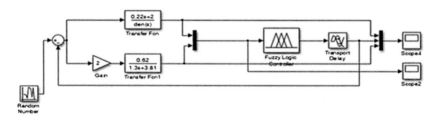

**Fig. 3.** Simulink of fuzzy logic Mamdani controller for PV sun tracking system.

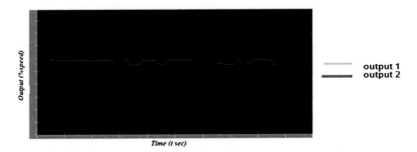

**Fig. 4.** Output signals to controlling two motors for sun tracking system.

## 5 Conclusion

The output fuzzy logic control is the signal that control two motors that drives, vertical and horizontal axis, for the Sun tracking system. The system responded to the variation of random signal to the fuzzy logic controller.

The spherical coordinate system is more applicable to the sun radiation, also the use of network sensor its more efficient than single sensor, the tracking system can be used for different application for monitoring, motion detection and security system.

## References

1. Tomova, A., Antchev, M., Petkova, M., Antchev, H.: Fuzzy logic hysteresis control of a single-phase on-grid inverter: computer investigation. Int. J. Power Electron. Drive Syst. (IJPEDS) 3(2), 179–184. ISSN: 2088-8694
2. Adewuyi, P.A.: Performance evaluation of Mamdani-type and Sugeno-type fuzzy inference system based controllers for computer fan. Int. J. Inf. Technol. Comput. Sci. 1, 26–36 (2013). Published Online December 2012 in MECS. http://www.mecs-press.org/; https://doi.org/10.5815/ijitcs.2013.01.03
3. Sona, N., Shantharama Rai, C.: Fuzzy logic controller for the speed control of an IC engine using Matlab\Simulink. Int. J. Recent. Technol. Eng. (IJRTE) 2(2) (2013). ISSN: 2277-3878
4. Ross, T.J.: Fuzzy Logic with Engineering Applications, 2nd edn. University of New Mexico, USA. World Appl. Sci. J. 6(1), 16–23 (2009)
5. Nhivekar, G.S., Nirmale, S.S., Mudholker, R.R.: Implementation of fuzzy logic control algorithm in embedded microcomputers for dedicated application. Int. J. Eng. Sci. Technol. 3(4), 276–283 (2011)
6. Mashhadi, S.K.M., Shokohinia, M., Reza, M.: Controlling the green-house irrigation system in fuzzy logic method. J. Math. Comput. Sci. 4(3), 361–370 (2012)
7. Naik, N.P.: Fuzzy Logic to Control Dam System for Irrigation and Flooding. Late Bhausaheb Hiray S.S. Trust's Institute of Computer Application, Government Colony, Bandra (East), Mumbai
8. Suresh, R., Gopinath, S., Govindaraju, K., Devika, T., SuthanthiraVanitha, N.: GSM based automated irrigation control using raingun irrigation system. Embedded System Technologies, and Electrical & Electronics Engineering Knowledge Institute of Technology, Salem, India. Int. J. Adv. Res. Comput. Commun. Eng. 3(2) (2014)

9. Foley, G.: Photovoltaic Application in Rural Area of Developing World. Technical report, World Bank, New York (1995)

10. Daut, I., Irwanto, M., Irwan, M.Y., Gomesh, N., Ahmed, S.N.: Three level single phase photovoltaic and wind power hybrid inverter. Energy Procedia **18**, 1307–1316 (2012)

11. Lopez, O., Freijedo, D.F., Yepes, G.A., Fernandez-comessa, P., Malvar, J., Teodorescu, R., Doval-Gandoy, J.: Eliminating ground current in a transformer less photovoltaic application. IEEE Trans. Energy Convers. **25**(1), 140–147 (2010)

# A Conceptual Model of Cyberterrorists' Rhetorical Structure in Protecting National Critical Infrastructure

Khairunnisa Osman[1], Ala Alarood[2], Zanariah Jano[3], Rabiah Ahmad[3], Azizah Abdul Manaf[2], and Marwan Mahmoud[4(✉)]

[1] Innovative and Sustainable Technical Education, Centre of Technopreneurship Development, University Teknikal Malaysia Melaka, Melaka, Malaysia
[2] Department of Computer Science, Faculty of Computing and Information Technology, University of Jeddah, Jeddah, Saudi Arabia
[3] Information Security and Networking Research Group, Center for Advanced Computing Technology, Universiti Teknikal Malaysia Melaka, Melaka, Malaysia
[4] King Abdulaziz University, Jeddah, Saudi Arabia
mmamahmoud@kau.edu.sa

**Abstract.** This study presents a conceptual paper on developing a model of cyberterrorists' rhetorical structure. The objectives of the study are to identify the type of rhetoric's in cyber terrorists' communication, analyze how the cyber terrorists utilize appeal to audience, stylistic devices and argumentation in their communication and to propose a model of cyber terrorists' rhetorical structure. A rhetorical analysis is conducted by examining recent cyberterrorists sites in popular media. Several categories are utilized; Appeal to the audience; Stylistic devices and Argumentation model. Drawing on a selected theoretical framework a combination of quantitative word-level analysis and qualitative coding will be used to assess the categories which are framed by cyber terrorists.

**Keywords:** Cyberterrorist · Argumentation model · Rhetorical analysis · Qualitative coding

## 1 Introduction

Literature on information and Technology security indicates dependency on technology. Yet, due to the fact that websites, nowadays, are attacked numerously with individuals who possess malicious intent, the dependency has become a liability, affecting the critical infrastructure and services that depend on technology. Hitherto, effective and efficient analysis of terrorist information on the Web are hampered by the problems of information overload and difficulty to obtain a comprehensive picture of terrorist activities. Much research at present focus on the ways in which organizations secure their networks and information in the supply chain, ignoring the ways in which organizations construct and understand cybersecurity risks. Moreover, the roles of rhetoricians and sociologists in cybersecurity realms are deemed essential given that the web is laden with terminological assumptions, violent metaphors, and ethical conflicts.

© Springer Nature Switzerland AG 2019
C. Benavente-Peces et al. (Eds.): SEAHF 2019, SIST 150, pp. 421–427, 2019.
https://doi.org/10.1007/978-3-030-22964-1_47

Hence, examining the discourse of cyber groups is deemed important in order to explore trends in security discourse from the past few years. The conversation and analysis from earlier years in the evolution of the computer science field is only partially relevant to today's challenges and does not address emergent topics like the cyberterrorists' corpus. The language-level studies that do exist for online security have focused almost exclusively on state and nation-level cyber-rhetoric and have not systematically examined the everyday business decisions that hackers (of all hat colors) are making—for example, studies like [11] inquiry into the growth of cyber-terrorism discourse as a reflection of physical acts of terrorism.

## 2 Literature Reviews

National critical infrastructure defined by many experts as artificial plant which locates resources that considered highly important for countries around the world. The term critical refer to how crucial of the components in country development. It is also representing a must component for life and therefore it's required a protection [1]. Recent technology connects critical infrastructure via communication network. As for example, countries rich with oil apply advanced communication technology to manage the distribution of that natural resources. Due to the rapid development of ICT, critical infrastructure is managed by a sophisticated computer system. With the growth of industry 4.0. and IoT maintaining critical infrastructure become more efficient but it increases demand on security and protection. Critical infrastructure become a major target for potential attackers. Traditionally physical attacks were a major technique in damaging the infrastructure. With the development of Internet, cyber become a medium to launch attack on critical infrastructure. Incidents like, use of ICT as a tool to create mass of destruction at public facilities and critical resources was reported elsewhere. This type of attacks known as terrorist attack. Extant literature highlights studies conducted on terrorists' use to the Web. Besides, studies on a rhetorical perspective are also featured.

### 2.1 Web Deploy Potential Terrorists

The main threat which concerns the world nowadays is the cyberterrorists' attack. The fact that they are using the web for channeling their communication to promote their ideology, deliver internal communications, attack their enemies, and conduct criminal activities is apparent. Terrorism is commonly defined as "the purposeful act or the threat of the act of violence to create fear and/or compliant behavior in a victim and/or audience of the act or threat" [20]. Cyber-terrorists simply mean those who have the intent of terrorism using technology. In short, terrorist groups are using advanced technology to accomplish their goals. Various warnings have been made on the fact that terrorists may launch attacks on critical infrastructure like major e-commerce sites and governmental networks [12]. Insurgents in Iraq have posted Web messages asking for munitions, financial support, and volunteers [6]. Hence, obtaining information from the terrorists' websites will permit better understanding and analysis of terrorist and extremist groups.

Recent studies have shown how terrorists use the Web to facilitate their activities. Weimann [24] analyzed the jihad group, Al-Qaeda and found that between 1998 and 2007 thousands of websites, online forums and chat rooms were utilized by terrorists and their sympathizers. Internet makes up about 90% of usage for their internal communication. The internet is seen as a virtual firewall to mask the identities of individuals, and subscribers have the chance to get in touch personally with terrorist representatives in order to ask for relevant information on anything including cyber-jihad. Weimann [25] asserted that communication through the Internet is relatively safe, cheap, easy and anonymous. Strategies by Al-Qaeda include searching for potential recruiters, international audiences and adversaries. Weimann [28] conducted a study using the "theater of terror perspective" and the theory of selective moral disengagement. While the "theater of terror perspective" ascertains that modern terrorism is attempting, through arranged messages, most media coverage available, the theory of selective moral disengagement is an analytical tool to investigate the terrorist rhetoric.

Weiman [26] found that Al-Qaeda widely utilizes the Internet to effect social change, instill fears or affect political decisions. Internet is also utilized for purposes such as fundraising, recruiting people, psychological warfare, propaganda and coordination of actions. Studies identified five categories of terrorist use of the Web [21]: propaganda (to disseminate radical messages); recruitment and training (to encourage people to join the Jihad and get online training); fundraising (to transfer funds, conduct credit card fraud and other money laundering activities); communications (to provide instruction, resources, and support via email, digital photographs, and chat session); and targeting (to conduct online surveillance and identify vulnerabilities of potential targets such as airports. Based on a study of 172 members of the global Salafi Jihad, Sageman [19] concluded that the Internet has instilled a close bond between individuals and a virtual religious community. His study ascertains that the Web appeals to lonely individuals by linking to people with common interests. Such virtual community offers a number of advantages to terrorists. It has become a neutral nation, emphasizing the facts that fighting against the far enemy (e.g., the United States) is the priority rather than the near enemy. Internet chat rooms instill extreme, abstract, but simplistic solutions in drawing most potential non-Islamic Jihad recruits because Internet cafés protect the anonymity of terrorists.

Since the 9/11 attacks, there is a growing interest in using the internet technology to curb terrorism. A study conducted by the U.S. Defense Advanced Research Projects Agency shows that their collaboration, modeling, and analysis tools speed analysis [17], but these tools are not customized in collecting and analyzing Web data. Although several Web mining technologies exist [9] and [13], there has not yet been a comprehensive methodology to address problems of collecting and analyzing terrorist data on the Web. Unfortunately, existing frameworks using data and text mining techniques [15] and [22] do not address issues specific to the Dark Web. Few studies have used advanced Web and data mining technologies to collect and analyze terrorist information on the internet despite these technologies being widely applied in such other domains as business and scientific research [10, 13] and [14] devise a methodology merging information collection, analysis, and visualization techniques from different Web information sources. However, qualitative measures are lacking such as persuasive appeals, rhetoric, and attribution of guilt to the Web site attributes.

## 2.2    A Rhetorical Perspective

Cybersecurity realms are relatively uncharted by rhetoricians and sociologists despite being laden with terminological assumptions, violent metaphors, and ethical conflicts. Hence, a rhetorical analysis is deemed appropriate to gain insights into cyber terrorists' realms of communication. Barton [4] argues that rhetoricians can use a methodical study of language to gain insights into cyber terrorists' realms. She states that.

"The fundamental insight that composition/rhetoric offers to the literature on ethics and bioethics is that decision-making with ethical dimensions is most often interactional and therefore rhetorical. In other words, such decision-making takes place between real people, in real time, in (semi-) ordinary language...".

Numerous scholars [5, 8] and [18] define the results of "in-the-wild" hacking, focusing on the matters like time, effort, and strategy involved to form functional exploits that destroy a country's infrastructure. A consensus among the scholars entails that speculation on large-scale, nation-level cyberwars is then not realistic. In addition, rhetorical study is essential in this area because "offensive" and "persuasive" discourse have been changing at a rapid pace. Thus, to explore trends in the cyberterrorists' communication from the past few years become essential. The results of earlier studies only contribute little insights to today's challenges and does not address cyberterrorists' corpus.

The online security studies mostly focus on state and nation-level cyber-rhetoric and do not examine the daily business decisions that hackers are making. Blank [7] highlights the possible consequences of terminological inflation, making comparison on trends in cybersecurity rhetoric to those following 9/11. She asserts that the "war on terror" discourse causes too much unchecked authority in the years following the September 11 attacks and admits her fear that the same could be true of hacking rhetoric. Hence, the term cyber-attack might trigger an aggressive response to cyber threats or cyber conduct, stretching or overstepping the relevant legal boundaries for authority.

# 3    Methods

This project proceeds through the following three main phases: Identifying the type of rhetorics in cyber terrorists' communication. Analyzing how cyber terrorists utilize various categories of the rhetorical analysis in their communication and proposing a model of cyber terrorists' rhetorical structure.

- **Phase 1**: Identifying the type of rhetorics used in potential cyber terrorists' communication.

  a. The 10 relevant websites and 10 blogs upon a recommendation by the experts will be gathered.
  b. The type of rhetorics is determined for each sample.

- **Phase 2**: Analyzing how cyber terrorists utilize various categories of the rhetorical analysis in their communication.

a. The phrases used in terms of appeal to audience, stylistic device and argumentation will be coded.
b. The coding schemes will be analyzed using ANOVA to ascertain the patterns which exist among the samples.

- **Phase 3**: Proposing a model of cyber terrorists' rhetorical structure.

a. Compiling the words and phrase as a model.
b. Samples.

Specialized corpora consist of 10 websites and 10 blogs of potential cyber terrorists are selected upon experts' recommendation. The samples are chosen based on their publication date between 2012 and 2015, the medium in which they are published and their relevance to the study at hand. Samples are taken from websites and blogs of potential cyber terrorists.

## 3.1  Sampling Method

The sampling method used is purposive as the listing of potential cyber terrorists are recommended by the experts. Using 'purposive sampling,' the population of cybersecurity discourse is separated into distinct and mutually exclusive categories or subgroups. Judgment is then exercised by the researchers to select samples from each subcategory according to predetermined proportions. In other words, selection of the data is non-random. The benefits of this method are that all relevant categories are covered and there is greater variability in the samples than random sampling can sometimes achieve. The study is primarily interested in whether rhetorics is being used by cyber terrorists' groups and in which ways.

## 3.2  Technique of Analysis

A content analytic approach is used. Systematic procedures are used to evaluate the content of communications. Content analysis analyses the frequency of certain items, symbols, or themes which appear in texts [29]. Furthermore, content analysis is applied to study attitudes or beliefs of one group, to understand characteristics of certain groups (e.g., the ideology of business elites) or changes in the group over time or space.

## 3.3  Theoretical Framework

This study applies a rhetorical analysis to characterize the rhetorical structure of the potential cyber terrorists' publication. Aristotle defines rhetoric as the art of public speaking to persuade [3]. Rhetoric refers to the art of logical discussion. In rhetoric, demonstration is used to persuade. Rhetoric is a combination of the science of logic and of the ethical branch of politics [3]. Based on Norreklit's study [16], several categories are utilized.

- Appeal to the audience—looking at audience's ethos or trust in the credibility of the source, to the audience's pathos or emotions, or to the audience's logos or logic [2]. The genre of text will typically influence the type of appeal used.

- Stylistic devices—analogies, metaphors, similes, metonymy, hyperbole, irony, antithesis, loaded adjectives and imprecise and intertextuality-based concepts.
- Argumentation model—involves three basic elements: a claim, data and a warrant [23]. The claim is the point of view the source wishes the audience to accept. Data is the evidence to support the claim. The warrant is an integration of claim and data [16].

## 4    Conclusion

In conclusion, the proposed model will benefit the relevant authorities in detecting potential cyber terrorists through their rhetorical structure. Ultimately, studying this unexplored area has important implications for policymakers creating new cybersecurity legislation, reporters attempting to accurately frame the debate, and information technology professionals whose livelihoods are affected by evolving social norms.

## References

1. Amoroso, E.G.: Cyber Attacks: Protecting National Infrastructure. Butterworth-Heinemann, Burlington, Massachusetts (2011)
2. Aristotle, G.A., Kennedy: Aristotle on Rhetoric a Theory of Civic Discourse (1991)
3. Aristotle: The Art of Rhetoric. Simile, vol. 1967, issue May 2009 p. li, [492] p.; Published by Harvard University Press; Heinemann (1926)
4. Barton, E.: Further contributions from the ethical turn in composition/rhetoric: analyzing ethics in interaction. Coll. Compos. Commun. **59**(4), 596–632 (2008)
5. Bendrath, R.: The cyber-war debate: perception and politics in U.S. critical infrastructure protection. Inf. Secur. **7**(1), 80–103 (2014)
6. Blakemore, B.: Web Posting May Provide Insight into Iraq Insurgency. ABC News (2004, November 23)
7. Blank, L.R.: Defining the battlefield in contemporary conflict and counterterrorism: understanding the parameters of the zone of combat. In: Georgia Journal of International and Comparative Law, vol. 39, no. 1, Emory Public Law Research Paper No. 10–139 (2010). Available at SSRN: https://ssrn.com/abstract=1677965
8. Brito, J., Watkins, T.: Loving the Cyber-Bomb? The Dangers of Threat Inflation in Cybersecurity Policy. SSRN Working Paper Series (2011)
9. Chen, H., Chau, M.: Web mining: machine learning for web applications. In: Williams, M.E. (ed.) Annual Review of Information Science and Technology (ARIST), vol. 38, pp. 289–329. Information Today Inc, Medford, NJ (2004)
10. Chen, H., Chung, W., Qin, J., Reid, E., Sageman, M., Weimann, G.: Uncovering the dark web: a case study of Jihad on the Web. J. Am. Soc. Inf. Sci. **59**, 1347–1359 (2008). https://doi.org/10.1002/asi.20838)
11. Dunn Cavelty, M.: Cyber-terror—looming threat or phantom menace? The framing of the US cyber-threat debate. J. Inf. Technol. Polit. **4**(1), 19–36 (2008)
12. Gellman, B.: Cyber-attacks by Al Qaeda feared. Washington Post (2002, June 27)
13. Last, M., Markov, A., Kandel, A.: Multi-lingual detection of terrorist content on the web. In: Paper Presented at the Proceedings of the PAKDD'06 International Workshop on Intelligence and Security Informatics, Singapore (2006)

14. Marshall, B., McDonald, D., Chen, H., Chung, W.: EBizPort: collecting and analyzing business intelligence information. J. Am. Soc. Inf. Sci. Technol. **55**(10), 873–891 (2004)
15. Nasukawa, T., Nagano, T.: Text analysis and knowledge mining system. IBM Syst. J. **40**(4), 967–984 (2001)
16. Norreklit, H.: The Balanced Scorecard: what is the score? A rhetorical analysis of the Balanced Scorecard. Account. Organ. Soc. **28**(6), 591–619 (2003)
17. Popp, R., Armour, T., Senator, T., Numrych, K.: Countering terrorism through information technology. Commun. ACM **47**(3), 36–43 (2004)
18. Rid, T.: "Think Again: Cyber-war." Foreign Policy. FP Group, 27 February 2012
19. Sageman, M.: Understanding Terror Networks. University of Pennsylvania Press, Philadelphia (2004)
20. Stohl, C., Stohl, M.: Networks of terror: theoretical assumptions and pragmatic consequences. Commun. Theory **17**, 93–124 (2007). https://doi.org/10.1111/j.1468-2885.2007.00289.x
21. Technical Analysis Group: Examining the cyber capabilities of Islamic terrorist groups. Institute for Security Technology Studies at Dartmouth College, Hanover, NH (2004)
22. Trybula, W.J.: Text mining. In: Williams, M.E. (eds.) Annual Review of Information Science and Technology, vol. 34, pp. 385–419. Information Today, Inc, Medford, NJ (1999)
23. Walton, D.N.: Argumentation Schemes for Presumptive Reasoning. L. Erlbaum Associates (1996)
24. Weimann, G.: The psychology of mass-mediated terrorism. Am. Behav. Sci. **52**(1), 69–86 (2008)
25. Weimann, G.: New Terrorism and New Media. Commons Lab of the Woodrow Wilson International Center for Scholars, Washington, DC (2014)
26. Weimann, G.: www.terror.net: How Modern Terrorism Uses the Internet. Special Report, United States Institute of Peace (116), pp. 1–12 (2004)
27. Weimann, G.: Lone wolves in cyberspace. J. Terror. Res. **3**(2) (2012). http://doi.org/10.15664/jtr.405
28. Weimann: Terror on the Internet. The New Arena, the New Challenges, 309 pp. United States Institute of Peace Press, Washington, DC (2006)
29. Williamson, J.B., Karp, D.A., Dalphin, J.R.: The Research Craft. An Introduction to Social Science Methods. Little Brown, Boston (1977)

# MP3 Steganalysis Based on Neural Networks

Marwan Mahmoud[1(✉)] and Alaa Abdulsalam Alarood[2]

[1] King Abdulaziz University, Jeddah, Saudi Arabia
mmamahmoud@kau.edu.sa
[2] Faculty of Computing and Information Technology, University of Jeddah,
Jeddah, Saudi Arabia

**Abstract.** In the history of human communication, the concept and need for secrecy between two parties has always been present. One way of achieving it is to modify the message so that it is readable only by the receiver, as in cryptography, for example. Hiding the message in an innocuous medium is called steganography whereas the counterpart of steganography, that is, discovering whether a message is hidden in a specific medium, is called steganalysis. In this paper, we propose a new model for steganalysis based on Artificial Neural Network (ANN) which is capable of detecting hidden messages within MP3 audio files. A three layer network with each layers interconnected with each other by three different transfer functions is used in this model. Statistical parameters calculated from dominant features extracted from stego MP3 files, are concatenated to form a single vector and subsequently is fed into the neural network. Gradient descent back propagation algorithm with adaptive learning rate is then used for training the network. Experimental results shows that the proposed model is robust to attacks, gives good stability and provides high accuracy in terms of minimum error rate.

**Keywords:** MP3 · Steganography · Steganalysis · Neural networks

## 1 Introduction

Steganography is a Greek word, meaning "covered writing". Which means hiding a secret message within another message, in such a way that no one can decode the message unless the master image voiced frame is found [1]. So it is possible to hide information in songs, movies or other information media [3]. Steganography is widely used on military and diplomatic issues [2].

The goal of steganalysis is to collect sufficient evidence about the presence of embedded message and to break the security of its carrier [4]. The importance of steganalytic techniques that can reliably detect the presence of hidden information in audio file is increasing. Steganalysis is use in computer forensics, cyber warfare, tracking criminal activities over the internet and gathering evidence for investigations particularly in case of anti-social elements [4]. In practice, a steganalyst is frequently interested in more than whether or not a secret message is present. The ultimate goal is to detect and extract the secret message.

© Springer Nature Switzerland AG 2019
C. Benavente-Peces et al. (Eds.): SEAHF 2019, SIST 150, pp. 428–435, 2019.
https://doi.org/10.1007/978-3-030-22964-1_48

There is a distinction between the concept of steganalysis and that of cryptanalysis in the same way that steganography is different from cryptography under cryptanalysis. The main goal here is in breaking the code with a view to decipher the contents of the encoded message. Steganalysis on the other is targeted only at determining the existence of a message rather than unraveling the content of the message hidden within enclosed medium. Consequently, it can be seen that the main intent of steganalysis is to answer whether a message exists or not, thereby providing a bivalent answer such as yes or no or true of false to the existence of a stego or that the message is genuine (Fig. 1).

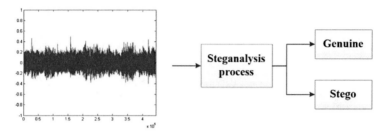

**Fig. 1.** Classical steganalysis process.

However, development of techniques and approaches for steganalysis of digital media afterwards was in conjunction with that of steganography, as seen in the preceding sections. For instance, a much related point is the search for the stego-key used in embedding a message will definitely require or involve cryptanalysis concepts [5]. It should be however noted that it is not a very straightforward task to highlight the various approaches to steganalysis in existence. An effort is made to discuss the various forms in the next frames.

## 2   Literature Review

Sharma and Bera proposed a steganalysis method based on the statistical moments of a wavelet characteristic function (SMWCF) and an artificial neural network (ANN) as a classifier [6]. In recent years, neural networks have proved their effectiveness in many applications, and ANN are recognized as powerful data analysis and modelling tool [7]. ANN consist of an interconnected collection of artificial neurons that change their structure based on the information that flows through the artificial network.

Qiao et al. proposed a scheme for detecting hidden messages in compressed audio files produced by MP3Stego. This scheme extract moment statistical features on the second derivatives, as well as Markov transition features and neighboring joint density of the MDCT coefficients based on each specific frequency band on MPEG-1 Audio Layer 3. For classification, a support vector machine is used on different feature sets. As shown by the experimental outcomes, this approach successfully discriminates MP3 covers and the steganograms produced with MP3Stego [8] (Fig. 2).

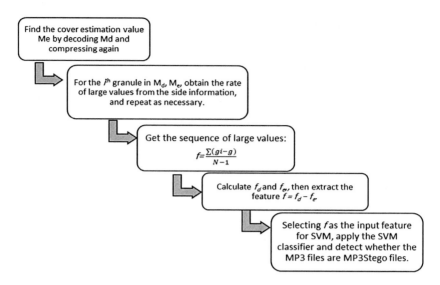

**Fig. 2.** Proposed steganalysis model [9].

Steganalysis algorithms appear fitting in the detection of detect Phase and Echo Steganalysis [10]. This method which comprises the use of statistical moments of peak frequency, proposed the use of Steganalysis algorithms for the detection of phase coding steganography according to the analysis of phase discontinuities and to detect the audio files. The phase steganalysis algorithm explores the situation that, phase coding causes corruption to the extrinsic continuities of unwrapped phase in every segment of audio. This leads to changes in the phase difference. In every audio segment, a statistical analysis of the phase difference is utilizable in monitoring the change and training the classifiers for differentiating an audio signal that is embedded from a clean audio signal. The echo Steganalysis algorithm statistically analyzes the peak frequency using short window extracting and then calculates the eighth high order center moments of peak frequency as feature vectors that are fed to a support vector machine, which is used as a classifier to differentiate between audio signals with and without data [10].

## 3   MP3 Dataset

MP3 is one of the most widely used techniques to encode and represent audio data [12], and of the reasons behind this popularity is its significant ability to reduce audio files sizes in comparison to other encoding techniques. The acronym MP3 is a short for Moving Picture Experts Group MPEG-1 Audio Layer 3 [11].

The used benchmark datasets for this study are adopted from [11], and consists of 10 s sample of 1886 MP3 files from the Garage band site. Garage band is a free and open source website that allows artists to upload their music and offers free download. There is access for visitors of the site to download the audio files, rate them, or write comments. So, people download the files and provide some Meta information about them. Analysis for each file. However, there is also personal classification schemes on those files described. The files belong to nine different genres namely: Pop, Rock, Folk, Funk, Jazz, Electronic, Blues, Rap/Hip, hop, and alternative. The number of songs in each genre varies. Each song is a 10-s MP3 file drawn from a random position of the corresponding song. Audio samples are encoded using mp3 method with a sampling rate of 44,100 Hz and a bit rate of 128 kbps.

## 4 Artificial Neural Networks

The Neural Networks have proven their effectiveness in numerous applications in the last few years. Artificial Neural Networks (ANNs) are regarded as powerful tools for data analysis and modeling. These methods can capture and correctly represent relationships both linear and nonlinear. They are valuable device for approximating functions, clustering data, and recognizing patterns that are imperceptible otherwise; Steganalysis often includes Neural Networks [13].

In our research we use ANN to predicate whether the object has a hidden message or not. The advantage of the usage of ANN for prediction is that they are able to learn from set of example only and that after their learning is finished, they are able to catch hidden and strongly non-linear dependencies, even when there is a significant noise in the training set.

## 5 Proposed Approach

Detection a hidden message in MP3 file is not an easy task in comparison with the image steganography. This presents a technique to detect the MP3 file and determine if it contains a stego message or not.

Figure 3 illustrates the block diagram of the proposed model. The model initially read the target MP3 file and gets its feature such as size, encoding, bit rate, sampling frequency, etc. These features are calculated then combined to form the input vector. The input vector propagated to the network to determine whether the input MP3 file is steganography file or not.

In training process, the network biases and weights are adjusted to build the final parameters of the network neurons. In the training process, historical input samples are entered to the network together with the target outputs. The training function is a function that takes the input and output from the historical samples and mp3 the weight and bias vectors which are associated to each network layer and neuron. In fact, each node in the network has weights vector and bias.

This variety of the learning samples supplies the neural network with all possibilities of the stego files, and such, the manipulated variable in the neural network propagation will be the consistency or inconsistency.

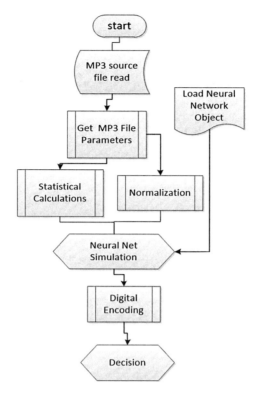

**Fig. 3.** Process of proposal approach.

## 5.1  Calculate Feature Extraction

The design of the neural network involves structure of network and vector of the network. In design process, many trials are used to select feature of MP3 files, after many experiment 10 feature was selected that are The peak signal to noise ratio (PSNR) of the MP3 file, Correlation coefficient, Mean, variance, standard deviation, The average of the total of MP3 file samples, The correlation between the first and second channel of the audio MP3 file, The mean square error of the MP3 file, The sum square error of the MP3 file, Sampling frequency, Number of bits. When selecting the feature parameter of the neural network structure we also selected the neural network type, the number of layers, number of neurons in each layer, training function or algorithm, and the activation function for each layer. For the network vector the entire feature are selected and stored in one vector to create the network that will be used in learning process.

In this work, we generated 50 MP3 file for experiments purposes 25 MP3 file are stego file and the other not stego. We utilized LSB algorithm to embedding a text message in MP3 files Table 1 shows some characteristics that have been extracted from files and utilized as input to the ANN. Each file represented as a vector of 10 fractures.

## 5.2   ANN Layers

As show in Fig. 4 the ANN has 2 hidden layers containing respectively 20 and 40 neurons and output layer to identify whether there is hidden stego or not. The used transfers ANN design are:

- Tan-sigmoid: activation functions for the input layer.
- Log-sigmoid network. The log-sigmoid transfer function was picked because its output range (0–1) is perfect for learning to output Boolean value
- Pure line activation functions for the output layer.

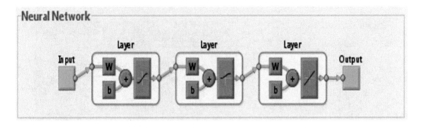

**Fig. 4.**  Layer of neural network.

Each layer is interconnected with each other tangent hyperbolic, logarithmic and linear transfer function respectively. The input statistical parameters are calculated and concatenated to shape the input vector, and then, it entered to the neural network. The neural network will start to propagate the input through the network layers and calculate the output value. The output value is 1 if the file is a stego MP3 file and 0 otherwise.

## 5.3   Training Function

We use a gradient descent with momentum and adaptive learning rate back propagation 'traingdx' is a network training function that updates weight and bias values according to gradient descent momentum and an adaptive learning rates shown in Fig. 5.

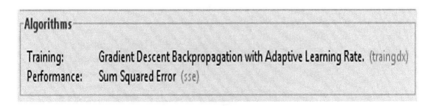

**Fig. 5.**  Training function.

Traingdx can train any network as long as its weight, net input, and transfer functions have derivative functions. Backpropagation is used to calculate derivatives of performance with respect to the weight and bias variables. Each variable is adjusted according to gradient descent with momentum.

## 5.4 Performance Function

We use Sum squared error performance function as a network performance function. It measures performance according to the sum of squared errors. Figure 5 illustrates the performance of the neural network training with respect to online training error.

The training behaviours are illustrated in the Fig. 7, where in such, a 10,000 epoch

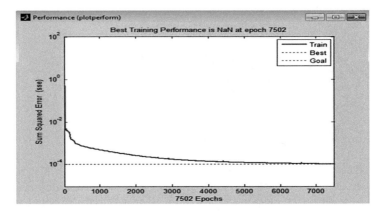

**Fig. 6.** Performance of training process for ANN learning steps.

was used. The epoch is a complete input learning iteration as illustrated before. The elapsed time for training is 51 s for each pass, where 100 pass were utilized. The final performance error is 0.0001 and the gradient of output change is $1 * 10^{-10}$. This means that, the learning of the network is stable. If the gradient is high, then, it's means that, the output is not stable, and the resulted error is not guaranteed (even though it's very small). But when accomplishing a very small gradient value, then, the output error result value will be guaranteed, and it is trusted value.

| Progress | | | |
|---|---|---|---|
| Epoch: | 0 | 7502 iterations | 10000 |
| Time: | | 0:00:51 | |
| Performance: | 16.6 | 0.000100 | 0.000100 |
| Gradient: | 1.00 | 0.00140 | 1.00e-10 |
| Validation Checks: | 0 | 0 | 6 |

**Fig. 7.** Progress of ANN.

# 6   Conclusion

In this paper we proposed utilization of a neural network for automated attack steganography on MP3 files. The MP3 file could be made steganography by inserting a hidden text message into it using LSB insertion which is the most known technique. An input vector is generated for ANN from the analyzed MP3 file. It consists of a set of dominant features those were extracted from the targeted MP3 file. The proposed algorithm achieved high accuracy attack results with numbers. In addition, it is capable to handle high quality steganography MP3 files with different encoding levels.

# References

1. Anguraj, S., Shantharajah, S.P., Murugan, R.A., Balaji, E., Maneesh, R., Prasath, S.: A fusion of A-B MAP cipher and ASET algorithms for the enhanced security and robustness in audio steganography. In: International Conference on Recent Trends in Information Technology (ICRTIT) (2011)
2. Shahadi, H.I., Jidin, R.: High Capacity and Inaudibility Audio Steganography Scheme. 978-1-4577-2155-7/11/$26.00. IEEE (2011)
3. Das, R., Tuithung, T.: A Novel Steganography Method for Image Based on Huffman Encoding. 978-1-4577-0748-3/12/$26.00. IEEE (2012)
4. Yadav, V.K., Chhikara, N., Gill, K., Dey, S., Singh, S., Yadav, S.: Three low molecular weight cysteine proteinase inhibitors of human seminal fluid: purification and enzyme kinetic properties. Biochimie 95(8), 1552–1559 (2013)
5. Chen, M., Sedighi, V., Boroumand, M., Fridrich, J.: JPEG-Phase-Aware Convolutional Neural Network for Steganalysis of JPEG Images (2017)
6. Sharma, M., Bera, S.: A review on blind still image steganalysis techniques using features extraction and Pattern classification method. International Journal of Computer Science. Eng. Inf. Technol. (IJCSEIT) 2, 117–135 (2012)
7. Usha, B., Srinath, D.N., Cauvery, D.N.: Data embedding technique in image steganography using neural network. Int. J. Adv. Res. Comput. Commun. Eng. 2, 2319–5940 (2013)
8. Jin, C., Wang, R., Yan, D.: Steganalysis of MP3Stego with low embedding-rate using Markov feature. Multimed. Tools Appl., 1–16 (2016)
9. Yan, D., Wang, R., Yu, X., Zhu, J.: Steganalysis for MP3Stego using differential statistics of quantization step. Digit. Signal Proc. 23, 1181–1185 (2013)
10. Zeng, W., Ai, H., Hu, R.: An algorithm of echo steganalysis based on power Cepstrum and pattern classification. In: Proceedings of the International Conference on Information and Automation, pp. 1667–1670 (2008)
11. Quackenbush, S.: MPEG Audio Compression Advances. The MPEG Representation of Digital Media. Springer, New York (2012)
12. Atoum, M.S., Ibrahim, S., Sulong, G., Ali, M.-A.: MP3 steganography: review. Int. J. Comput. Sci. Issues 9(6), 3 (2012)
13. Hashemi, S.A., Monadjemi, S.A.H., et al.: Price index forecasting using BP neural network and wavelet neural networks. Asian J. Res. Bank. Financ. 4(3), 105–116 (2014)

# Author Index

**A**
Abad, David, 358
Abdelkrim, Afef, 63
Abdulsalam Alarood, Alaa, 428
Adnane, Cherif, 55, 258, 274
Adnen, Cherif, 311
Ahmad, Rabiah, 421
Alarood, Ala, 421
Alghamdi, Anas, 11
Al-Hadeethi, Farqad F., 114
Alhaidari, Fahd Abdulsalam, 11, 24
Alhiyafi, Jamal, 24
ALraddadi, Abdullah, 414
Alsowayegh, Najat, 91
Alsyouri, Hatem M., 114
Amor, Nesrine, 347
Amores, Ernesto, 358
Aparicio, Daniel, 305
Atta-ur-Rahman, 11, 24

**B**
Bacha, Faouzi, 401
Bahl, Vaishali, 211
Bartolini, David, 296
Benavente-Peces, César, 221, 296
Benmbarek, M., 71
Benzergua, F., 71, 333
Bustamante, Alexis, 305

**C**
Caponetto, R., 245
Chaker, A., 71, 333
Chebbi, Souad, 347
Cherif, Adnen, 190

Cherki, I., 333
Chokri, Bouchoucha, 311
Chung, Tae-Sub, 270
Clemente-Jul, Carmen, 358
Corselli, Cesare, 98

**D**
Dash, Sujata, 11
Djellouli, Abderrahmane, 1
Djidar, Z., 333
Dwairi, Sohaib A., 114

**E**
Elngar Ahmed, A., 274

**F**
Fathi, Troudi, 393

**G**
Garba, Ibrahim, 91
Gargouri, Faiez, 289
Garsallah, Ali, 119
Giacobbe, Maurizio, 369

**H**
Hajaiej, Zied, 119, 190
Heer, Mandeep Singh, 35
Helali, Wafa, 190
Heo, Taek-Soo, 270
Hu, Ao, 221
Hussein, Emad Kamil, 85

**I**
Iqbal, Tahir, 169

© Springer Nature Switzerland AG 2019
C. Benavente-Peces et al. (Eds.): SEAHF 2019, SIST 150, pp. 437–438, 2019.
https://doi.org/10.1007/978-3-030-22964-1

**J**
Jano, Zanariah, 421
Janeska-Sarkanjac, Smilka, 165
Joshi, Rahul, 196
Jouini, Houda, 393

**K**
Kakkar, Ketan, 196
Kaur, Gurleen, 211
Kaur, Parminder, 42
Khaireddine, Zarai, 55
Khalfellah, N., 333
Khan, Mohammed Aftab, 24
Khanna, Disha, 266
Khorchani, Abdelghaffar, 98
Kies, Fatima, 98
Kim, Jung-Sook, 270
Kumar, Ravinder, 211
Kumar, Sunil, 150

**L**
Lakdja, Fatiha, 1
Loussaief, Sehla, 63

**M**
Mahmood, A., 386
Mahmood, Muhammad Habib, 251
Mahmoud, Marwan, 421, 428
Mami, Abdelkader, 393
Manaf, Azizah Abdul, 421
Marrouchi, Sahbi, 347
Minevska, Marija, 165
Mohamed, Mankour, 125
Moussa, Sonia, 119
Muthulakshmi, B., 318

**N**
Najam-l-Islam, M., 386
Nar, Mandip Kumar, 144
Nasim, Mohammad, 282
Nawaz, Rehan, 251
Nouha, Mansouri, 311

**O**
Osman, Khairunnisa, 421

**P**
Piña, Andhers, 305
Puliafito, Antonio, 369

**R**
Rachid, Meziane, 1
Rafique, Samina, 386
Ramaraju, G V, 282
Ratti, Nisha, 42
Rizzo, F., 245
Rodríguez, Lourdes, 358
Russotti, L., 245

**S**
Saeed, Ali Hamouda Ali, 414
Sahbani, Abdelaziz, 376, 414
Sami, Ben Slama, 125, 258, 274
Sami, BenSlama, 55
Sánchez, Mónica, 358
Sandeep, 231
Shahid, Muhammad Awais, 251
Sharda, Anuranjan, 196
Sharma, Vikrant, 35, 211, 231
Sharma, Yogesh Kumar, 196
Shekhar, Jayant, 150
Shishido, Hirofumi, 306
Sihem, Nasri, 258, 274
Singh, Bhanu Pratap, 150
Singh, Paprinder, 144
Singh, Prabhjeet, 144

**T**
Tena-Ramos, David, 221

**V**
Vaid, Neetu, 266
Vasudeva, Rajiv, 231
Venkatesulu, M., 318
Vir, Rajinder, 231

**X**
Xibilia, M. G., 245
Xibilia, Maria Gabriella, 369

**Y**
Yoshioka, Rentaro, 306
Yousef, Malik, 180

**Z**
Zafar, Bassam, 258, 274
Zahaf, Sahbi, 289
Zekri, Manel, 289
Zoghlami, Maha, 401

PGSTL